Atomic Physics
in Hot Plasmas

DAVID SALZMANN

New York   Oxford
Oxford University Press
1998

Oxford University Press

Oxford   New York
Athens   Auckland   Bangkok   Bogota   Bombay
Buenos Aires   Calcutta   Cape Town   Dar es Salaam
Delhi   Florence   Hong Kong   Istanbul   Karachi
Kuala Lumpur   Madras   Madrid   Melbourne
Mexico City   Nairobi   Paris   Singapore
Taipei   Tokyo   Toronto   Warsaw

and associated companies in
Berlin   Ibadan

Library of Congress Cataloging-in-Publication Data
Salzmann, David, 1938–
Atomic physics in hot plasmas / David Salzmann.
p.   cm.—(International series of monographs on physics)
Includes bibliographical references and index.
ISBN 0-19-510930-9
1. Plasma spectroscopy.   2. High temperature plasmas.   3. Atoms.
4. Ions.   I. Title.   II. Series: International series of monographs
on physics (Oxford, England).
QC718.5.S6S35   1998   97-28763

1 3 5 7 9 8 6 4 2
Printed in the United States of America
on acid-free paper

Preface

In recent years, with the advent of new applications for x-ray radiation from hot plasmas, the field of *atomic physics in hot plasmas*, also called *plasma spectroscopy*, has received accelerated importance. The list of new applications includes the high tech and industrial prospects of x-ray lasers, x-ray lithography, and microscopy. It also includes new methods for the traditional use of spectroscopy for plasma diagnostics purposes, which are important in laboratory and astrophysical research. Finally, some aspects of plasma spectroscopy are routinely used by the rocket and aircraft industries, as well as by environmental and other applied research fields which use remote sensors. The aim of this book is to provide the reader with both the basics and the recent developments in the field of plasma spectroscopy. The structure of the book enables its use both as a textbook for students and as a reference book for professionals in the field.

In contrast to the rapid progress in this field, there has been no parallel coverage in the literature to follow these developments. The most important treatise is still H. Griem's thirty-year-old book (Griem, 1964). Mihalas's book (Mihalas, 1970) contains important material, but is intended for other purposes. The more recent book by Sobelman (Sobelman, 1981) gives an excellent theoretical background, but more limited material which can be used directly by a group involved in plasma experiments or simulations. A series of shorter review articles focus mainly on partial aspects of the field, and a few volumes of conference proceedings present research papers on highly specialized subjects. There seems to be a need for a new comprehensive book, for both tutorial and professional use, which describes the subject in a coherently organized way, and which can be used by both students and the community of professionals active in this field. In fact, the idea of publishing a book on this subject came after discussions with colleagues in the United States, France, Germany, and Japan, in which countries I spent a few months during 1993, while on sabbatical leave. In particular, in the Institute of Laser Engineering, University of Osaka, Japan, I gave a series of eight seminars for the staff and graduate students on this topic. The notes of these seminars were the starting point of this book.

Plasma spectroscopy is a multidisciplinary field, which has roots in several other fields of physics. As such, it is impossible to describe from basic principles all the ingredients required for the understanding of this field in one book. It is, therefore, assumed that the reader is familiar with the basics of the underlying fields. First of all, it is assumed that the reader has a basic knowledge of quantum theory and atomic spectroscopy, so that the terminology of the notations and quantum numbers of simple and complex atoms, as well as the angular momentum coupling schemes (LS, jj, and the corresponding $3j$, $6j$, $9j$ symbols) are known. Second, although in Chapter 1 we give a brief recapitulation of the basic formulas of statistical physics that are used in the book, we assume that the reader understands the origin and the meaning of these formulas. Finally, in several places we mention advanced methods of approximations or computations without giving any further explanation. These are, in most cases, advanced topics, and the reader interested in more detail will find them in other references.

I take the opportunity to express my thanks to several of my colleagues who helped me in the preparation of this book. First, special thanks to Dr. Aaron Krumbein, my friend and colleague, with whom I have had the privilege working for several years. Aaron read the manuscript of this book and helped me in many of its aspects, including the organization and the explanation of the material, and even the style.

I would like to thank Professor H. Takabe, who was the organizer of the course in ILE, Osaka, and with whom I was also in close collaboration on several more specific subjects of plasma spectroscopy and laser plasma interactions. He helped me with many major and minor daily problems during my stay in Japan. He was also my chief source of information about Japan, and his explanations covered subjects from the research fields in ILE, through the shapes of the Kanji letters, to the traditions of the Japanese way of life. I would like to acknowledge very interesting professional and nonprofessional discussions with Professor K. Mima, the present director of ILE. I would also like to thank Professor S. Nakai, the director of ILE, for his invitation and generous hospitality. Without his help, my visit to Japan, which finally resulted in the writing of this book, would not have been possible.

My thanks are also given to the management of Soreq NRC, Israel, and particularly to Dr. U. Halavy, the director of Soreq, who encouraged me in writing this book and provided the help of the institute in several technical aspects, such as preparing the figures and library help in the search for some older research papers. I also take the opportunity to thank about 40 of my colleagues (their names appear in the references) who responded to my letter and sent me their recent research papers (altogether approximately 300 of them). Their responses helped me to advance the quality of this book, and at the same time to update myself on the recent achievements in the field. The page limit, however, allowed me to include only a part of this material in the book.

David Salzmann

Soreq NRC,
Yavne, June 1997

Contents

ATOMIC PHYSICS
IN HOT PLASMAS

Introductory Remarks, Notations, and Units

1.1 The Scope of This Book

The field of atomic physics in hot plasmas, also called *plasma spectroscopy*, is the study of the properties of the electrons and ions in an ionized medium and the electromagnetic radiation emitted from this medium. The plasma is assumed to have a temperature high enough so that the greater part of the atoms are ionized and all the molecular bonds are broken.

Classical textbooks on plasma physics regard the ions and electrons as point-like structureless objects, and consider mainly the mutual electric and magnetic interactions inside the plasma. Most of the texts describe first the charged particle motion under the influence of these fields and then the collective macroscopic properties of a plasma. The presence of electromagnetic radiation is considered only as long as it has collective effects on the plasma particles, which constrains the treatment generally to long wavelength radiation in the radio or microwave regions. In this regard, plasma physics is a macroscopic theory.

In contrast, the field of atomic physics in hot plasmas zooms into the microscopic atomic structure. It is interested in three general topics. The first is the influence of the plasma environment on the atomic/ionic potential, and thereby on the ionic bound electrons wavefunctions and energy levels. The second central topic is the study of the electron–ion and ion–ion collisional processes inside the plasma, their cross-sections and rates. These processes, which are responsible for the ionization and excitation of the ions, determine the charge and excited states distributions. Finally, this field includes also the subject of the emission and absorption spectra of the plasma.

Atomic physics in hot plasmas, like other areas of plasma physics, can be roughly divided into two regions: low density plasmas, up to approximately 10^{17} ions/cm^3 and high density plasmas, above 10^{19} ions/cm^3. Plasmas in the lower density regime are the subject of research in astrophysics, tokamaks, and magnetic confinement devices. In these plasmas the central interest is the behavior of the atomic processes, charge state distributions, the average charge ($\bar{Z}$), and emission spectra. At higher densities, which generally include inertial confinement plasmas and matter in star interiors, there are also direct effects of the plasma environment on the ions, accompanied with several phenomena such as energy level and emission line shifts, line profiles, forbidden transitions, and others, which are of no importance in low density plasmas.

In this book we have tried to include updated research results. Although the active research in the field goes on at an accelerated pace, we hope that the material included will provide a sufficient basis for everybody who intends to enter this field.

1.2 The Basic Plasma Parameters

A plasma consists of three kinds of particles: ions, electrons, and photons. The electrons can be further divided into *bound electrons*, namely, electrons that occupy negative-energy quantum states bound to a single ion, and *free electrons*, which are positive-energy electrons moving freely in the plasma.

Consider a homogeneous plasma of ions of *atomic number* Z, which is also the electric charge of the nucleus, and denote the ion density (that is, the average number of ions per unit volume) by n_i. Generally, the ions are in various *ionization* or *charge states*, depending on the number of electrons missing from each atom. Denote by ζ the charge of the ion, which is also the number of electrons missing from the atom. The average partial density of a charge state ζ over the ensemble will be denoted by N_ζ, and obviously

$$\sum_{\zeta=0}^{Z} N_\zeta = n_i \tag{1.2.1}$$

Although the summation includes, in principle, all of the possible charge states from neutral ($\zeta = 0$) up to fully ionized ($\zeta = Z$) species, significant contributions to the sum in equation (1.2.1) come only from a limited range of charge states which have nonvanishing partial densities in the plasma.

As a charge state ζ contributes ζ electrons to the population of the free electrons in the plasma, one gets for the electron density, n_e,

$$n_e = \sum_{\zeta=0}^{Z} \zeta N_\zeta \tag{1.2.2}$$

Equation (1.2.2) is equivalent to the requirement of *charge neutrality* in the plasma. This relationship is correct only as an average over the whole ensemble. Locally, the electron density does not exactly neutralize the local positive charge at every point.

The average charge state, denoted by $\bar{Z}$, is defined as

$$\bar{Z} = \frac{\sum_{\zeta=0}^{Z} \zeta N_\zeta}{\sum_{\zeta=0}^{Z} N_\zeta} = \frac{n_e}{n_i} \tag{1.2.3}$$

or

$$n_e = \bar{Z} n_i \qquad (1.2.4)$$

This last equation is perhaps the most frequently used relationship in the theory of atoms in hot plasmas. The charge state distribution as well as $\bar{Z}$ depend on both the temperature and the density. The calculation of $\bar{Z}$ is one of the central topics of plasma spectroscopy.

Using a formula similar to equation (1.2.3), one can also calculate higher moments of the charge state distribution. For example, one can calculate the second moment, $\overline{Z^2} = \sum_{\zeta=0}^{Z} \zeta^2 N_\zeta / n_i$, which is important in calculating the standard deviation,

$$\sigma_Z^2 = \overline{Z^2} - \bar{Z}^2 \qquad (1.2.5)$$

σ_Z provides a criterion about the range of charge states which have nonvanishing density in the plasma.

In a plasma of ion density n_i, the average volume available for every ion is $V_i = 1/n_i$. A more frequently used quantity is the *ion sphere*, which is a sphere of radius R_i, and which has the same average volume V_i,

$$V_i = \frac{4\pi}{3} R_i^3 = \frac{1}{n_i} \qquad (1.2.6)$$

R_i is called the *ion sphere radius*,

$$R_i = \left(\frac{3}{4\pi n_i} \right)^{1/3} \qquad (1.2.7)$$

referred to in some publications also as the *Wigner–Seitz radius*. Since an ion sphere contains, on the average, one ion of charge $\bar{Z}$, the requirement of plasma neutrality implies that there are also, on the average, $\bar{Z}$ free electrons inside the ion sphere. This, however, is true only as an average statement, and in a real plasma there are significant fluctuations around this average.

1.3 Statistics, Temperature, Velocity, and Energy Distributions

The Partition Function and the Helmholtz Free Energy

Although the main intention of this book is not a study of statistical mechanics, we will use the statistical properties of plasmas to deduce several important results. We assume that the reader is familiar with the general ideas of statistical mechanics and thermodynamics, but a reminder of the corresponding definitions and formulas is always welcome. We list here a few formulas and definitions that will rather frequently be applied in the following chapters.

Our starting point is the *Gibbs distribution* or the *canonical distribution*, which is one of the basic quantities of statistical mechanics. It provides the formula for the distribution of a macroscopical subsystem in a large closed system that is in equilibrium (see Landau, 1959). Gibbs deduced that the probability to find this subsystem in a state of energy E_n is proportional to the exponential factor,

$$\text{Probability} \propto g_n e^{-E_n/T} \tag{1.3.1}$$

where g_n is the statistical weight of this state, and T is the temperature of the system. Here and in the following we write the temperature in energy units, namely, the Boltzmann constant, k_B, is unity, and we write only T where books on statistical mechanics would write $k_B T$.

The normalization condition of the probability distribution function requires that

$$\text{Probability} = \frac{g_n e^{-E_n/T}}{Z(T)} \tag{1.3.2}$$

where

$$Z(T) = \sum_n g_n e^{-E_n/T} \tag{1.3.3}$$

is the *partition function*. The summation in equation (1.3.3) should be understood as a simple sum over discrete energy states, or an integration for a continuity of energy states. In the second case, g_n has to be replaced by $\rho(E)\,dE$, where $\rho(E)$ is the density of quantum states per unit energy.

The partition function is a basic quantity from which, in principle, all the other parameters of a statistical system can be deduced. We first calculate its value for a free electron in the plasma. The energy of a free electron has only one component, namely, its kinetic energy,

$$E_n = \frac{1}{2m_e}(p_x^2 + p_y^2 + p_z^2)$$

where $\vec{p} = (p_z, p_y, p_z)$ is the electron's momentum. The statistical weight of electrons having this momentum, which are located in the volume element dV is (Landau and Lifschitz, 1959),

$$\frac{dp_x\,dp_y\,dp_z\,dV}{(2\pi\hbar)^3}$$

By virtue of equation (1.3.3), the partition function of a free electron is

$$z_e(T) = 2\int dV \int\int\int \exp\left(-\frac{1}{2m_e T}(p_x^2 + p_y^2 + p_z^2)\right)\frac{dp_x\,dp_y\,dp_z}{(2\pi\hbar)^3}$$

$$= 2V\left(\frac{m_e c^2 T}{2\pi(\hbar c)^2}\right)^{3/2} \tag{1.3.4}$$

where V is the volume of the system and the factor 2 comes from the two possible spin states of the electron.

The total energy of the ions consists of two parts: the kinetic and the potential energies: $E = E_k + E_n$, where the kinetic energy has the same form as for the electrons, with the ion mass replacing the electron mass in equation (1.3.4), and E_n is the binding energy of the ionic state. The partition function of an ion of charge ζ splits, therefore, into two multiplicative factors,

$$z_\zeta(T) = z_{\zeta,k} \cdot z_{\zeta,b}$$

$$= \sum_n \int dV \int\int\int \exp\left[-\frac{1}{T}\left(\frac{1}{2m_e}(p_x^2 + p_y^2 + p_z^2) + E_{\zeta,n}\right)\right] \frac{dp_x \, dp_y \, dp_z}{(2\pi\hbar)^3}$$

$$= V\left(\frac{m_i c^2 T}{2\pi(\hbar c)^2}\right)^{3/2} \sum_n \exp\left(-\frac{E_{\zeta,n}}{T}\right) \tag{1.3.5}$$

Here, m_i is the ionic mass, $z_{\zeta,k}$ is the kinetic energy part, and $z_{\zeta,b}$ the *internal* or *excitational* energy part of the partition function. It should be mentioned explicitly that this factorization of the partition function into multiplicative components according to the mutually independent degrees of freedom of the particle is a general property of the partition function.

Assume a system of M indistinguishable particles, and assume that the partition function of each particle is $z(T)$. The partition function of the whole system is calculated by means of a binomial probability function

$$Z(T) = \frac{1}{M!}[z(T)]^M \tag{1.3.6}$$

In this formula the power M stems from the multiplicative property of the partition function for independent degrees of freedom, see above, whereas the origin of $M!$ in the denominator comes from the fact that the particles are indistinguishable. Equation (1.3.6) can be further reduced, by using Stirling's formula, $M! \approx (M/e)^M$, $M \to \infty$, to

$$Z(T) \approx \left(\frac{ez}{M}\right)^M \tag{1.3.7}$$

where $e = \sum_n 1/n! = 2.718\ldots$ is the base of the natural logarithm. For large number of particles this is a very good approximation, and for a real plasma one can regard it, for all practical purposes, as an equality.

Another important thermodynamic function is the *free energy*, F, also called the *Helmholtz free energy*, whose definition is

$$F = -T \log[Z(T)] = -MT \log\left(\frac{ez}{M}\right) \tag{1.3.8}$$

The second equation is correct only for a one-component plasma.

Using the general relationship of thermodynamics it can be shown that the derivatives of the free energy with respect to its variables give some of the most important physical parameters of the system. For instance, its partial derivative with respect to the temperature at constant volume is the system's entropy, whereas its partial derivative with respect to the volume at constant temperature yields the pressure (Landau and Lifshitz, 1959),

$$S = -\left(\frac{\partial F}{\partial T}\right)_V, \qquad P = -\left(\frac{\partial F}{\partial V}\right)_T \qquad (1.3.9)$$

In chapter 5 we shall be interested in the partial derivative of F with respect to the number of particles. From equation (1.3.8) it follows that

$$\frac{\partial F}{\partial M} = -T\left[\log\left(\frac{\epsilon z}{M}\right) - 1\right] = -T \log\left(\frac{z(T)}{M}\right) \qquad (1.3.10)$$

Energy and Velocity Distribution Functions

A system of electrons and ions in a plasma is not necessarily in equilibrium. Even when not in equilibrium, the particles have some energy and velocity distributions,

$$f_{\vec{v}}(v_x, v_y, v_z)\, dv_x\, dv_y\, dv_z = f_E(E)\, dE \qquad (1.3.11)$$

$f_{\vec{v}}(v_x, v_y, v_z)\, dv_x\, dv_y\, dv_z$ is the density of electrons or ions whose velocity components are within the limits $[v_x, v_x + dv_x]$, $[v_y, v_y + dv_y]$, $[v_z, v_z + dv_z]$ and $f_E(E)\, dE$ is the density of electrons or ions whose energies are in the range $[E, E + dE]$. If the plasma is isotropic, without a preferential direction, one uses the total velocity distribution function,

$$f_v(v)\, dv = \int\int\int_{v_x^2 + v_y^2 + v_z^2 = v^2} f_{\vec{v}}(v_x, v_y, v_z)\, dv_x\, dv_y\, dv_z$$

$$= \int_{\theta=0}^{\pi} \int_{\phi=0}^{2\pi} f_{\vec{v}}(v) v^2\, dv \sin\theta\, d\theta\, d\phi = 4\pi v^2 f_{\vec{v}}(v)\, dv \qquad (1.3.12)$$

which is the density of electrons/ions whose total velocity is between $[v, v + dv]$, regardless of the direction of their motion.

We will first treat the electron subsystem. If the system is in equilibrium, then by virtue of equations (1.3.2) and (1.3.4), the energy and velocity distributions acquire the form

$$f_v(v)\, dv = n_e \left(\frac{m_e}{2\pi T_e}\right)^{3/2} 4\pi v^2 \exp\left(-\frac{mv^2}{2T_e}\right) dv$$

$$f_E(E)\, dE = \frac{2}{\sqrt{\pi}} n_e \left(\frac{E}{T_e}\right)^{1/2} \exp\left(-\frac{E}{T_e}\right) \frac{dE}{T_e} \qquad (1.3.13)$$

These are called the *Boltzmann–Maxwell velocity* and *energy* distributions. In equation (1.3.13) $f_v(v)\, dv$ is the *density of electrons* whose velocity is between v and $v + dv$, $f_E(E)\, dE$ is the *density of electrons* whose kinetic energy is between E and $E + dE$, and T_e is the *electron temperature* (in energy units).

When equation (1.3.13) holds true, it is possible to characterize the whole distribution by one parameter, the temperature. For equation (1.3.13) to be correct, several conditions must be fulfilled. First, it is assumed that the plasma is nonrelativistic, namely, the electron velocities are well below the speed of light (although generalizations for relativistic plasmas can be found in the literature). Second, it is also assumed that the system has had enough time to thermalize, that is, to attain thermal equilibrium as defined by a Boltzman–Maxwell velocity distribution. The thermalization times of electrons are given by the electron self-collision time (Spitzer, 1962),

$$t_c = 0.2896 \frac{(mc^2)^{1/2} T_e^{3/2}}{n_e c e^4 \log \Lambda} \tag{1.3.14}$$

where $\log \Lambda$ is the *Coulomb logarithm* (Spitzer, 1962),

$$\Lambda = \frac{3}{2\bar{Z}^2} \left(\frac{T_e^3}{\pi e^6 n_e} \right)^{1/2} \tag{1.3.15}$$

$\log \Lambda$ is a slowly varying function of the temperature and density whose values are tabulated in Spitzer's book. The values of $\log \Lambda$ are generally between 5 and 15. Numerically this formula predicts a thermalization time of

$$t_c = 3.3 \times 10^{-13} \left(\frac{T_e}{100 \, \text{eV}} \right)^{3/2} \left(\frac{10^{21} \, \text{cm}^{-3}}{n_i \log \Lambda} \right) \quad \text{s,} \tag{1.3.16}$$

(complying with the conventions of the SI system, we write s for seconds). For a $T_e = 100 \, \text{eV}$ plasma this last equation predicts $t_c = 20 \, \mu\text{s}$ at $n_i = 10^{12} \, \text{cm}^{-3}$, $t_e = 30 \, \text{ns}$ at $n_i = 10^{15} \, \text{cm}^3$, $t_e = 40 \, \text{ps}$ at $n_i = 10^{18} \, \text{cm}^{-3}$ and $t_c = 60 \, \text{fs}$ at $n_i = 10^{21} \, \text{cm}^{-3}$. These times are generally very short relative to the plasma evolution times at the given densities, so that in laboratory and, of course, for astrophysical plasmas one can safely assume that the electron temperature, T_e, is a well defined quantity. Examples of plasmas in which this condition is clearly incorrect are the femtosecond laser-generated plasmas that are presently under intensive research. In these plasmas, the plasma evolution time is too short for the electrons to thermalize during the laser pulse duration. In this case one must consider non-Maxwellian energy and velocity distribution functions (Rousse, et al. 1994).

Finally, equation (1.3.13) is incorrect also when the electron density is very high and exchange effects play an important role. At such high electron densities the Fermi–Dirac distribution, rather than a Boltzmann–Maxwell one, is the correct statistical method to describe the electron distribution. One obtains for the momentum and energy distributions (see e.g., Eliezer, 1986; Landau and Lifshitz, 1959)

$$f_p(p) \, dp = \frac{1}{\pi^2 \hbar^3} \frac{p^2 \, dp}{1 + \exp[(p^2/2m - \mu)/T_e]}$$

$$f_E(E) \, dE = \frac{\sqrt{2}}{\pi^2} \frac{(mc^2)^{3/2}}{(\hbar c)^3} \frac{E^{1/2} \, dE}{1 + \exp[(E - \mu)/T_e]} \tag{1.3.17}$$

Here, as in equation (1.3.13), T_e is the electron temperature, and μ is the *chemical potential* or the *Fermi energy* of the electrons. It can be shown that for a gas of nondegenerate free electrons

$$\mu = T_e \log \left[\frac{n_e}{2} \left(\frac{2\pi(\hbar c)^2}{mc^2 T_e} \right)^{3/2} \right] \qquad (1.3.18)$$

In the case of Fermi–Dirac statistics the total electron density is

$$n_e = 2 \frac{(mc^2 T_e)^{3/2}}{\sqrt{2\pi^2}(\hbar c)^3} F_{1/2} \left(\frac{\mu}{T_e} \right) \qquad (1.3.19)$$

where

$$F_j(x) = \int_0^\infty \frac{y^j\, dy}{1 + \exp(y - x)} \qquad (1.3.20)$$

is the *complete Fermi–Dirac integral*. Given the electron density n_e and the electron temperature T_e, equation (1.3.19) is an equation from which the Fermi energy μ is calculated. When $\mu \to -\infty$ the Fermi–Dirac distribution reduces to the Boltzmann–Maxwell one. The criterion for this to happen is

$$\frac{n_e}{2} \left(\frac{2\pi(\hbar c)^2}{mc^2 T_e} \right)^{3/2} \ll 1 \quad \text{or} \quad \bar{Z} \left(\frac{n_i}{10^{21}\,\text{cm}^{-3}} \right) \left(\frac{100\,\text{eV}}{T_e} \right)^{3/2} \ll 6 \times 10^3 \quad (1.3.21)$$

Although in this book we will be interested mainly in high temperature plasmas, for the sake of completeness we also write the functional behavior of the Fermi–Dirac statistics at the $T \to 0$ limit. In fact, this version of the statistics is important in solid state and other low temperature fields of physics. In the $T \to 0$ limit equations (1.3.17) reduce to the form

$$f_p(p)\, dp = \begin{cases} 0 & p^2/2m > \mu \\ p^2\, dp/\pi^2\hbar^3 & p^2/2m < \mu \end{cases}$$

$$f_E(E)\, dE = \begin{cases} 0 & E > \mu \\ (\sqrt{2}/\pi^2)[(mc^2)^{3/2}/(\hbar c)^3]E^{1/2}\, dE & E < \mu \end{cases} \qquad (1.3.22)$$

The electron density is obtained by integrating the second of these equations over all possible energies,

$$n_e = \frac{\sqrt{2}}{\pi^2} \left(\frac{mc^2}{(\hbar c)^2} \right)^{3/2} \int_0^\mu E^{1/2}\, dE = \frac{2}{3} \frac{\sqrt{2}}{\pi^2} \left(\frac{mc^2}{(\hbar c)^2} \right)^{3/2} \mu^{3/2} \qquad (1.3.23)$$

wherefrom one obtains the low temperature limit of the chemical potential,

$$\mu = \frac{(\hbar c)^2}{2mc^2} (3\pi^2 n_e)^{2/3} \qquad (1.3.24)$$

Numerically, this formula is written as

$$\mu = 0.3646 \left(\frac{n_e}{10^{21}\,\text{cm}^{-3}} \right)^{2/3} \text{eV} \qquad (1.3.25)$$

At this low temperature limit the average energy density (energy per unit volume) is

$$\rho_E = \int_0^\mu E f_E(E)\, dE = \frac{\sqrt{2}}{\pi^2} \left(\frac{mc^2}{(\hbar c)^2}\right)^{3/2} \int_0^\mu E^{3/2}\, dE = \frac{2}{5} \frac{\sqrt{2}}{\pi^2} \left(\frac{mc^2}{(\hbar c)^2}\right)^{3/2} \mu^{5/2} \quad (1.3.26)$$

and the average energy per particle is

$$\frac{\rho_E}{n_e} = \frac{3}{5}\mu \quad (1.3.27)$$

The ions, too, have a Boltzmann-type energy distribution,

$$f_E(E)\, dE = n_i \left(\frac{m_i}{2\pi T_i}\right)^{3/2} \exp\left(-\frac{E}{T_i}\right) \quad (1.3.28)$$

where T_i is the ion temperature. Being particles much more massive than the electrons, the degeneracy effects in ions are expected to show up only at extremely high densities, which are beyond any present experimental technique. A Fermi–Dirac statistics for the ions is, therefore, never needed in plasma spectroscopy.

The ion temperature does not necessarily equal the electron temperature T_e. When the ion and the electron temperatures are equal, we shall simply denote the plasma temperature by $T (= T_e = T_i)$. The electron–ion energy equipartition time is given by (Spitzer, 1962)

$$t_{eq} = \frac{3m_e c^2 m_i c^2}{8(2\pi)^{1/2} n_i (Ze^2)^2 c \log \Lambda} \left(\frac{T_e}{m_e c^2} + \frac{T_i}{m_i c^2}\right)^{3/2}$$

$$\approx 3.16 \times 10^{-10} \frac{A}{Z^2} \left(\frac{T_e}{100\,\text{eV}}\right)^{3/2} \left(\frac{10^{21}\,\text{cm}^{-3}}{n_i \log \Lambda}\right) \quad \text{s} \quad (1.3.29)$$

where A is the atomic weight of the ions. For a hydrogen plasma this is about 1000 times longer than the electron–electron thermalization time, t_c, equations (1.3.14) and (1.3.16), and decreases for higher-Z plasmas. The temperature difference between the electrons and the ions plays an important role, for instance, in plasmas generated by nanosecond duration laser pulses in which the laser energy is absorbed mainly by the free electrons. This energy is transferred to the ions only later, within a few tenths of nanoseconds, by means of collisions between the electrons and the ions. At the first stages of the evolution of such plasmas, the two temperatures are quite different, and they equalize only on a timescale of one nanosecond or so.

1.4 Variations in Space and Time

All the preceding equations are true only in the average sense. Large fluctuations in space and time may occur in all the above quantities. For example, one can speak about the *time-dependent* and *space-dependent* charge state densities, $N_\zeta(\vec{r}, t)$, and similarly about the *local instantaneous* ion density and electron temperature, $n_i(\vec{r}, t)$, $T_e(\vec{r}, t)$. In the following we denote the space- and time-dependent quantities by explicitly showing the $\vec{r}$- and/or t-dependence in parentheses, whereas quantities averaged over the ensemble will be denoted by the same letter but without any extra notation. We hope that this will not cause confusion. When viewed locally and instantaneously, equation (1.2.1) is rewritten as

$$\sum_{\zeta=0}^{Z} N_\zeta(\vec{r}, t) = n_i(\vec{r}, t) \tag{1.4.1}$$

One cannot, however, write equation (1.2.2) in a space-resolved form, because the electron density does not necessarily neutralize the ionic positive charge at every point in space.

Assume that an ion is located at $r = 0$. In a homogeneous isotropic steady state plasma the average electron and ion densities are independent of time, do not have a preferential direction, and depend only on the radius. One can write $n_e(\vec{r}, t) = n_e(r)$, $n_i(\vec{r}, t) = n_i(r)$. Quantities related to the local electron and ion densities are the *ion–ion* and *ion–electron radial distribution functions*, defined as

$$g_i(r) = \frac{n_i(r)}{n_i}, \qquad g_e(r) = \frac{n_e(r)}{n_i} \tag{1.4.2}$$

$g_i(r)$ vanishes near the central ion, at $r = 0$, due to the mutual rejection of positively charged ions, and approaches unity asymptotically for large distances from the origin. At very high densities the distribution function approaches unity in an oscillatory manner, reflecting the buildup of a lattice type structure in the plasma. The *ion–electron pair distribution function*, $g_e(r)$, measures the polarization of the electrons around the central ion. This function gets its maximum near the nucleus, due to the ion–electron electrostatic attraction, and tends to unity asymptotically beyond the outer peripheries of the ion. These two quantities are frequently used in the computation of the spatial distributions of the ions and the electrons in the vicinity of a given ion.

Some care should be taken about the meanings of the space- and time-dependent quantities, because in real plasma there are several scale lengths and characteristic times. Regarding the variations in space, there is first the scale length of the plasma gradients, generally defined by

$$L_{plasma} = \left| \frac{n_i}{\nabla n_i} \right| \quad \text{or} \quad L_{plasma} = \left| \frac{T_e}{\nabla T_e} \right| \tag{1.4.3}$$

which is a measure of the characteristic distance along which the *average* quantities change substantially. Unless stated explicitly, throughout this book we will assume a homogeneous isotropic plasma, and will not be interested in this parameter.

On the other hand, one can zoom into shorter distances in the plasma, and inquire about density variations on the scale length of the ion sphere radius, or even within the ionic volume. In fact, the nucleus of an ion, the only positive charge, occupies only a small region at the center of the ion sphere. The bound electrons with their characteristic charge distribution occupy the rest of the ionic volume, thereby generating a charge distribution which varies on the scale of the *ionic radius*. The free electrons span the outer parts of the ion sphere. Altogether, there are significant variations in the electric charge density on the scale of the ionic as well as the ion sphere radii. These scale lengths are the subject of several chapters in this book.

There are several characteristic times in hot plasmas. The longer one is related to the time of the plasma evolution,

$$t_{plasma} = \left| \frac{n_i}{\partial n_i / \partial t} \right| \quad \text{or} \quad t_{plasma} = \left| \frac{T}{\partial T / \partial t} \right| \tag{1.4.4}$$

Except where otherwise mentioned, throughout this book we assume a stationary plasma and will not be interested in this time scale. Three other time scales are, however, of greater interest on the atomic scale in hot plasmas. The first of these is the time connected to the plasma frequency (Spitzer, 1962),

$$\tau_p = \frac{2\pi}{\omega_p}, \quad \omega_p = \left[\frac{4\pi e^2 n_e}{m_e} \left(1 + Z \frac{m_e}{m_i} \right) \right]^{1/2} \approx \left(\frac{4\pi e^2 n_e}{m_e} \right)^{1/2} \tag{1.4.5}$$

This time scale determines the shortest time at which the free electron cloud can adjust to any change in the local ionic pattern. This time scale depends only on the electron density, and becomes longer for lower electron densities. For purposes of comparison we cast this formula into the form

$$\tau_p = 3.52 \times 10^{-15} \left(\frac{10^{21} \, \text{cm}^{-3}}{n_e} \right)^{1/2} \text{s} \tag{1.4.6}$$

The second time scale is connected to the atomic processes rates,

$$\tau_a = \frac{1}{n_e \langle \sigma v \rangle} \tag{1.4.7}$$

Here σ is the cross section of the most frequent atomic reaction in the plasma, and v is the electron velocity. The angle brackets in the denominator indicate averaging over the velocity distribution of the electrons. This parameter is the subject of the discussion in chapter 4. This time scale, too, becomes longer as the density drops. In contrast to τ_p, τ_a depends, through the velocity distribution, also on the temperature. In fact, this parameter indicates the average time between two collisions of an ion with other plasma particles which cause a change of the excitation or ionization state of the ion. Due to the complex behavior of the atomic cross sections, one cannot put equation (1.4.7) into a simple numerical form, as in the case of τ_p.

Finally, there is the atomic time scale, which is the time of revolution of the bound electrons in their orbitals. This is given approximately by the Kepler–Bohr formula,

$$\tau_B = \frac{\pi \bar{Z} \hbar}{E_H} \left(\frac{E_H}{|E|} \right)^{3/2} \tag{1.4.8}$$

where $E_H = 13.6$ eV is the hydrogen atom ground state energy, see table 1.1, and $|E|$ is the energy of the electron. This quantity is independent of the plasma density or temperature, and depends solely on the ionic charge and the energy of the ionic state under consideration. Numerically, equation (1.4.8) is rewritten as

$$\tau_B = 1.52 \times 10^{-16} \bar{Z} \left(\frac{E_H}{|E|} \right)^{3/2} \quad \text{s} \tag{1.4.9}$$

It may be worthwhile to emphasize that the scale lengths and characteristic times on the atomic scale depend on the plasma densities only, but not on the temperatures. The concept of temperature is intrinsically a statistical quantity which is obtained by averaging over the distributions of a large ensemble of particles. A local temperature, $T(\vec{r})$, therefore has meaning only when considering a portion of a plasma which has a size of the order of L_{plasma}, during a time period of the order of t_{plasma}. There is, of course, no meaning to temperature on the scale length of the ion sphere or during the period of the atomic time scale.

1.5 Units

A few words about the units used in this book are in order. The general experience suggests that a system based on the three basic units eV, cm, and s is most useful when working with atomic physics in hot plasmas. Only macroscopic energies and derived quantities, such as total radiation rates, are expressed in cgs or MKS units such as erg, joule, ergs/s or watt. Energies on the atomic scale, such as temperature (in the sense of average kinetic energy per electron or ion), energy levels, and photon energies are expressed in eV.

It turns out that most of the formulas relevant to the field can be expressed in terms of a few atomic constants, which we list in the above units in table 1.1. In fact, the reader could note that in this chapter we have already expressed the

Table 1.1 Some Constants Frequently Used in Plasma Spectroscopy

Constant	Symbol	Value and units
Fine structure constant	$\alpha = e^2/\hbar c$	$1/137.035\,9895$
Planck constant/2π	$\hbar$	$6.582\,1220 \times 10^{-16}$ eV s
	$\hbar c$	$1.973\,271 \times 10^{-5}$ eV cm
Mass of the electron	mc^2	$5.109\,99 \times 10^5$ eV
	$\hbar^2/m = (\hbar c)^2/mc^2$	$7.619\,973 \times 10^{-16}$ eV cm^2
(Charge of the electron)^2	e^2	$1.439\,9652 \times 10^{-7}$ eV cm
Bohr radius	$a_0 = (\hbar c)^2/mc^2 e^2$	$5.291\,772\,49 \times 10^{-9}$ cm
Hydrogen atom ground state energy	$E_H = e^2/2a_0$	$13.605\,698$ eV
	$= (1/2)\alpha^2 mc^2$	
ergs/eV		$1.602\,177\,33 \times 10^{-12}$ erg/eV

formulas by means of these elementary constants. For example, in equations (1.3.8–9) powers of c, the speed of light, could be reduced, but we preferred to keep these formulas in a form in which the coefficients are given in terms of the combinations of mc^2 and $\hbar c$, which can be expressed in units of eV and eV cm, respectively.

Modeling of the Atomic Potential
in Hot Plasmas

2.1 General Properties of the Models

To get a full picture of the interactions of an ion immersed in a hot plasma with all the other plasma particles, one should in principle solve 10^{23} coupled Schrödinger equations with 10^{23} unknown wavefunctions. This is certainly beyond the capacity of present-day computers, so one needs the help of various models, which approximate the plasma influence on the ionic potential.

Although the full solution of a large scale problem is impossible, one attempt was made by S. Younger and colleagues (Younger et al., 1988, 1989) to find such a solution on a smaller scale. For this purpose they developed a self-consistent-field molecular dynamic type computer code, which simulates the evolution in time of a system of 30 ground state neutral helium atoms at densities between 0.1 and 1.5 g/cm^3 (1.5×10^{22}–2.25×10^{23} atoms/cm^3) and temperatures between 1 and 5 eV. Their code takes into account the motion of the atoms due to the inter-atomic forces, using the Hellmann–Feynman theorem (Younger et al., 1988). For each configuration of the atoms the electronic wavefunctions were calculated by means of a self-consistent-field method. The central conclusion from their studies is that many-atom screening effects become increasingly important in higher density plasmas.

In their studies Younger and colleagues have identified four regimes for the electronic behavior with density (see figure 2.1). At low densities the atoms are far apart compared to their mean radii and they interact relatively weakly. At this limit the influence of the plasma environment on the atomic parameters is small, except perhaps for the outermost excited states, which are, however, only seldom populated. As the density is increased neighboring atomic potentials overlap, resulting in screening of the atomic potential by the free electrons. At still higher densities, neighboring ions share the outermost electronic charge density by covalent bonding. Inner electrons are localized within the potential well of a single ion. In this quasimolecular regime several atomic potentials combine to form a potential deep and wide enough to tightly bind the electrons. At higher densities, the wavefunctions of electrons in lower quantum states overlap an increasing number

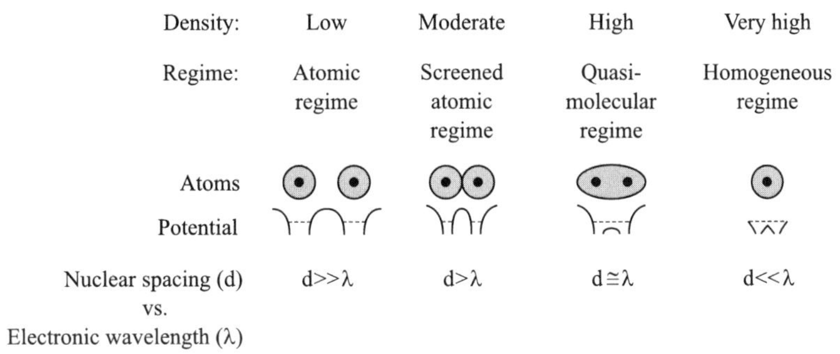

Density:	Low	Moderate	High	Very high
Regime:	Atomic regime	Screened atomic regime	Quasi-molecular regime	Homogeneous regime
Atoms				
Potential				
Nuclear spacing (d) vs. Electronic wavelength (λ)	$d \gg \lambda$	$d > \lambda$	$d \cong \lambda$	$d \ll \lambda$

Figure 2.1 The basic models of the ionic potential. (From Younger, 1988.)

of adjacent ions. At very high densities, the potential wells are too closely spaced to support bound states, and a homogeneous negative-energy electron gas is formed.

In spite of its limitations, this work is very important in giving some general ideas and directions to the basic physical picture that should underlie the various atomic models that aim to describe the physics of atoms in hot plasmas. Although the inclusion of more atoms or the extension of the model to charged or excited ions at higher densities or temperatures seems to be too difficult, it gives very important hints as to the correct picture for modeling ionic potential at other density and temperature regimes as well.

All the models of the ionic potential in hot plasmas share some common concepts and terminology. The most important is the screening of the nuclear electrostatic potential by the free and the bound electrons. The quantity that measures the deviation of the screened potential from the Coulombic one is the *screening factor*, $S(r)$, defined by

$$V(r) = \frac{Ze}{r} S(r) \qquad (2.1.1)$$

where Z is the charge of the nucleus. In other words, the screening factor is the ratio between the combined nuclear + electronic potential and the bare nuclear Coulomb potential. The main contribution to the total potential near the nucleus comes from the nuclear Coulomb one; therefore one expects that for small r: $S(r \to 0) = 1$. On the other hand, at large distances the cloud of the free and bound electrons completely screens the nuclear potential, so that far from the nucleus $S(r \to \infty) = 0$, which means that the total potential diminishes more rapidly than the Coulomb one. This statement also means that in fact, only the location and the structure of the few neighboring ions have appreciable influence on a given ion. The effects of ions farther away is compensated by the influence of the electron background in their vicinity.

The differences between the various models of the ionic potential are focused on several questions. The first is the accuracy to which the model treats the bound electrons. Are they treated with the full apparatus of quantum mechanics or only by means of statistical methods? Second, what statistical method is used to describe the influence of the free electrons: a Boltzmann–Maxwell, or Fermi–Dirac, or a full quantum mechanical treatment? Finally, to what distance are the neighboring ions treated as separate objects whose interactions with the given ion should be accounted for in detail and from what distance on can the potential be assumed to behave as statistical continuous background.

We now introduce the models used in the various temperature and density regimes.

2.2 The Debye–Hückel Theory

Initially, the Debye–Hückel (DH) theory was devised to account for the polarization of polar molecules of the solvent around the ions of dissociated molecules in a solution. Only later was it recognized that it is also suitable to model the local electric potential in plasmas. Although it has a limited range of validity, the DH model is a very good example to illustrate the ingredients of modeling the ionic potential in hot plasmas.

Assume that a structureless pointlike ion with nuclear charge ζ_0 is located at $r = 0$. The first equation that has to be satisfied by the electric potential around this ion is, of course, the Poisson equation,

$$\nabla^2 V(\vec{r}) = -4\pi e \left(\sum_{\zeta=0}^{Z} \zeta N_\zeta(\vec{r}) - n_e(\vec{r}) \right) \tag{2.2.1}$$

where $N_\zeta(\vec{r})$ is the density of ions having charge ζ, and $n_e(\vec{r})$ is the electron density. The first term in equation (2.2.1) describes the positive and the second the negative charge density. If the spatial distributions of all the ionic charge states and the electrons are known, then from this equation one can solve for the electric potential $V(\vec{r})$.

A second set of equations, which form the basis of the DH theory, is the Boltzmann statistical distribution for the ions and electrons,

$$N_\zeta(\vec{r}) = N_\zeta \exp\left(-\frac{e\zeta V(\vec{r})}{T}\right)$$

$$n_e(\vec{r}) = n_e \exp\left(+\frac{eV(\vec{r})}{T}\right) \tag{2.2.2}$$

Here N_ζ and n_e, without explicit arguments, are the *average densities* over the whole plasma. This set of equations indicates that by knowing the electric potential one can find the local charge state and the electron distributions at every point inside the plasma.

In principle, one can insert equation (2.2.2) into equation (2.2.1) to obtain one differential equation from which the potential can be inferred. This, however, yields a highly nonlinear equation. A simple approximate solution can be obtained when

$$\frac{eV(\vec{r})}{T} \ll 1 \tag{2.2.3}$$

A full explanation of the physical meaning of this assumption will be the subject of the next section. When equation (2.2.3) holds true, the exponentials in equation (2.2.2) can be expanded into power series. Keeping first order terms only, equation (2.2.1) can be linearized to the following rather simple form,

$$\nabla^2 V(\vec{r}) = 4\pi e \left[\sum_{\zeta-0}^{Z} \zeta N_\zeta \left(1 - \frac{e\zeta V(\vec{r})}{T} \right) - n_e \left(1 + \frac{eV(\vec{r})}{T} \right) \right]$$

$$= -4\pi \frac{e^2 V(\vec{r})}{T} \left[-\sum_{\zeta=0}^{Z} \zeta^2 N_\zeta - n_e \right]$$

$$= \frac{4\pi e^2 (\overline{Z^2} + \bar{Z})}{T} n_i V(\vec{r}) = \frac{1}{D^2} V(\vec{r}) \tag{2.2.4}$$

where

$$D = \sqrt{\frac{T}{4\pi e^2 n_i (\overline{Z^2} + \bar{Z})}} \tag{2.2.5}$$

is the *Debye screening length* and the notation $\overline{Z^2} = \sum \zeta^2 N_\zeta / n_i$ is used. Equation (1.2.4) was used in the transition from the second to the third equation of (2.2.4).

If the potential has spherical symmetry, the derivatives with respect to θ and ϕ drop out, and the Poisson operator reduces to

$$\nabla^2 V(r) = \frac{1}{r^2} \frac{\partial}{\partial r} \left(r^2 \frac{\partial V}{\partial r} \right) \tag{2.2.6}$$

Further, replacing $V(r)$ by the screening factor $S(r) = rV(r)/Ze$ reduces equation (2.2.4) to an even simpler form,

$$\frac{d^2 S}{dr^2} = \frac{1}{D^2} S(r) \tag{2.2.7}$$

whose solution, which has the correct behavior at $r = 0$ and $r \to \infty$, is

$$S(r) = e^{-r/D} \tag{2.2.8}$$

Inserting this result into equation (2.2.1) yields the *Debye–Hückel potential*,

$$V(r) = \frac{Ze}{r} e^{-r/D} \tag{2.2.9}$$

The DH model predicts an exponentially decreasing screening factor, $S(r) = \exp(-r/D)$, with a characteristic screening distance which equals the Debye screening length. The *Debye sphere* is defined as the sphere around the central ion whose radius is D. The influence of the central ion extends out only to ions that are included in the Debye sphere, and, conversely, this ion is influenced only by ions inside this sphere. The number of ions, M, inside the Debye sphere is

$$M = \frac{4\pi}{3} n_i D^3 = \left(\frac{D}{R_i}\right)^3 \tag{2.2.10}$$

where R_i is the ion sphere radius, equation (1.2.7).

The Debye radius decreases as the ion density goes up. At some high enough density, the Debye radius may become smaller than the ion sphere radius. Obviously, at such high densities the DH theory is no longer valid, because the Debye sphere contains, on the average, less than one ion and the statistical treatment of equation (2.2.2) cannot be justified.

The DH theory is also not valid at too small distances, when

$$r \leq \frac{Ze^2}{T} \tag{2.2.11}$$

because then equation (2.2.3) becomes incorrect. It is shown in chapter 5 that, as a rule of thumb, the ionization and recombination processes in a plasma of temperature T come to a steady state when the ionization energy of the outermost bound electron of the most abundant charge state is about 2 to 5 times larger than the temperature. The validity condition (2.2.3) is incorrect, therefore, at distances for which $r \leq 5Ze^2/\chi_\zeta$ where χ_ζ is the ionization energy of the most abundant charge state. But, $\chi_\zeta \approx Ze^2/2R^*$, where R^* is the average radius of the outermost ionic state. It turns out then that the DH theory is valid only for $r \geq 10R^*$, that is, at most to the outer peripheries of the ion but not inside the ionic volume.

It follows that the basic assumption of the Debye–Hückel theory is valid in low density high temperature plasmas where the average interionic distance is large and the interaction of a given ion with the other plasma particles is relatively small. Under these conditions, the internal electronic structure of the ion does not influence the ion–ion or the ion–electron interactions, and the ion can be treated as a pointlike structureless object.

The DH potential, equation (2.2.9), can be substituted back into equations (2.2.2) to obtain spatial ion and electron distributions. These equations show that the electrons are strongly polarized around the central ion, their density increasing as $\exp(Ze^2/rT)$ when r goes to zero. Such a distribution is highly unreasonable because it predicts that the number of electrons in any small sphere around the nucleus is infinite. But, as we have already seen, one should not really expect the theory to be true in the ions' interiors.

On the other hand, the positively charged ion species are rejected from the central ion, their concentration diminishing as $\exp(-\zeta_0\zeta e^2/rT)$ for small r. Both the polarization of the electrons and the rejection of the ions vanish rapidly beyond distances of the order of Z^2e^2/T and the distributions tend to their average values.

To summarize, the DH model consists of two ingredients. The first is the Poisson equation which reflects the electrostatic nature of the plasma particles interactions. The second is the statistical distribution of the electrons and ions in the electric potential generated by all the plasma particles. In the case of the DH model, the Boltzmann statistical distribution was used. The use of this statistics limits the validity of the DH model to relatively low density plasmas. In the following, we will see that all the other models also use the same two ingredients, but use different statistical models for the electron and ion densities, thereby moving their validity to other domains in the density–temperature plane.

2.3 The Plasma Coupling Constant

The DH theory can be applied only as long as condition (2.2.3) is correct, namely, at high temperature and low ion–ion interactions. In such a plasma the ions and electrons are moving almost freely in space and the motion of one of these particles affects only weakly the motion of nearby ones. This is a *weakly interacting* or *uncorrelated* or *weakly coupled* plasma—all three of these terminologies can be found in the literature. Complying with recent trends, in the following we shall use the third of these expressions.

At the other extreme, *strongly coupled plasma*, the density is high, temperature is low, and the mutual ion–ion interactions are strong. In this limit, the shift of one ion from its position immediately influences, through their electrostatic interactions, the motion of nearby particles, in much the same way as in a fluid the motions of the atoms are correlated.

The parameter that reflects the above behavior is the *plasma coupling constant* (Ichimaru, 1982), defined by

$$\Gamma_{ii} = \frac{\bar{Z}^2 e^2}{R_i T} = \left(\frac{4\pi n_i}{3}\right)^{1/3} \frac{\bar{Z}^2 e^2}{T} \qquad (2.3.1)$$

It is proportional to the ratio between the average potential energy, $\bar{Z}^2 e^2 / R_i$, of two ions that are at the average distance R_i from each other and the average kinetic energy, which is proportional to T. The quantity in equation (2.3.1) is also referred to as the *ion–ion coupling constant*. A similar definition holds for the *electron–ion coupling constant*,

$$\Gamma_{ei} = \frac{\bar{Z} e^2}{R_i T} \qquad (2.3.2)$$

This last quantity and a similar *electron–electron coupling constant* are, however, only seldom used in the literature, and in the following by the term "coupling constant," Γ, we shall mean the ion–ion one (equation 2.3.1).

Regarding the coupling between the plasma particles, plasmas can be divided into three general regions: weakly coupled plasmas for which $\Gamma \leq 0.1$, strongly coupled plasmas for which $\Gamma \geq 10$, and an intermediate region in which $0.1 \leq \Gamma \leq 10$. In figure 2.2 the lines of constant Γ are plotted for an aluminum plasma on a density–temperature plane with some indications of regions of experimental interest. In figure 2.3 the same plot is shown for a hydrogen plasma. As $\Gamma \sim n_i^{1/3}$, the intermediate region, which spans 2 orders of magnitude difference in Γ, corresponds to 6 orders of magnitude difference in the ion densities, exactly around the domain where present-day laser–plasma experiments are carried out.

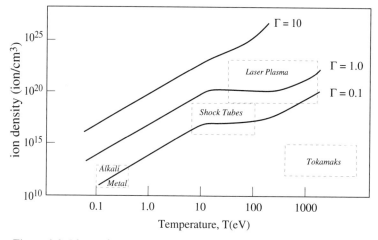

Figure 2.2 Lines of constant Γ_{ii} for an aluminum plasma.

It is interesting to note that the model of the one-component plasma (see section 2.7) predicts that at $\Gamma \geq 172$, which corresponds to very strong correlations, the ions of a hydrogen plasma settle down in a body-centered cubic structure (Ichimaru, 1982), thereby producing a solid state plasma. Such a lattice has a lower free energy than a plasma in which the ions and the electrons move randomly. This may happen either at very low temperatures or at extremely high densities. For a plasma at room temperature ($T = 300\,\mathrm{K} = 0.026\,\mathrm{eV}$), with $\bar{Z} = 1$, this should occur around $n_i \sim 7 \times 10^{21}$ ions/cm³. In contrast, a hot fully ionized

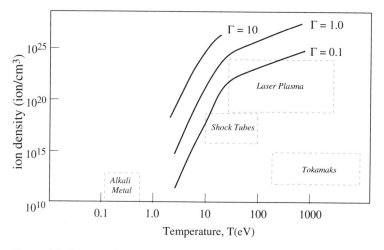

Figure 2.3 Same as figure 2.2, for a hydrogen plasma.

hydrogen plasma of $T = 100\,\text{eV}$ becomes a solid state plasma only at the tremendous ion density of about $4 \times 10^{32}\,\text{ions/cm}^3$.

In terms of the plasma coupling constant, one can reformulate the validity domain of the DH model (equation 2.2.3) by the condition $\Gamma \ll 1$. For higher Γ's one needs different methods of modeling.

2.4 The Thomas–Fermi Statistical Model

The Basic Equations of the Thomas–Fermi Model

The Thomas–Fermi (TF) model was initially developed to study the potential of high-Z neutral atoms, in which, due to their large number of electrons, a statistical approach can be a plausible approximation. Only much later was it adapted to strongly coupled high density plasmas where the number of electrons within the ion sphere is relatively large. The earliest versions of the TF model, which were developed mainly for the zero temperature case, are presented in most of the standard textbooks. The model is widely used in solid state physics, astrophysics, and equation-of-state computations. It is particularly useful for getting information about the behavior of matter under extreme conditions of high temperature and high density that are not amenable to experiment. We introduce the model through its finite temperature version, which is the suitable form for high temperature plasmas.

A large body of literature exists on the TF model. An excellent review is given in an article by Gombas (1956), which summarizes the earlier works for atoms and ions. A comprehensive description of the model for the purposes of equation-of-state, pressure, and similar thermodynamic quantities is given in the book by Eliezer, Ghatak and Hora (1986). Spruch (1991) wrote "Pedagogical notes on TF theory" with applications for atoms, astrophysics, and stability of matter.

The TF model is widely used to simulate the ionic potential in hot plasmas, particularly those of high-Z material. It is relatively simple for computations, and provides reasonably accurate results that were tested experimentally for some thermodynamic quantities of high-Z materials, such as equation-of-state and pressure. The basic TF equation has undergone a gamut of corrections to improve its predictions, some of which will be discussed below.

The problem solved by the TF theory is of a nucleus of charge Z positioned at $r = 0$, and Z electrons (bound and free) confined to the ion sphere. Together they generate charge neutrality within this sphere. The total potential, nuclear + electronic, is zero on and beyond the boundaries of the ion sphere. Implicitly this model assumes that the ion–ion correlations reject other ions to beyond the boundaries of this sphere. It is also assumed implicitly that beyond the ion sphere there is a continuous background of electrons that exactly neutralize, at every point, the positive ion charge spatial distribution.

The TF model, similarly to the Debye–Hückel model, has two ingredients. The first is the Poisson equation,

$$\nabla^2 V(\vec{r}) = -4\pi e[Z\delta(\vec{r}) - n_e(\vec{r})] \tag{2.4.1}$$

This potential can be divided to a nuclear part,

$$V_N(r) = \frac{Ze}{r} \qquad (2.4.2)$$

and an electronic part, $V_e(\vec{r})$, which satisfies

$$\nabla^2 V_e(\vec{r}) = 4\pi e n_e(\vec{r}) \qquad (2.4.3)$$

The solution of this last equation is readily written as

$$V_e(\vec{r}) = -\int_V \frac{e n_e(\vec{r})}{|\vec{r} - \vec{r}'|} d^3 r' \qquad (2.4.4)$$

where the integration is over the volume of the ion sphere.

The TF model assumes spherical symmetry around the nucleus. In this case the electron density is isotropic, $n_e(\vec{r}) = n_e(r)$, and equation (2.4.4) is reduced to a simpler form in the following way: first, the denominator is expanded into a series of Legendre polynomials (Abramowitz and Stegun, 1965)

$$|\vec{r} - \vec{r}'|^{-1} = \sum_{k=0}^{\infty} \frac{r_<^k}{r_>^{k+1}} P_k(\cos\theta) \qquad (2.4.5)$$

where $r_<$ is the smaller and $r_>$ is the larger of r and r'. Inserting this expansion into equation (2.4.4), one obtains

$$V_e(r) = -e \sum_{k=0}^{\infty} \int_0^{\infty} r'^2 dr' \int_0^{\pi} \sin\theta\, d\theta \int_0^{2\pi} d\phi\, n_e(r') \frac{r_<^k}{r_>^{k+1}} P_k(\cos\theta) \qquad (2.4.6)$$

The integrations over ϕ trivially gives 2π. The integration over θ can be carried out using the orthogonality conditions of the Legendre polynomials,

$$\int_0^{\pi} d\theta \sin\theta\, P_k(\cos\theta) = \int_{-1}^{1} dx\, P_k(x) = \int_{-1}^{1} dx\, P_k(x) P_0(x)$$

$$= \frac{2}{2k+1} \delta_{k,0} = 2\delta_{k,0} \qquad (2.4.7)$$

where we have used the definition $P_0(x) = 1$. Substituting this result into the integrals, one gets

$$V_e(r) = -2\pi e \sum_{k=0}^{\infty} 2\delta_{k,0} \int_0^{\infty} r'^2 dr'\, n_e(r') \frac{r_<^k}{r_>^{k+1}} = -4\pi e \int_0^{\infty} n_e(r') \frac{r'^2 dr'}{r_>}$$

$$= -4\pi e \left(\frac{1}{r} \int_0^r n_e(r') r'^2 dr' + \int_r^{R_i} n_e(r') r'\, dr' \right) \qquad (2.4.8)$$

and the total potential is

$$V(r) = V_N(r) + V_e(r) \qquad (2.4.9)$$

On the boundaries of the ion sphere, $r = R_i$, the second integral in equation (2.4.8) vanishes, while 4π times the first integral gives exactly the number of electrons inside the ion sphere, which is Z. It turns out that $V_e(R_i) = -Ze/R_i$, is equal in value and opposite in sign to $V_N(R_i) = Ze/R_i$, so that the total potential vanishes on the ion sphere boundaries,

$$V(R_i) = 0 \qquad (2.4.10)$$

The second set of equations of the TF model, similarly to the DH model, are the equations of statistical distributions. These equations relate the electron and ion distributions to the local potential. In the high density high Γ domain at which the TF model is aimed, the Pauli exclusion principle and the electron degeneracy effects are important and therefore the Fermi–Dirac statistics is appropriate.

The Fermi–Dirac electron momentum distribution in a high electron temperature plasma, taking into account the presence of a local electric microfield, is obtained from equation (1.3.17),

$$f_e(r, p)\, dp = \frac{1}{\pi^2 \hbar^3}\, \frac{p^2\, dp}{1 + \exp\{[p^2/2m - eV(r) - \mu]/T_e\}} \qquad (2.4.11)$$

We recall that T_e is in energy units, in other words, we write simply T_e instead of $k_B T_e$, (k_B is the Boltzmann constant). The local electron density is obtained by integrating this distribution over the momenta. Denoting $y = [eV(r) + \mu]/T_e$, and changing the variable of the integration to $x = p^2/2mT_e$, one gets for the electron density,

$$n_e(r) = \int dp\, f_e(r, p) = \frac{(2mT_e)^{3/2}}{2\pi^2 \hbar^3} \int_0^\infty \frac{x^{1/2}\, dx}{1 + \exp(x - y)}$$

$$= \frac{1}{2\pi^2} \left(\frac{2mc^2 T_e}{(\hbar c)^2} \right)^{3/2} F_{1/2}\left(\frac{\mu + eV(r)}{T_e} \right) \qquad (2.4.12)$$

where $F_{1/2}$ is the Fermi–Dirac integral (1.3.20). The Fermi energy, μ, is determined from the charge neutrality requirement,

$$Z = \int_0^{R_i} n_e(r, \mu)\, d^3 r \qquad (2.4.13)$$

Equations (2.4.2), (2.4.8–9), and (2.4.12–13) form the basic formulas of the TF model. These equations are generally solved using an iterative procedure. The results of such a computation are the local electrostatic potential, $V(r)$, the electron spatial distribution, $n_e(r)$, and the Fermi energy, μ.

The Fermi–Dirac integral appears in a variety of fields in physics, such as thermionic emission, theory of metals and semiconductors, and many others. Great effort was invested to obtain reliable, high accuracy approximation formulas for the Fermi–Dirac integral that are appropriate for use in computer codes. High accuracy ($\epsilon \sim 10^{-10}$), easily computable formulas were developed by Fernandez-Velicia (1986) for positive arguments, $x > 0$, and any j, and in particular for $j = 1/2$ and $3/2$. The results are too lengthy to be reproduced here, and the interested reader is encouraged to go back to the original paper. Interpolation formulas, valid also for $x \leq 0$, but having lower accuracy ($\epsilon \sim 10^{-5}$) can be found in a paper by Arpigny (1963). It should be mentioned that in mathematics-oriented papers the Fermi–Dirac integral is defined slightly differently, by dividing equation (1.3.20) by $\Gamma(j+1)$, the Gamma function with argument $j+1$.

The inverse problem also received much attention: Given the Fermi–Dirac integral, $F_{1/2}(\mu)$, what is the Fermi energy μ? This problem has importance in the analysis of experimental results, because the electron density, being proportional to $F_{1/2}(\mu)$, is the parameter that is measurable in experiments. Aguilera-Navarro et al. (1988) developed a rather simple algorithm that gives μ as function of $F_{1/2}$. Their algorithm is accurate to within 0.6% for $-\infty < \mu \le 170$, and 2.5% for $\mu = 260$. However, their approximation formula does not have the correct asymptotic behavior, $\mu \sim [F_{1/2}]^{2/3}$, for $[F_{1/2}] \to \infty$. Chang and Izabelle (1989) developed a different approximation formula, which has the correct behavior at both limits. They claim an accuracy of better than 10^{-4}.

The TF model does not distinguish between bound and free electrons and treats them together self-consistently. Nevertheless, free electrons are characterized by having positive energy,

$$\frac{p^2}{2m} - eV(r) \ge 0 \qquad (2.4.14)$$

and their partial density, $n_{e,f}$, is calculated by

$$n_{e,f}(r) = \frac{1}{2\pi^2}\left(\frac{2mc^2 T_e}{(\hbar c)^2}\right)^{3/2} F_{1/2}\left(\frac{\mu + eV(r)}{T_e}; \left|\frac{eV(r)}{T_e}\right|\right) \qquad (2.4.15)$$

where $F_{1/2}(x; \beta)$ is the *incomplete Fermi–Dirac integral*,

$$F_j(x; \beta) = \int_\beta^\infty \frac{y^j\, dy}{1 + \exp(y - x)} \qquad (2.4.16)$$

In view of equation (2.4.16), the integration in equation (2.4.15) is carried out only over the positive-energy free electrons. The bound electron partial density is obtained as the complement of the above result,

$$n_{e,b}(r) = \frac{1}{2\pi^2}\left(\frac{2mc^2 T_e}{(\hbar c)^2}\right)^{3/2}\left[F_{1/2}\left(\frac{\mu + eV(r)}{T_e}\right) - F_{1/2}\left(\frac{\mu + eV(r)}{T_e}; \left|\frac{eV(r)}{T_e}\right|\right)\right] \qquad (2.4.17)$$

To get some understanding about the predictions of the TF model we will investigate the electron distribution under several extreme conditions. First, the case of $\mu \to -\infty$. As we have seen in chapter 1, in this case the Fermi–Dirac distribution approaches the Boltzmann distribution, and the two basic equations of the TF model become, in fact, the two basic equations of the DH model. The boundary conditions are, however, not the same. In the TF model the ionic potential vanishes on the ion sphere boundaries, whereas in the DH model there is no limitation of this kind.

Second, near the nucleus the potential is approximately Ze^2/r. At very small r, this term can be much larger than μ and the argument of the Fermi–Dirac function becomes $\sim Ze^2/(rT) \to \infty$. To get the asymptotic behavior of the Fermi–Dirac integral for large argument, we divide the integral in equation (1.3.20) into two parts, (i) $y - x \geq 0$, in which case $\exp(y - x) > 1$, and (ii) $y - x < 0$, in which case $\exp(y - x) \leq 1$. Accordingly, one obtains

$$F_j(x) = \int_0^\infty \frac{y^j \, dy}{1 + \exp(y - x)} = \left(\int_0^x + \int_x^\infty \right) \frac{y^j \, dy}{1 + \exp(y - x)} = J_1 + J_2 \qquad (2.4.18)$$

In J_1 the exponential factor in the denominator is small relative to 1, so that this integral is approximated by

$$J_1 = \int_0^x \frac{y^j \, dy}{1 + \exp(y - x)} \approx \int_0^x y^j \, dy = \frac{x^{j+1}}{j + 1} \qquad (2.4.19)$$

Regarding J_2, in this case one can neglect 1 relative to the exponential factor,

$$J_2 = \int_x^\infty \frac{y^j \, dy}{1 + \exp(y - x)} \approx e^x \int_x^\infty e^{-y} y^j \, dy = e^x \Gamma(j + 1; x) \qquad (2.4.20)$$

where $\Gamma(j + 1; x)$ is the *incomplete Gamma function*. For large x it can be shown that this function behaves as $e^{-x} x^j$, see equation 6.5.32 in Abramowitz and Stegun (1965), so that for large x, $J_2 \sim x^j$, is much smaller than J_1. Substituting these approximate values into equation (2.4.12), one obtains for small r,

$$n_e(r) \approx \frac{1}{2\pi^2} \left(\frac{2mc^2 T_e}{(\hbar c)^2} \right)^{3/2} \frac{2}{3} \left(\frac{Ze^2}{rT_e} \right)^{3/2} \propto r^{-3/2} \qquad (2.4.21)$$

i.e., the electrons are strongly polarized around the nucleus, their density increasing to infinity proportionally to $r^{-3/2}$. The number of electrons in any small sphere of radius r is, however, finite, decreasing to zero as the sphere radius r constricts to zero. An infinite electron density in the vicinity of the nucleus is in contradiction to the quantum mechanical estimates, which predict a finite electron density near the nucleus.

The average degree of ionization, $\bar{Z}$, which is the number of free electrons per ion within the ion sphere, is given by

$$\bar{Z}(T_e, n_i) = \int_V n_{e,f}(r, \mu) \, d^3 r \qquad (2.4.22)$$

It will be shown below that $\bar{Z}(T_e, n_i)$ has a useful scaling property,

$$\bar{Z}(T_e, n_i) = Z \, g\left(\frac{T_e}{Z^{4/3}}; \frac{n_i}{Z} \right) \qquad (2.4.23)$$

where g is a universal function common to all ions.

R. More has developed an algorithm that approximates the TF values of $\bar{Z}$ for all materials to an accuracy of a few percent; see, for example More (1981). The algorithm is given in table 2.1. It gives remarkably good estimate for the average Z as computed within the framework of the TF theory, and can be used to get an estimate of the expected values of $\bar{Z}$ for any given electron temperature and ion density.

Table 2.1 The average ionization according to the Thomas–Fermi model

---

ρ = mass density in g/cm^3
T = temperature in eV
Z = atomic number
$\mathcal{A}$ = atomic mass

Let

$$R = \rho/(Z\mathcal{A}), \qquad T_0 = T/Z^{4/3}$$
$$\alpha = 14.3139, \qquad \beta = 0.6624$$
$$f(x) = x/(1 + x + \sqrt{1 + 2x})$$

Then

(A) **T = 0**

$$\frac{\partial \bar{Z}}{\partial \rho} = \frac{\bar{Z}}{\rho}\left(\frac{\beta}{\sqrt{1+2x}}\right), \qquad x = \alpha R^{\beta}$$
$$\bar{Z} = f(x)Z$$

(B) **T > 0**

$$T_F = \frac{T_0}{1 + T_0}$$

$$A = a_1 T_0^{a_2} + a_3 T_0^{a_4}$$

$$a_1 = 3.323 \times 10^{-3}, \qquad a_2 = 0.971\,832, \qquad a_3 = 9.26\,148 \times 10^{-5},$$

$$a_4 = 3.101\,65$$

$$B = -\exp(b_0 + b_1 T_F + b_2 T_F^7)$$

$$b_0 = -1.7630, \qquad b_1 = 1.431\,75, \qquad b_2 = 0.315\,463$$

$$C = c_1 T_F + c_2$$

$$c_1 = -0.366\,667, \qquad c_2 = 0.983\,333$$

$$Q_1 = AR^B, \qquad Q = (R^C + Q_1^C)^{1/C}, \qquad x = \alpha Q^{\beta} \quad (\alpha, \beta \text{ see above})$$

$$\bar{Z} = f(x)Z \quad (f(x) \text{ see above})$$

Before discussing further the various properties of the TF model, a short summary is in order.

- The TF model treats the bound and free electrons statistically in a self-consistent potential generated by both of them. The Pauli exclusion is taken into account through the Fermi–Dirac statistics.
- The model accounts for the plasma correlation effects by confining the ion, together with Z bound + free electrons, in an ion sphere enclosure, assuming that no other ion can penetrate into this sphere.
- The TF model assumes that the plasma consists of one kind of ions only, which represents the average ionic species. Charge state distribution, as well as the distribution of the electrons among the various excited states must be computed by other means.
- The TF theory does not account for atomic shell effects, energy quantization and other quantum mechanical details of the electronic configurations.

These should be calculated by other theories, which may use the TF total potential as their starting point.

- The electrons are strongly polarized around the nucleus. For $r \to 0$, $n_e(r)$ is divergent,

$$\lim_{r \to 0} n_e(r) \propto r^{-3/2}$$

see equation (2.4.21). Nevertheless, the number of electrons near the nucleus is always finite. The divergence of $n_e(r)$ near the nucleus does not agree with quantum mechanical calculations, which predict finite results.

Solutions of the Thomas–Fermi Model

There is no analytical solution for the TF equations. One can get some important information, however, even without getting a full solution.

We first define the *Thomas–Fermi radius*, R_{TF}, as

$$R_{TF} = \frac{1}{2} \left(\frac{3\pi}{4} \right)^{2/3} \frac{a_0}{Z^{1/3}} = 0.885\,341 \ldots \frac{a_0}{Z^{1/3}} \tag{2.4.24}$$

The significance of this definition will become clear when we speak about the TF model at $T \to 0$. Using R_{TF} one can rewrite the second term in the argument of the Fermi–Dirac integral, equation (2.4.12), as

$$\frac{eV(r)}{T} = \frac{Ze^2}{Tr} S(r) = \frac{Ze^2}{R_{TF} T} \frac{S(x)}{x} = \alpha \frac{S(x)}{x} \tag{2.4.25}$$

where $x = r/R_{TF}$, $S(r)$ is the screening factor (equation 2.1.1), and

$$\alpha = \frac{Ze^2}{R_{TF} T} = 2 \left(\frac{4}{3\pi} \right)^{2/3} \frac{e^2}{a_0} \frac{Z^{4/3}}{T} \tag{2.4.26}$$

In equations (2.4.24–26) we have assumed that the electron temperature equals the ion temperature, and their common value is denoted by T. The Poisson equation (2.4.3) can also be expressed in terms of R_{TF}. In spherical coordinates one obtains

$$\nabla^2(eV(r)) = \nabla^2 \left(\frac{Ze^2}{r} S(r) \right) = \frac{Ze^2}{r} \frac{d^2 S(r)}{dr^2} = \frac{Ze^2}{R_{TF}^3} \frac{1}{x} \frac{d^2 S(x)}{dx^2} \tag{2.4.27}$$

Substituting these results into equations (2.4.3), one gets

$$\frac{Ze^2}{R_{TF}^3} \frac{1}{x} \frac{d^2 S(x)}{dx^2} = 4\pi e^2 n_e(r) = \frac{4\pi e^2}{2\pi^2} \left(\frac{2mc^2 T}{(\hbar c)^2} \right)^{3/2} F_{1/2} \left(\frac{\mu}{T} + \alpha \frac{S(x)}{x} \right) \tag{2.4.28}$$

After reordering the various terms,

$$\frac{d^2 S(x)}{dx^2} = \frac{9\pi}{64} \left(\frac{T}{Z^{4/3} E_H} \right)^{3/2} x F_{1/2} \left(\frac{\mu}{T} + \alpha \frac{S(x)}{x} \right) \tag{2.4.29}$$

This is an equation in the screening factor $S(x)$. The initial conditions corresponding to this equation are the following: First, close to the nucleus,

$$S(0) = 1 \tag{2.4.30}$$

see the discussion following equation (2.1.1). Second, due to the charge neutrality
inside the ion sphere the electrostatic field on the ion sphere boundaries is zero,

$$e \frac{dV(r)}{dr}\bigg|_{r=R_i} = \frac{Ze^2}{R_{TF}^2} \frac{d}{dx}\left(\frac{S(x)}{x}\right)\bigg|_{x=X_i} = 0 \tag{2.4.31}$$

or

$$X_i S'(X_i) - S(X_i) = 0 \tag{2.4.32}$$

where $X_i = R_i/R_{TF}$ is the ion sphere radius in units of the TF radius.

Although we are still far from getting an analytical solution for the TF equation, rewriting the equations in the form of equation (2.4.29) can already give
some insight into its results. First, one can see that both the coefficient on the right
hand side and α, in the argument of the Fermi–Dirac integral, depend on the
temperature only through the combination $Z^{4/3}/T$. This means that the screening
factor and the potential, as well as all the derived quantities, such as the free and
bound electron densities, $\bar{Z}$, pressure and so on, are all dependent within the
framework of the Thomas–Fermi theory only on this combination. This is a
very important scaling property of the TF theory, which we have already mentioned in conjunction with equation (2.4.23).

Second, the ion density affects the equations through the boundary conditions,
where it appears in the combination $X_i = R_i/R_{TF} \propto (n_i/Z)^{1/3}$. This provides a
scaling law of the TF model with respect to the ion density. This scaling law was
also mentioned in equation (2.4.23).

A different way to write the TF equations is to denote the argument of the
Fermi–Dirac integral by

$$\Psi(\xi) = \frac{\mu + eV(r)}{T} \zeta \tag{2.4.33}$$

where $\xi = r/R_i$. The advantage of this parametrization will be seen shortly. In
terms of Ψ, the Laplacian of the potential is rewritten as

$$\nabla^2[eV(r)] = \nabla^2\left(T\frac{\Psi(\xi)}{\xi} - \mu\right) = T\frac{1}{r^2}\frac{d}{dr}\left[r^2\frac{d}{dr}\left(\frac{\Psi(\xi)}{\xi}\right)\right]$$

$$= \frac{T}{R_i^2}\frac{1}{\xi^2}\frac{d}{d\xi}(\xi\Psi'(\xi) - \Psi(\xi)) = \frac{T}{R_i^2}\frac{\Psi''(\xi)}{\xi} \tag{2.4.34}$$

Substituting back into equation (2.4.1) and using equation (2.4.12), one obtains a
second order nonlinear differential equation for $\Psi(x)$ (Latter, 1955) which, however, has a simple structure,

$$\Psi''(\xi) = A\xi F_{1/2}\left(\frac{\Psi(\xi)}{\xi}\right) \tag{2.4.35}$$

A is a dimensionless constant,

$$A = \frac{R_i^2}{T} \frac{4\pi e^2}{2\pi^2} \left(\frac{2mc^2 T}{(\hbar c)^2}\right)^{3/2} = \frac{4}{\pi}\left(\frac{3}{4\pi}\right)^{2.3}\left(\frac{T}{E_H}\right)^{1/2}\frac{1}{(n_i a_0^3)^{2/3}} \tag{2.4.36}$$

The boundary conditions (2.4.30–32) are transformed into

$$\Psi(0) = \frac{Ze^2}{R_i T} = \frac{1}{Z}\Gamma, \qquad \Psi'(1) = \Psi(1) \tag{2.4.37}$$

where Γ is the plasma coupling constant, equation (2.3.1). The advantage of this formalism is that in terms of $\Psi(\xi)$ the forms of several plasma parameters appear in a relatively simple way. For example, the kinetic energy per unit volume is

$$\epsilon_{kin} = \int_0^\infty \frac{p^2}{2m} f_e(r,p)\, dp = \frac{1}{2\pi^2}\left(\frac{2mc^2 T}{(\hbar c)^2}\right)^{3/2} TF_{3/2}\left(\frac{\Psi(\xi)}{\xi}\right) \tag{2.4.38}$$

where we have used equation (2.4.11) for $f_e(p)$, and substituted $x = p^2/2mT$ to carry out the integration. The total kinetic energy within the ion sphere and the average kinetic energy per electron are

$$E_{kin} = \int_0^{R_i} 4\pi r^2\, dr\, \epsilon_{kin} = \frac{3}{2\pi^2}\frac{T}{n_i}\left(\frac{2mc^2 T}{(\hbar c)^2}\right)^{3/2}\int_0^1 \xi^2\, d\xi\, F_{3/2}\left(\frac{\Psi(\xi)}{\xi}\right) \tag{2.4.39}$$

$$\langle E_{kin}\rangle = \frac{E_{kin}}{Z} = \frac{3}{2\pi^2}\frac{T}{n_e}\left(\frac{2mc^2 T}{(\hbar c)^2}\right)^{3/2}\int_0^1 \xi^2\, d\xi\, F_{3/2}\left(\frac{\Psi(\xi)}{\xi}\right)$$

The difference between these two apparently similar formulas is the appearance of the ion density, n_i, in the denominator of the first and the electron density, n_e, in the second of these equations. When the solution of equation (2.4.35) is substituted for $\Psi(\xi)$, the integration can be carried out on a computer with relative ease. Using similar technique, one finds for the electron–nucleus potential energy,

$$E_{e-n} = -\frac{4\pi R_i^3}{2\pi^2}\left(\frac{2mc^2 T}{(\hbar c)^2}\right)^{3/2}\int_0^1 \xi^2\, d\xi\, \frac{Ze^2}{R_i\xi} F_{1/2}\left(\frac{\Psi(\xi)}{\xi}\right)$$

$$= -\frac{3}{2\pi^2}\frac{Ze^2}{R_i n_i}\left(\frac{2mc^2 T}{(\hbar c)^2}\right)^{3/2}\int_0^1 \xi\, d\xi\, F_{1/2}\left(\frac{\Psi(\xi)}{\xi}\right) \tag{2.4.40}$$

The electron–electron potential energy is carried out in a similar way, although needs slightly more algebra. The potential generated by the electrons is

$$V_e(r) = V(r) - \frac{Ze}{r} = \frac{T}{\xi}\Psi(\xi) + \mu - \frac{Ze}{r} \tag{2.4.41}$$

wherefrom one gets

$$E_{e-e} = \frac{3}{4\pi^2} \frac{1}{n_i} \left(\frac{2mc^2 T}{(\hbar c)^2}\right)^{3/2} \int_0^1 d\xi\, \xi^2 \left(\frac{T}{\xi} \Psi(\xi) + \mu - \frac{Ze}{R_i} \frac{1}{\xi}\right) F_{1/2}\left(\frac{\Psi(\xi)}{\xi}\right) \quad (2.4.42)$$

Similar formulas are obtained for other thermodynamic parameters, such as the pressure, entropy, and others, that are essential for computations of the equation of state (Eliezer et al., 1986).

Some Useful Results

The TF model can be used to obtain useful scaling laws for several atomic properties. For example, the scaling of the ionization potential of a given ion can be derived using the following rather simple reasoning: The number of bound electrons per ion is $Z_b = Z - \bar{Z}$. Assuming that these electrons fill up Bohr orbitals up to some principal quantum number n_p, their number can also be estimated by

$$Z_b = Z - \bar{Z} \cong \int_0^{n_p} 2n^2\, dn = \frac{2}{3} n_p^3 \quad (2.4.43)$$

The ionization energy, I_{TF}, of a bound electron from this last bound state into the continuum is, approximately, $I_{TF} = \bar{Z}^2 E_H/n_p^2$. Substituting n_p from equation (2.4.43) into this estimate, one gets

$$I_{TF} = \left(\frac{2}{3}\right)^{2/3} E_H \frac{\bar{Z}^2}{Z_b^{2/3}} = \left(\frac{2}{3}\right)^{2/3} E_H Z^{4/3} \frac{(\bar{Z}/Z)^2}{[1-(\bar{Z}/Z)^{2/3}]} = Z^{4/3} I_0\left(\frac{\bar{Z}}{Z}\right) \quad (2.4.44)$$

where $I_0(x)$ is the same function for all the ions. By virtue of equation (2.4.23), one finds that

$$\frac{I_{TF}}{Z^{4/3}} = I_0\left[g\left(\frac{T}{Z^{4/3}}; \frac{n_i}{Z}\right)\right] = g^*\left(\frac{T}{Z^{4/3}}; \frac{n_i}{Z}\right) \quad (2.4.45)$$

Equation (2.4.44), with all its limitations, can provide some crude estimate of the ionization potential of ionized species. Figure 2.4 gives a comparison between the predictions of equation (2.4.44) and more precise computations of the ionization potential. Moreover, equation (2.4.45) connects the ionization potential of the most abundant species of a plasma of ions with atomic number Z at a given temperature to the ionization potential of a plasma with a different Z and T. Such scaling laws may be useful in some cases.

The Thomas–Fermi Theory in Low Temperature Plasmas

Originally the TF model was devised for high-Z atoms at zero temperature. Only later was it adjusted for high temperature plasmas (Metropolis et al., 1953). We have inverted the order of the presentation, because the high temperature version of the model fits better the goals of this book. Nevertheless, we shall devote a short subsection to the low temperature version of the TF model, first to show

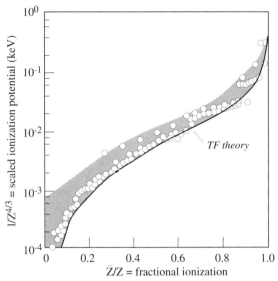

Figure 2.4 The ionization potential as calculated by equation (2.4.44) (solid line), and a more accurate Hartree–Fock computer code (circles). (From More, 1981.) (Courtesy of the University of California, Lawrence Livermore National Laboratory and the Department of Energy.)

how this model approaches the low temperature limit, and, second, because this approximation can depict the conditions in some low density plasmas as well.

Assume a very low density plasma consisting of ions that have a sufficiently large number of bound electrons that it is justified to use a statistical approach. Since the ion density is low, so is the free electron density, and one can assume that the number of free electrons within the bound electron orbitals is negligibly small. The screening and other free electron effects can therefore be neglected. Moreover, in low density plasmas the electron–ion collision rate, and therefore the excitation rate of the bound electrons, is small, so that the ions are, with rather high probability, in their ground state. It turns out that in this case a zero temperature isolated ion treatment can be a plausible approximation.

The zero-temperature TF theory is described in detail in textbooks on quantum mechanics. See the article of Gombas (1956) for an excellent overview of the theory for this case. At zero temperature the starting points of the theory are the Poisson equation (2.4.1) and the zero temperature Fermi–Dirac statistics, equations (1.3.22–25). Defining $x = r/R_{TF}$, the two equations are combined into one,

$$\frac{d^2 S(x)}{dx^2} = \frac{[S(x)]^{3/2}}{x^{1/2}} \tag{2.4.46}$$

(It should be mentioned that in the standard texts on the TF theory the screening factor is denoted by $\chi(x)$. To standardize our notations, we shall continue to use $S(x)$ for the screening factor.)

The requirements that the potential is Coulombic near the nucleus, and that the electric potential and field will both vanish on the boundaries of the ion sphere, $r = R_i$, $X_i = R_i/R_{TF}$, are expressed by the conditions

$$S(0) = 1, \qquad S(X_i) = 0, \qquad X_i S'(X_i) = -\frac{\bar{Z}}{Z} \qquad (2.4.47)$$

All isolated $(X_i = \infty)$ and neutral $(\bar{Z} = 0)$ atoms have the same screening function, which, however, still depends implicitly on Z through the dependence of x and R_{TF}. This neutral atom screening function, which will be denoted by $S_0(x)$, is tabulated in several places in the literature, see for instance Gombas (1956).

A formal solution of equation (2.4.46) for $x < 1$ can be obtained by means of a power series,

$$S_0(x) = 1 + \sum_{k=1}^{\infty} a_k x^{k/2} \qquad (2.4.48)$$

where we have already used the boundary condition at $r = 0$. This sum is inserted back into equation (2.4.46). Raising the series to the power $3/2$ is carried out by the standard methods, and the double derivation is straightforward. Equating the coefficients of equal power of x on both sides of the equation, one finds

$$a_1 = 0, \qquad a_3 = \frac{4}{3}, \qquad a_4 = 0, \qquad a_5 = \frac{2}{5} a_2, \qquad a_6 = \frac{1}{3},$$

$$a_7 = \frac{3}{70} a_2, \qquad a_8 = \frac{2}{15} a_2, \qquad \cdots \qquad (2.4.49)$$

The solution depends on one parameter, $a_2 = S'(0)$, which should be determined by the boundary conditions at $x = X_i$. The value that corresponds to a neutral isolated atom, $X_i = \infty$, is $a_2 = -1.588\,07 \ldots$.

As already mentioned, the expansion (2.4.48) is valid for $x < 1$. One may try to find a solution for large x by expanding the solution around $x \to \infty$. Direct substitution into equation (2.4.46) shows that the leading term of the expression at infinity is

$$S_0(x) \to \frac{144}{x^3}, \qquad x \to \infty \qquad (2.4.50)$$

The full expansion has the form

$$S_0(x) \to \frac{144}{x^3} \sum_{k=0}^{\infty} c_k (F x^\lambda)^k, \qquad x \to \infty \qquad (2.4.51)$$

where

$$F = 13.270\,973\,848, \qquad \lambda = -\frac{\sqrt{72} - 7}{2} = -0.772 \qquad (2.4.52)$$

and the coefficients c_k are listed in Eliezer et al. (1986) up to $k = 30$.

Formulas (2.4.48) and (2.4.51) correspond to an isolated neutral atom. The extension of these formulas to "isolated" ions was suggested by Fermi (1930),

$$S(x) = S_0(x) + k(\bar{Z})\eta(x) \tag{2.4.53}$$

where $k(\bar{Z})\eta(x)$ is a correction term in which $k(\bar{Z})$ depends only on $\bar{Z}$ and $\eta(x)$ only on x. A full treatment is described in Gombas (1956). An approximate form suggested for $k(\bar{Z})$ by Fermi and Amaldi is

$$k(\bar{Z}) = -0.083 \left(\frac{\bar{Z}}{Z}\right)^3 \tag{2.4.54}$$

whereas Sommerfeld, using asymptotic expansions, proposed that $\eta(x)$ has the form

$$\eta(x) = \left[1 + \left(\frac{x^3}{144}\right)^{\lambda/3}\right]^n \tag{2.4.55}$$

Here λ is defined in equation (2.4.52) and n is chosen so that the boundary conditions (2.4.47) are satisfied. Combining the last two expressions, his final formula is

$$S(x) = S_0(x)\left[1 - \left(\frac{1+z}{1+z_0}\right)^{6/\lambda^2}\right] = \frac{1}{(1+z)^{3/\lambda}}\left[1 - \left(\frac{1+z}{1+z_0}\right)^{6/\lambda^2}\right] \tag{2.4.56}$$

where $z = (x^3/144)^{\lambda/3}$ and $z_0 = (X_i^3/144)^{\lambda/3}$. For $x < 10$, equation (2.4.56) gives results that are accurate to better than 10%. Thus, this simple formula can provide reasonably accurate ionic potentials when the assumptions leading to its derivations are fulfilled.

A useful relationship that can be deduced from the zero-temperature TF model for isolated ions is the total energy,

$$E = \frac{12}{7}\left(\frac{2}{9\pi^2}\right)^{1/3}\left[\frac{dS(0)}{dx} + \frac{1}{x_0}\left(\frac{\bar{Z}}{Z}\right)^2\right]Z^{7/3}\frac{e^2}{a_0} \tag{2.4.57}$$

This formula was first obtained by Hulthen (1935) through minimizing the Gibbs chemical potential. For neutral atoms ($\bar{Z} = 0$) it reduces to a remarkably simple formula,

$$E = \frac{12}{7}\left(\frac{2}{9\pi^2}\right)^{1/3}\frac{dS(0)}{dx}Z^{7/3}\frac{e^2}{a_0} = -20.93Z^{7/3} \quad \text{eV} \tag{2.4.58}$$

This result is in good agreement with Hartree–Fock self-consistent-field type calculations. Schwinger proposed a multiplicative correction factor $F = 1 - 0.6504Z^{-1/3} + 0.346Z^{-2/3}$, which improves the agreement with the Hartree–Fock calculations.

Formal Derivation of the TF Theory

The TF model can be derived formally using the variation principle. According to this principle, the system of the plasma particles will take positions and energy distribution in the plasma which minimize their free energy. This simple require-ment should be complemented by two constraints: the first set is the requirement of the plasma neutrality, equation (2.4.13), and the second is the requirement that the chemical potential, μ, being a macroscopic quantity, is independent of the position r. In the TF theory these conditions should be fulfilled inside every ion sphere.

The functional of the free energy, $\mathcal{F}[n_e(r)]$, of an inhomogeneous electron gas at finite temperature, with local electron density $n_e(r)$, and pointlike nuclei posi-tioned at points $\vec{R}_j$, has the general form

$$\mathcal{F}[n_e(r)] = \int n_e(r)\epsilon[n_e(r), T]\, d^3r + \frac{1}{2}\sum_{i \neq j} \frac{Z^2 e^2}{|\vec{R}_i - \vec{R}_j|}$$

$$- \sum_j \int \frac{Ze^2}{|\vec{r} - \vec{R}_j|} n_e(e)\, d^3r + \frac{1}{2}\int\int d^3r\, d^3r'\, \frac{e^2 n_e(r) n_e(r')}{|\vec{r} - \vec{r}'|} \qquad (2.4.59)$$

The first term of equation (2.4.59) is the kinetic energy term (Brachman, 1951; March, 1953, 1957). The other three terms in equation (2.4.59) describe, respec-tively, the ion–ion, electron–ion, and electron–electron potential energy. A varia-tion of the free energy functional, $\mathcal{F}[n_e(r)]$, with respect to the electron density reproduces the original Thomas–Fermi equations. We shall not repeat this deri-vation here because it would take us too far away from the mainstream of our subject. The interested reader is referred to March (1957), Eliezer et al. (1986) or Gombas (1956).

Corrections to the Thomas–Fermi Theory

A series of corrections have been introduced into the TF model to improve its predictions. These are generally given as additional terms to the free energy functional (2.4.59). The equations for the electrostatic potential and the electron density are then derived by means of the variational principle.

Exchange and Correlation Terms

The corrections due to the exchange interaction of the electrons were introduced into the TF model by Dirac, and later by Jensen. The resulting theory is called the Thomas–Fermi–Dirac (TFD) model. Recently, this model has been treated together with the electron–electron correlation effects (Gupta and Rajagopal, 1980) which also have some influence at high electron densities.

The exchange interactions stem from Pauli's exclusion principle, which states that the wavefunctions of electrons in the same quantum state should be antisym-metric. In other words, electrons in the same quantum state move apart from each other not only because of their electrostatic repulsion, but also due to this principle. The average exchange interaction is expressed in terms of a repulsive potential. It is represented in the form of the Slater exchange potential,

$$\mathcal{F}_{xc}[n_e(r)] = -\frac{3e^2}{4\pi}[3\pi^2 n_e(r)]^{1/3} n_e(r) \tag{2.4.60}$$

Here an extra factor of $n_e(r)$ was added to get the exchange energy per unit volume. A factor of 2/3 should be inserted if a Kohn–Sham (Kohn and Sham, 1965) exchange potential is used. A more refined exchange potential was proposed by Lindgren and Rosen (Rosen and Lindgren 1972, Cowan 1981), which gives results intermediate between the Slater and the Kohn–Sham formulas. For most purposes equation (2.4.60) and its corrections are adequate, even at higher temperatures.

The electron correlation correction term originates from the fact that the mutual electrostatic potential between the nucleus and the electrons, as well as among the electrons themselves, tend to position the electrons in space so that the interparticle distance will be as large as possible. Such an arrangement reduces the free energy to the minimum. This correlation potential is also represented in the form of a repulsive free energy functional. At low temperatures this functional was first calculated by Gell-Mann and Brueckner, and is also called sometimes as the *Brueckner correction*, or the TFDB theory. The Gell-Mann–Brueckner term at $T = 0$ is

$$\mathcal{F}_{corr}[n_e(r)] = -\frac{e^2}{2a_0}\left[\frac{2}{3\pi^2}(1 - \log 2)\log\left(\frac{4\pi}{3}a_0^3 n_e(r)\right) + 0.096\right] \tag{2.4.61}$$

Both the exchange and the correlation terms tend to reduce the total energy of the system, that is, they add a repulsive component to the total potential.

The high temperature, high density formulas for both $\mathcal{F}_{xc}$ and $\mathcal{F}_{corr}$ were developed by Dharma-wardana and Taylor (1981) and by Gupta and Rajagopal (1980). Both results are too complicated to be reproduced here, and the interested reader is referred to the original papers for more information. An approximate expression, which has the correct low and high temperature asymptotic behavior, was suggested by Perrot (1979) for $\mathcal{F}_{xc}$,

$$\mathcal{F}_{xc}[n_e(r)] = -\frac{\pi e^4 a_0 [n_e(r)]^2}{2T}\tanh\left(ay^{-2/3} + by^{-2}\right) \tag{2.4.62}$$

where

$$y = \frac{\pi^2}{\sqrt{2}}\left(\frac{e^2}{a_0 T}\right)^{3/2} a_0^3 n_e(r) \tag{2.4.63}$$

is a dimensionless variable, $a = (3\sqrt{2})^{4/3}/4$ and $b = 27/16$ are numerical constants. This formula reproduces equation (2.4.60) for $T \to 0$, whereas for $T \to \infty$ it predicts, in accordance with other calculations, a $\mathcal{F}_{xc}[n(r)] \sim T^2$ behavior.

The high temperature asymptotic behavior of the free energy functional of the correlation energy is given in the paper of Gupta and Rajagopal (1980):

$$\mathcal{F}_{corr} \rightarrow -\frac{2}{3}\left(\pi\,\frac{e^6 n_e^3(r)}{T}\right)^{1/2} \qquad \text{for} \qquad T \rightarrow \infty \qquad (2.4.64)$$

Density Gradient Corrections

The quantum mechanical description of the electron as a wave necessitates another correction to the original TF theory. It stems from the fact that the electron is not a pointlike particle and the average electron density may vary within the volume of its wavefunction. The local quantum mechanical kinetic energy of an electron is proportional to the gradient of its wavefunction.

$$E_k(r) = \frac{\hbar^2}{2m}[\nabla\psi(r)]^2 \qquad (2.4.65)$$

Since $n_e(r) = |\psi(r)|^2$, it turns out that this correction term is important in regions of high electron density gradients. This point was recognized first by von Weizsäcker in the context of the calculations of the nuclear mass. Later Kalitkin (1960) and Kirzhnitz (1959) introduced it into the TF theory. The model which incorporates this correction is called the TFK or the TFC model, and the correction is referred to as the *Weizsäcker* or the *density gradient correction*.

The general form of the free energy corresponding to this correction is

$$\mathcal{F}_{grad}[n_e(r)] = \frac{\hbar^2}{2m}\int h[n_e(r)]\frac{[\nabla n_e(r)]^2}{n_e(r)}\,d^3r \qquad (2.4.66)$$

There has been some confusion in the literature as to the function $h[n_e]$. It has changed values for $r = 0$ from $1/9$ in the original paper of von Weizsäcker, to $1/72$ when the correct gradient expansion was used. Perrot has computed this function for y (see equation (2.4.63), from zero to infinity, and presented his results in a power series whose accuracy is better than 10^{-3} (Perrot, 1979).

The Model of Parr and Ghosh

In an attempt to correct the divergence of the electron density near the nucleus, Parr and Ghosh (1986) amended the TF model in a manner that produces continuity of the electron density at $r \rightarrow 0$. They treated only the $T = 0$ case. They imposed an additional constraint on the TF model, namely, that the integral

$$\int d^3r\, e^{-2kr}\nabla^2 n_e(r) \qquad (2.4.67)$$

should be finite for some constant k. They found that this condition is satisfied when

$$k = \left(\frac{1}{6}(3\pi)^{2/3} n_e(0)\right)^{1/2} \tag{2.4.68}$$

This choice yields finite electron density near the nucleus. Moreover, with this choice one obtains that

$$\int \nabla^2 n_e \, d^3 r = 0 \tag{2.4.69}$$

in contrast to the original TF theory where this integral is nonintegrable. The electron density near the nucleus tends to $n_e(r) = n_e(0) \exp(-2Zr/a_0)$, which is the correct behavior. It is interesting to note that the total energy in their version becomes

$$E = -20.91 Z^{7/3} (1 - 0.5360 Z^{1/3}) \quad \text{eV} \tag{2.4.70}$$

This is in agreement with equation (2.4.58), with Schwinger's correction factor, but equation (2.4.70) has a firmer foundation.

2.5 Ion Sphere Models

Under the terminology *ion sphere models* we gather all the models that assume charge neutrality inside the ion sphere. In other words, this term corresponds to all the models that consider a nucleus of charge Z, and the same number, Z, of electrons inside the ion sphere volume. This assumption ensures that the electrostatic potential and field are exactly zero on the ion sphere boundaries, and it is implicitly assumed that the charge distributions of the ions and electrons beyond the boundaries exactly cancel, resulting in a zero potential and field beyond the ion sphere. In this sense the Thomas–Fermi model also belongs to this category of models; however, due to its special status among these models, we prefer to devote a special section to it.

The greatest limitation of the TF theory is the fact that it cannot account for the detailed electron shell structure and other quantum mechanical properties of the ion. It is suitable only for finding average quantities, such as the potential or the average electron density. For spectroscopic purposes, more detailed quantum mechanical description of the ionic structure is required. The ion sphere model tries to overcome this difficulty by solving the accurate Schrödinger or Dirac equations for the bound electrons, while the free electrons are treated statistically, generally by means of the Fermi–Dirac distribution.

The ion sphere model assumes an ion having Z_b bound electrons, positioned at $r = 0$ and Z_f free electrons $(Z_b + Z_f = Z)$ occupies the rest of the ion sphere volume. The plasma effects are taken into account by confining the ion and the Z_f free electrons inside the ion sphere. Z_b and Z_f are integers, and the ion can be either in its ground state or in one of its excited states. The partial fractions of the populations of the various species cannot be calculated directly and require other methods, see chapter 5.

The basic equations of the ion sphere models are the following. First, again denoting by $V_N(r)$ and $V_e(r)$ the nuclear and electronic parts of the potential, the total local potential is

$$V(r) = \begin{cases} V_N(r) + V_{e,b}(r) + V_{e,f}(r) + V_{xc}(r) & r < R_i \\ 0 & r \geq R_i \end{cases} \tag{2.5.1}$$

Here $V_N(r) = Ze/r$ is the potential of the nucleus, and $V_{e,b}(r)$ and $V_{e,f}(r)$ are the potentials generated by the bound and the free electrons, respectively. The total potential produced by the electrons is $V_e(r) = V_{e,b}(r) + V_{e,f}(r)$, and $V_{xc}(r)$ is the exchange potential, which is added to obtain higher accuracy energy levels and wavefunctions for the bound electrons. Similarly to the derivation of equation (2.4.10) it can be shown that the two regions of the potential join continuously on the ion sphere boundaries.

The electronic part of the potential must satisfy the Poisson equation,

$$\nabla^2 V_{e,b}(r) = 4\pi e n_{e,b}(r), \quad \nabla^2 V_{e,f}(r) = 4\pi e n_{e,f}(r) \tag{2.5.2}$$

whereby $n_{e,b}(r)$ and $n_{e,f}(r)$ are the local bound and free electron densities. When the electron densities are known, then, in practice, the potentials $V_{e,b}(r)$ and $V_{e,f}(r)$ are computed by means of equation (2.4.8). For the exchange potential one may use the Slater type,

$$(-e)V_{xc}(r) = -\frac{3e^2}{2\pi}[3\pi^2 n_e(r)]^{1/3} \tag{2.5.3}$$

or the Kohn–Sham (Kohn and Sham, 1965) type exchange potential (which is obtained by multiplying equation (2.5.3) by 2/3). A more refined shape of the exchange potential was proposed by Lindgren and Rosen (Rosen and Lindgren, 1972; Cowan, 1981), and gives values between the Slater and the Kohn–Sham formulas.

In the ion sphere model, like all the other models, the Poisson equation is complemented by the equations for the bound and free electron densities. To obtain the bound electron density, one first solves the Schrödinger equation or, even better, the Dirac equation for each of the Z_b bound electrons (Rozsnyai, 1972, 1982; Skupsky, 1980; Davis and Blaha, 1982, Salzmann and Szichman, 1987). In the central field approximation the equation for the radial part is

$$\left[-\frac{\hbar^2}{2m} \frac{1}{r^2} \frac{\partial}{\partial r}\left(r^2 \frac{\partial}{\partial r}\right) - eV(r) + \frac{l(l+1)}{r^2} \right] R_{nl}(r) = \epsilon_{nl} R_{nl}(r) \tag{2.5.4}$$

Here $V(r)$ is the total electrostatic potential equation (2.5.1). The boundary condition for the solution of equation (2.5.4) is that the electrons' wavefunctions approach zero at infinity, $R_{nl}(r) \to 0$ for $r \to \infty$, rather than on the ion sphere boundaries, thereby allowing the tunneling of the bound electrons beyond the ion sphere. This feature of the model is very important in reflecting the physical situation in which the wavefunctions of electrons in highly excited states can overlap two or more ions, thereby generating a short-lived instantaneous quasi-molecule in the plasma. After obtaining the wavefunctions and eigenvalues of all the bound electrons, their density is calculated from

$$n_{e,b}(r) = \sum_{n,l}(2l + 1)|R_{nl}(r)|^2 \tag{2.5.5}$$

where the summation runs over all the occupied ionic states.

The free electron distribution is treated statistically and is computed from the incomplete Fermi–Dirac integral, equation (2.4.15),

$$n_{e,f}(r) = \frac{1}{2\pi^2}\left(\frac{2mc^2 T_e}{(\hbar c)^2}\right)^{3/2} F_{1/2}\left(\frac{\mu + eV(r)}{T_e}; \left|\frac{eV(r)}{T_e}\right|\right) \tag{2.4.15}$$

with the auxiliary requirement

$$Z_f = \int_0^{R_i} n_{e,f}(r, \mu)\, d^3r \tag{2.5.6}$$

from which the chemical potential, μ, is solved.

Inserting $n_{e,b}(r)$ and $n_{e,f}(r)$ from equations (2.5.5) and (2.4.15) into (2.5.1–2) results in one closed equation which can be solved for the potential. In practice equations (2.5.1), (2.5.4–5), (2.4.15), and (2.5.6) are solved by an iterative procedure. This is done in the following steps.

1. Assume initial approximate bound and free electron densities and a chemical potential μ. This can be done by taking the electron density of the isolated ion as $n_{e,b}(r)$, and a homogeneous distribution, $n_{e,f}(r) = Z_f/(4\pi R_i^3/3)$, for the free electrons. A first estimate of the chemical potential can be obtained from equation (1.3.25).
2. Calculate the potential generated by the bound + free electrons, $V_e(r)$, as a function of r using equation (2.4.8).
3. Find the total potential, $V(r)$, nuclear + electronic + exchange, equation (2.5.1).
4. Solve the Schrödinger equation in the potential $V(r)$ for all the bound electrons and calculate the wavefunctions, $\psi_{nl}(r)$, and energies, E_{nl}, of all the occupied bound states.
5. Get an improved bound electron density as function of r, from equation (2.5.5).
6. Find a better free electron spatial distribution by means of equation (2.4.15) using the total potential $V(r)$.
7. Solve equation (2.5.6) to find the chemical potential μ.
8. Check if the results converge. If not, then return to step (2) to start a new iteration.

The results of this procedure incorporate the electrostatic potential, the binding energies and wavefunctions of all the bound ionic states, and the chemical potential, as well as the bound and free electron spatial distributions.

The plasma density influences the results of these calculations through the boundary conditions imposed on the potential, that is, through the neutrality conditions of the ion sphere. The plasma temperature effects are taken into account through the free electron spatial distribution, equation (2.4.15). It must be emphasized that the total potential is produced by both the bound and the free electrons and in this sense the solution is self-consistent in both.

The great advantage of the ion sphere models is the fact that the bound states are treated in an exact quantum mechanical way, and only the effects of the surrounding plasma are incorporated in an approximate form. In fact, in very low density plasmas, where the screening of the free electrons is small, this procedure reproduces the Hartree–Slater self-consistent-field method. The ion sphere models therefore produce results of high accuracy which can be used for calculations of emission lines for spectroscopic purposes, and for detailed studies of the influence of the plasma environment on the atomic energy levels and wavefunctions. More about this point will be discussed in chapter 3.

Before entering more deeply into the ion sphere model, a short summary of the basic principles is in order.

- The ion sphere model treats the bound electrons by accurate quantum mechanical means, that is by solving the Schrödinger or Dirac equations for each bound electron in a potential generated by the nucleus and the bound and the free electrons. The Pauli exclusion principle is taken into account by adding a Slater (or other) exchange potential.
- The plasma correlation effects are accounted for by confining the ion together with $Z = Z_b + Z_f$ bound + free electrons inside the ion sphere, assuming that there are no other ions in this sphere.
- The free electrons are treated statistically by Fermi–Dirac statistics, thereby accounting for the possibility of polarization of the free electrons near the nucleus.
- The ion sphere model does not assume that the plasma consists of one species only. However, the charge state distribution and the excited states populations cannot be solved directly from the model, and these have to be calculated by other methods. These methods will be discussed in chapter 5.

Versions of the Ion Sphere Model

The natural way to improve the ion sphere model can proceed in two directions, in both the bound and the free electron parts. The bound electron part can be improved by applying more sophisticated and accurate methods to calculate the wavefunctions and densities. The relativistic Dirac equation was used (Liberman 1979; Salzmann and Szichman, 1987) in some works, thereby taking into account relativistic effects on the bound electrons, which may be of importance for high-Z materials. A very efficient computer package, RELDIR, was developed to include the plasma effects along the lines explained above (Salzmann and Szichman, 1987). This code is based on the HEX computer program, (Cowan, 1981), which is widely used to solve the Dirac equation for isolated ions. The exchange interactions are taken into account by adding a Slater-type exchange potential, see equation (2.5.3), to the total potential. The Kohn–Sham version (Kohn and Sham, 1965) or the more detailed Lindgren–Rosen type exchange interactions (Rosen and Lindgren, 1972) can be applied as well.

In principle, there is no difficulty in using the more sophisticated Hartree–Fock (HF) or even the multiconfiguration Dirac–Fock methods to calculate the bound electrons' wavefunctions and binding energies, rather than the simple single-particle central-field approximation. Although such high accuracy methods have still not been applied for plasma spectroscopy purposes, several questions in plasma spectroscopy may require the use of such special techniques.

To make improvements in the free electron part of the model, one can choose either of two directions: One may make an approximation to get simple analytical formulas, or improve the accuracy by using more sophisticated methods to estimate the free electron density. Use of simple analytical formulas may become adequate in most plasma spectroscopy applications, while saving significant amounts of computation time. A widely used approximation assumes that the free electron polarization around the nucleus has only negligible effect on the potential, and that the free electrons are distributed in space more or less uniformly,

$$n_{ef}(r) = n_{ef} = \frac{Z_f}{4\pi R_i^3/3} \tag{2.5.7}$$

This assumption is sufficiently accurate for low and intermediate density plasmas, when the number of the free electrons inside the bound electron orbitals is relatively small. When this assumption holds true, the electrostatic potential generated by the free electrons can be found by means of equation (2.4.8),

$$V_{ef}(r) = -4\pi e n_{ef} \left(\frac{1}{r} \int_0^r r'^2 \, dr' + \int_r^{R_i} r' \, dr' \right) = -\frac{3}{2} \frac{Z_f e}{R_i} \left[1 - \frac{1}{3} \left(\frac{r}{R_i} \right)^2 \right] \tag{2.5.8}$$

This form of the potential is rather frequently used when one is looking for some analytical, though approximate, result.

The treatment of the free electrons can be made more accurate by using detailed wavefunctions for the determination of the local free electron density. This was used in a paper by Davis and Blaha (1982). The free electron radial Schrödinger equation, in a manner similar to equation (2.5.4), can be written as

$$\left[-\frac{\hbar^2}{2m} \frac{1}{r^2} \frac{\partial}{\partial r} \left(r^2 \frac{\partial}{\partial r} \right) - eV(r) + \frac{l(l+1)}{r^2} - eV_{xc} - eV_{corr} \right] R_{pl}(r) = \frac{p^2}{2m} R_{pl}(r) \tag{2.5.9}$$

where R_{pl} is the wavefunction of a free electron with momentum p and angular momentum $\sqrt{l(l+1)}\hbar$. V, V_{xc}, and V_{corr} are the electrostatic, the exchange, and the correlation potentials, respectively. The asymptotic behavior of R_{pl} is an outgoing spherical wave,

$$R_{pl} \sim \sin\left(kr - l\frac{\pi}{2} + \delta_l(k) \right)/kr \tag{2.5.10}$$

where $k = p/\hbar$ is the wavenumber, and $\delta(k)$ is the phase shift of the wavefunction at infinity. The total free electron density is obtained by weighting the densities of the individual electrons with the Fermi–Dirac distribution function,

$$n_{e,f}(r) = \int_0^\infty dp\, f_e(r,p) \sum_{l=0}^\infty |R_{p,l}(r)|^2 \qquad (2.5.11)$$

where $f_e(r,p)$ is given by equation (2.4.11). The chemical potential is calculated by means of equation (2.5.6). The number of partial waves which should be taken into the l-sum is determined by the finite temperature version of the Friedel sum rule,

$$Z = Z_b + \frac{2}{\pi}\frac{1}{mT}\int_0^\infty p\, dp\, f_0(p)[1 - f_0(p)] \sum_{l=0}^\infty (2l+1)\delta_l(k) \qquad (2.5.12)$$

where now

$$f_0(p) = \frac{1}{1 + \exp[(p^2/2m - \mu)/T]} \qquad (2.5.13)$$

is a factor which is proportional to the free space Fermi–Dirac distribution function. Davis and Blaha (1982) reported that 43 terms were included in the l-sum until reasonable convergence of the results was obtained. Dharma-wardana and Perrot (1982) mentioned that the Friedel sum rule can be satisfied by as few as 8 to 10 l-terms.

Nearest Neighbor Effects in the Ion Sphere Model

The effects of the ion–ion correlations are treated by the ion sphere model through the confinement of the ion together with Z bound + free electrons in the ion sphere enclosure, and the assumption that no other ion or electron can penetrate into this volume. This picture can serve as a reasonable initial approximation. In a real plasma, however, such a static picture is an oversimplification that is very successful for the computation of several plasma quantities but needs refinement when some other quantities are required. In particular, one would not expect that the ion sphere model in its basic formulation can provide a sufficiently accurate picture for highly excited states whose orbitals are close to the ion sphere boundaries. In a real plasma, the local microfield near these boundaries is the superposition of the electrostatic fields generated by several nearby ions, so that the assumption of a spherically symmetric electrostatic potential is no longer valid. Highly excited states therefore are those which are most affected by the nonsphericity of the fields.

The greatest perturbation is, of course, due to the nearest neighbor. More distant ions have smaller influence on the bound states of the central ion. A natural extension of the ion sphere starts, therefore, with the incorporation of the nearest neighbor effects into the ion sphere model.

There are several effects by which the nonsphericity of the potential manifests itself. The first is the formation of an instantaneous quasi-molecule in which an excited electron's wavefunction tunnels out from its original ion and overlaps a second nearby ion. The original energy and wavefunction of the electron are modified accordingly. This modification of the energy of the excited state strongly depends on the interionic separation, which, however, fluctuates in time whenever the nearest neighbor is approaching or moving away from the central ion. Thus, the energy of the state fluctuates as well. Moreover, such a highly excited state is strongly affected also by other nearby ions, which enhance the fluctuations in the state's energy. All in all, the state cannot be regarded as having a constant well-defined energy, but it behaves more like an energy band. It turns out that un-localized excited states, which overlap two or more ions, form a continuum of negative-energy bound states, which are bound to the plasma as a whole rather than to any ion separately.

Localized excited states are affected by the nearby ions even when they are below the electrostatic barrier. This comes from the fact that the ionic potential is deformed by the nearest neighbor, particularly in the region halfway between the two nuclei, where the spherical symmetry is distorted into a cylindrical symmetry. This point has several consequences. The most important is the fact that the orbital quantum number, l, is no longer a good quantum number for highly excited states, and the state becomes a mixture of several ls. Transitions that are strictly forbidden for isolated ions can now proceed through the mixed state. For example, a $4s - 1s$ transition is, of course, forbidden for ions in spherically symmetric potential. However, when the ion is immersed in a high density plasma, so that the upper $4s$ state is significantly distorted by the nearby ions, this $4s$ state becomes mixed with the $4p$ and, to a small extent, also with the $4d$ and $4f$ states; that is, its wavefunction becomes of the form $\psi = \alpha\psi_{4s} + \beta\psi_{4p} + \gamma\psi_{4d} + \epsilon\psi_{4f}$, with $|\alpha|^2 + |\beta|^2 + |\gamma|^2 + |\epsilon|^2 = 1$. The transition can now proceed through the $4p$ state, and its intensity will be proportional to $|\beta|^2$, the mixing coefficient between the $4p$ and the $4s$ states.

The first work that attempted to get some information about these effects was carried out by S. Rose, who calculated the $1s–2p$ transitions in $H^+–H$ collisions in plasmas containing H and H^+ atoms (Rose, 1983). Rose solved accurately the Schrödinger equation for the $H–H^+$ transient quasi-molecule as function of the internuclear separation for bonding and antibonding structures. He found high oscillator strengths for the forbidden $1s\,\sigma_g–2p\,\sigma_u$ transition when the ions were close, which would not show up in a spherically symmetric case. Also, the average oscillator strength of an allowed transition, $1s\,\sigma_g–2p\,\pi_u$, for ions in close encounter is significantly altered relative to the isolated ion case.

A more complete study was carried out by J. Stein and D. Salzmann (Stein et al., 1989, Salzmann et al., 1991) for a plasma consisting of identical ions. For this purpose they developed a cylindrically symmetric ion sphere model that takes into account the electrostatic potential generated by both the central ion and its nearest neighbor. In their model the ion sphere deforms into two truncated spheres (figure 2.5), each with a constant volume $V_i = 1/n_i$. The boundary conditions are the following: (i) the potential is zero on and beyond the truncated spheres boundaries, and (ii) the normal component of the electrostatic field (the component parallel to the line connecting the two nuclei) is zero on the surface that separates the two spheres. This condition is a result of the reflection symmetry around this surface. The Born–Oppenheimer approximation is assumed, namely, that the electron cloud adjusts itself rapidly to any change in the inter-

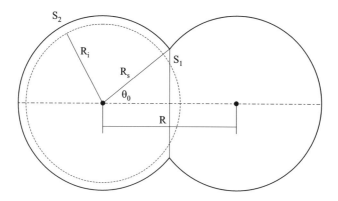

Figure 2.5 Approximate shape of the quasi-molecule. (From Stein, 1989.)

ionic distance, so that at every moment the electrons' spatial distribution is the same as in the case of static ions positioned at the same distance.

Their method includes a multipole expansion of the electrostatic potential inside the truncated spheres for a range of interionic separations. For each separation the total energy and the free energy of the ion are computed. Using this free energy as a weighting factor, a Monte Carlo program is then used to find the probability that the distance to the nearest neighbor is between R and $R + dR$. The last step in their method is to solve the Schrödinger equation for one of the ions in the multipole field for a range of probable interionic separations, and to convolute the results with the above probability function. A fast method for the calculation of the multipole components of the electron distribution and the electrostatic potential was proposed in a separate paper (Salzmann, 1994).

Representative results are shown in figure 2.6 for the equipotentials, and in figure 2.7 for the probability function of the interionic distance. In figure 2.8 the

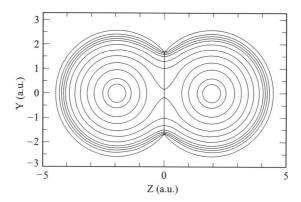

Figure 2.6 The equipotentials of the quasi-molecule. (From Stein, 1989.)

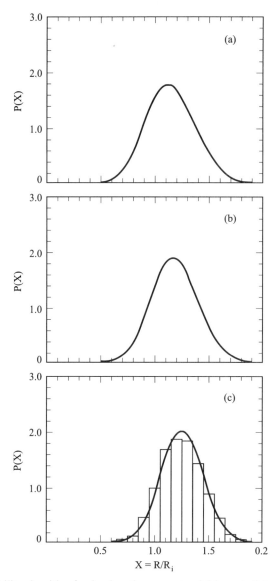

Figure 2.7 Probability densities for having the nearest neighbor at distance R in a Li-like krypton plasma at a temperature of 1100 eV and ion densities of (a) 6×10^{21} cm^{-3}, (b) 3×10^{22} cm^{-3}, and (c) 1×10^{23} cm^{-3}. (From Stein, 1989.) The histogram represents the Monte Carlo results from which the probability density moments were derived.

computed emission spectra from a $T = 1400$ eV krypton plasma are shown at various densities. There are several interesting features in these spectra, such as the shift of the lines toward lower energies (red shift) as the ion density goes up, which will be discussed in detail in chapter 3, and the line broadening at the higher densities, discussed in chapter 7. For the purposes of the present chapter the most

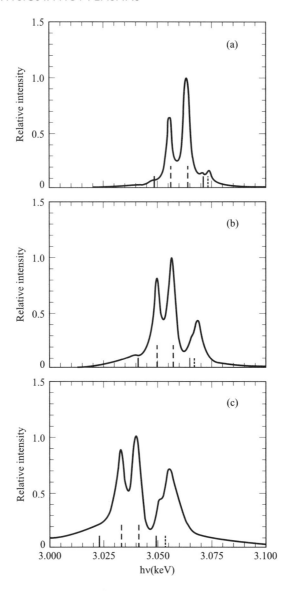

Figure 2.8 Computed spectra for the $n = 4 \rightarrow 2s_{1/2}$
transition. (a) 6×10^{21} cm^{-3}, (b) 3×10^{22} cm^{-3}, and
(c) 1×10^{23} cm^{-3}. (From Stein, 1989.)

striking feature is the growth of the forbidden $4d$–$1s$ and $4s$–$1s$ bands at both sides
of the allowed $4p_{1/2}$–$1s$ and $4p_{3/2}$–$1s$ lines. These bands could not be predicted in a
spherically symmetric model. Recently, such forbidden bands have been observed
experimentally (Leboucher-Dalimier et al., 1993) in a highly ionized fluorine
plasma.

Other attempts to calculate the nearest neighbor effects in dense plasmas were made by Malnoult, d'Etat, and Nguyen (1989) who have carried out the calculation of a similar two-centered system using elliptical coordinates. In their calculations they used a homogeneous electron density, equation (2.5.7). They defined the molecular envelope as the equipotential that satisfies the electrical neutrality condition. This choice assures, through Gauss's law, that the normal electric field to this surface vanishes (Dirichlet boundary condition). To first order, the molecule is, therefore, decoupled from the other regions in the plasma. In their paper they present the equipotentials and the shape of the quasi molecule, which seem to be less artificial than for the truncated sphere shape. They present results for the spectral line shapes and shifts in a hydrogenlike neon plasma.

A two-center problem with slightly different assumptions was solved by Furutani et al. (1993) within the framework of the Thomas–Fermi–Dirac model with the Weizsäcker correction. They, too, have used elliptical coordinates, and their definition of the shape of the quasi molecule is similar to that of Malnoult et al. Their results include the equipotentials and contours of the electron density as functions of the temperature for H–H, N–N and C–H quasi molecules.

2.6 Ion Correlation Models

The ion correlation models (ICM) may be regarded as intermediate models between the Debye–Hückel theory and the ion sphere models, thereby closing the gap between weakly and strongly coupled plasmas. These models consider an interaction volume that is larger than the ion sphere, thereby taking into account the influence of several nearby ions on the potential around the central ion. In this sense, these models are extension of the ion sphere models, which consider the central ion only. Moreover, these models use Fermi–Dirac statistics, thereby making them superior at high densities to the Debye–Hückel theory, which, although it treats the vicinity of the ion beyond the ion sphere in a similar manner, is valid for weakly coupled plasmas only.

Of the models in this category, the density functional theory (DFT), developed by M. W. C. Dharma-wardana and F. Perrot (1982) has the firmest foundations. The basic picture solved by the DFT, as in all the previously introduced models, consist of a central nucleus having charge Ze in a medium containing electrons and ions. The electrons pile up around the central ion, some forming negative-energy bound states, others occupying positive-energy states (Dharma-wardana and Perrot, 1982). The ions form a charge depletion around the central ion. Their distribution is not assumed to have a sharp discontinuity on the ion sphere boundaries, as in the ion sphere models, but rather increases continuously as the distance from the central ion increases. The ion density attains its asymptotic value at a distance of several ion sphere radii from the central ion.

The DFT considers a volume Ω_c which is sufficiently large that the ion–ion and the electron–ion radial distribution function, $g_i(r)$ and $g_e(r)$, see equation (1.4.2), are both essentially unity on its boundary. In other words, the ions and the electrons on the boundaries of Ω_c are no longer correlated to the central ion. It is shown by Dharma–wardana and Perrot (1982) that in the limit of weak coupling the correlation sphere becomes identical with the Debye sphere, whereas in strongly coupled plasmas the correlation sphere constricts to the ion sphere. This feature of DFT makes it a very attractive candidate as a generalization of all the previously introduced models.

The formal derivation of the DFT equations is similar to the derivation of the TF theory, see section 3.3: One sets up the free energy of the system and uses the variational principle with respect to the ion and electron densities to minimize the free energy. Detailed description of the formalism and the equations is given in Dharma-wardana and Perrot (1982). See also Rozsnyai (1991). The final equations are similar to those of the ion sphere model. For instance, equations (2.5.2–5) for the potential and the bound electrons are valid also in DFT, and equations (2.5.9–13) are valid for the free electrons. The neutrality condition, equations (2.5.6) or (2.4.13), is, however, replaced by

$$Z = \int_{0+}^{R_c} [n_e(r) - \bar{Z}n_i(r)] \, d^3r = n_e \int_{0+}^{R_c} [g_e(r) - g_i(r)]4\pi r^2 \, dr \qquad (2.6.1)$$

where the lower limit of the integral is chosen so that the integral does not include the nucleus of the central ion, and now, in contrast to the ion sphere or the TF models, the upper limit of the integral extends to *the correlation sphere radius*, R_c. This equation requires the knowledge of the spatial distribution of the ions, $n_i(r)$, around the central ion, or equivalently, the ion–ion correlations in the plasma. These are calculated in DFT by means of the hypernetted-chain equation, see section 2.7, which gives a good estimate of the ion radial distribution function. When $n_i(r)$ is known, one can calculate the local charge density, $n_e(r) - \bar{Z}n_i(r)$ and, in particular, the distance at which the local charge density vanishes (to within a desired accuracy). This distance is defined as the correlation sphere radius, R_c.

The DFT model was applied by its developers to a series of problems: the determination of effective proton–proton interaction, the examination of the onset of a bound state in a hydrogen plasma, as well as the optical properties and the electrical resistivity of Fe and Xe plasmas. Other questions that were calculated by the DFT are the dynamic structure factors of a hydrogen plasma and the electric microfield distributions. Accordingly, the DFT seems to be able to give answers to a wide variety of questions on atomic physics in plasmas.

The approximations and the accuracy of the DFT were discussed in a paper by Chihara (1991). He has shown that DFT "breaks down" when it treats dense plasmas of ions that have many bound electrons, while it gives reasonably accurate results when the bound electron contribution to the plasma free energy is relatively small. For more details about DFT, the reader is referred to the original papers.

2.7 Statistical Theories

Under the term "statistical theories" we include a series of theories that use the methods of statistical mechanics to calculate properties of dense plasmas. Generally, these theories have been developed not for studying atomic or ionic properties but rather for some statistical parameters, such as the ion–ion radial distribution function, equation (1.4.2). They treat the ions and electrons as point-like structureless objects. A treatise on atomic behavior in hot plasmas cannot enter too deeply into these theories, which ignore the internal structure of the ions. The brief survey below is introduced only for two purposes. First, the terminology and concepts of these models are sometimes used within the field of atomic physics in hot plasmas and it is important to explain the meaning of this terminology to researchers in the field of plasma spectroscopy. Second, these models give information about the radial distribution function, equation (1.4.2), that is needed in high density plasmas for the determination of the local microfield and the calculation of spectral line shapes.

As a first step we derive a few quantities of statistical nature for weakly coupled plasmas. These results are important also as benchmarks for comparison with the behavior of the same quantities at higher densities.

Weakly Coupled Plasmas, $\Gamma \ll 1$

Assume a homogeneous and uniform weakly coupled plasma that has a constant distribution for both the electrons and ions, $g_i(r) = g_e(r) = 1$, where $g_i(r)$ and $g_e(r)$ are the ion–ion and ion–electron radial distribution functions defined in equation (1.4.2). Also assume that an ion is located at $r = 0$. We first derive the probability density function that its nearest neighbor is at a distance between R and $R + dR$. If N_0 is the number ions confined in a large volume V_0, the average ion density is $n_i = N_0/V_0$. Divide V_0 into a large number, k, of equal size cells each of volume $\Delta V = V_0/k$. Choose k to be much larger than N_0, so that most of the cells are empty and the probability of finding two particles in one cell is negligibly small. The probability of finding an ion within one of these cells is $P = N_0/k = N_0 \Delta V/V_0 = n_i \Delta V$. The probability that the cell is empty is the complementary probability $\tilde{P} = 1 - n_i \Delta V$.

The probability of finding the nearest neighbor at a distance R is the product of the probability of not having any ion within a sphere of radius R with the probability of having at least one ion in a spherical shell of volume $4\pi R^2 dR$. The probability that the sphere of radius R is empty is the product of the probabilities of the separate cells inside this sphere, $\tilde{P}(V) = \prod_{m=1}^{M}(1 - n_i \Delta V) = (1 - n_i \Delta V)^M$, where M is the number of cells within the sphere, $M = (V/V_0)k$, and $V = (4\pi/3)R^3$ is the sphere's volume. One therefore gets

$$\tilde{P}(V) = (1 - n_i \Delta V)^M = \left(1 - \frac{N_0}{V_0}\frac{V_0}{k}\right)^{k(V/V_0)}$$

$$= \left[\left(1 - \frac{N_0}{k}\right)^k\right]^{V/V_0} \rightarrow [\exp(-N_0)]^{V/V_0} = e^{-n_i V}$$

The probability that the nearest neighbor is at a distance R is therefore

$$P_{NN}(R)\, dR = 4\pi n_i R^2\, dR \exp\left(-n_i \frac{4\pi}{3} R^3\right) \tag{2.7.1}$$

The corresponding formula in strongly coupled plasmas, in which $g_i(r)$ and $g_e(r)$ are not constant, is

$$P_{NN}(r)\, dR = 4\pi n_i R^2 g_i(R)\, dR\, \exp\left(-4\pi n_i \int_0^R R'^2 g_i(R')\, dR'\right) \qquad (2.7.2)$$

It is easy to show that both equations (2.7.1) and (2.7.2) satisfy the normalization condition $\int_0^\infty P_{NN}(r)\, dR = 1$. Using equation (2.7.1), one can derive the moments of the distance of the nearest neighbor,

$$\langle R_{NN}^k \rangle = \int_0^\infty R^k P_{NN}(R)\, dR = \int_0^\infty R^k 4\pi n_i R^2\, dR\, \exp\left(-n_i \frac{4\pi}{3} R^3\right) \qquad (2.7.3)$$

Transforming the variable of integration to $m = 4\pi R^3 n_i / 3 = (R/R_i)^3$ (m is the average number of ions in a sphere of radius R), or $R = R_i m^{1/3}$, $dR = (1/3)m^{-2/3} R_i\, dm$, one gets

$$\langle R_{NN}^k \rangle = R_i^k \int_0^\infty m^{k/3} e^{-m}\, dm = R_i^k \Gamma\left(1 + \frac{k}{3}\right) \qquad (2.7.4)$$

In particular, the average distance to the nearest neighbor is

$$\langle R_{NN} \rangle = R_i \Gamma\left(\frac{4}{3}\right) = 0.892\,97\ldots R_i \qquad (2.7.5)$$

Equation (2.7.5) tells us that for weakly coupled plasmas the average distance to the nearest neighbor is about 89% of the ion sphere radius, substantially less than assumed by the ion sphere models.

The second moment is

$$\langle R_{NN}^2 \rangle = R_i^2 \Gamma\left(\frac{5}{3}\right) \qquad (2.7.6)$$

wherefrom one gets the standard deviation of the fluctuations of the distance around the average,

$$\sigma(R_{NN}) = [\langle R_{NN}^2 \rangle - \langle R_{NN} \rangle^2]^{1/2} = 0.325 R_i \qquad (2.7.7)$$

and the relative fluctuations are

$$\frac{\sigma(R_{NN})}{\langle R_{NN} \rangle} = 0.364 \qquad (2.7.8)$$

Thus, in weakly coupled plasmas there are large fluctuations in the interionic distance and, therefore, also in the local electrostatic microfield.

One Component Plasmas (OCP)

We next turn to the methods used to study the properties of strongly coupled plasmas. Of these, the *one component plasma* (OCP) is, perhaps, the most widely used method to simulate the conditions of matter in extremely highly coupled plasmas where no bound electrons or atomic properties remain, such as in inertial confinement fusion plasma and in the interior of the main-sequence stars.

The OCP, in its basis, is a Monte Carlo type method that is used to calculate the internal energy, pressure, static and dynamic structure factors, and other macroscopic thermodynamic quantities by means of computer simulations. It can also be used to compute the ion radial distribution function, and this is its main importance for our purposes.

The OCP assumes pointlike structureless ions of charge Z immersed in a continuous uniform negatively charged background produced by the electrons. Charge neutrality is assumed within any large volume that contains many ions, in other words, *the local charge density* of the continuous electron background exactly equals the *average charge density* of the ions. Polarization of the electrons around the positive ions is neglected. The ions interact with each other through their Coulomb fields, and the total free energy is computed by the Monte Carlo simulation.

A very large body of literature has been published on results of OCP. Setsuo Ichimaru et al. published a series of papers (see Ichimaru et al., 1985) that summarize various aspects of this subject from both the theoretical and the computational sides of the topic. An earlier summary is presented in Ichimaru (1982). Computational results were published by J. Hansen, H. DeWitt, W. L. Slattery, W. B. Hubbard and coworkers (see, for instance, Hansen, 1973; Hubbard and Slattery 1971; DeWitt 1976; Slattery et al., 1980, 1982 and references therein).

The OCP model is particularly useful for calculating the ion radial distribution function at high-Γ (plasma coupling constant) conditions. In figure 2.9 this function is shown (Ichimaru 1982) between $\Gamma = 0.1$ up to $\Gamma = 140$. It can be seen from the figure that the neighboring ions are rejected by the central ion, thereby forming a depletion of the positive charge near the ion. For $\Gamma \leq 5$ the radial distribution function shows a monotonically increasing behavior as a function of r. Such behavior corresponds to an unordered distribution of the ions in space. For $\Gamma \geq 10$, the radial distribution function exhibits an oscillatory behavior, indicating that there are some distances that are more likely to being occupied by the neighboring ions. In other words, a lattice-type structure starts to build up in the plasma.

For $\Gamma \geq 175(\pm 5)$ the OCP model predicts that the free energy is minimum when the ions settle down in a body-centered cubic lattice, thereby generating a solid state high temperature plasma. Some recent simulations indicate that approaching this freezing point from a lower Γ (higher temperature) state one can obtain under suitable conditions a *supercooled liquid*, inasmuch as one obtains it in room temperature laboratory experiments with regular liquids. Similarly, approaching the $\Gamma = 175$ melting point from higher Γ (lower temperature) conditions, one may expect an *overheated solid*, which at a critical Γ undergoes a phase transition to a disordered plasma state.

One of the most important predictions of OCP is for energy density in high density strongly coupled plasmas. The results of Monte Carlo runs for the ion energy density were fitted to the following form of the plasma coupling constant, Γ (Slattery et al., 1980),

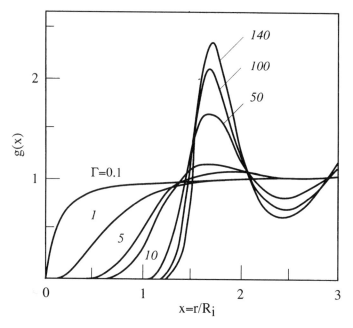

Figure 2.9 The radial distribution function for various values of Γ. (From Ichimaru, 1982.)

$$\frac{E}{n_i T} = a\Gamma + b\Gamma^{1/4} + c\Gamma^{-1/4} + d \tag{2.7.9}$$

where a, b, c, and d are numerical constants (Ichimaru, 1982),

$$a = -0.897\,52, \qquad b = 0.945\,44$$
$$c = 0.179\,54, \qquad d = -0.800\,49$$

Equation (2.7.9) can reproduce the results of the Monte Carlo runs to an accuracy of 10^{-5}.

In addition to the quantities already mentioned, OCP also gives useful information about the electrostatic microfield distribution and other related quantities that describe the collective and macroscopic thermodynamic properties of the plasma, such as the structure factor, dielectric function, electric susceptibility and optical reflectivity, to mention a few.

Hypernetted Chain Model (HNC)

The hypernetted chain model (HNC), similarly to OCP, treats the plasma particles as pointlike structureless objects that interact with each other through their Coulomb fields. The HNC model was developed initially to study the statistical properties of the spatial distribution and autocorrelation functions of the molecules in liquids. It was adapted into plasma physics without major modifications.

The HNC model counts the interactions of a given ion with all the other ions in its vicinity, and organizes them in space so that the free energy will be minimum. This requires that adjacent ions are neither too close nor too far from the central ion. In the second step, the model also takes into account the interactions among the neighboring ions themselves, arranging these, too, in a way that minimizes the free energy. The model then takes into account the feedback of this ordering of the neighboring ions on the central ion. In the next steps, more and more distant ions are included in this self-organizing procedure, and the influence of their ordering on the central ion is taken into account.

The general formalism uses the *Mayer cluster expansion* technique (see, e.g., Feynman 1974) to account for the particle correlations and to describe the interactions in a graphic method. The full analysis shows an interesting relationship between the mutual interaction of groups of particles (Meeron, 1960), which is expressed by means of the *Ornstein–Zernicke equation* (Ornstein and Zernicke, 1914),

$$h(\vec{r}) = c(\vec{r}) + n_i \int d^3 r' \, c(|\vec{r} - \vec{r}'|) h \vec{r}') \qquad (2.7.10)$$

where $h(r) = g_i(r) - 1$ is the excess relative local density of the ions, and $c(r)$, called the *direct correlation function* is computed from the basic two-particle potential. Generally, the computation of $c(r)$ is very difficult, so that (2.7.10) is still not in a form that facilitates its immediate use. The HNC model consists of neglecting a certain class from the mutual interaction picture described above. When the so-called *bridge diagrams* are neglected from the above iterative procedure, one gets the characteristic approximation of the HNC model (see equations (5.4.18) and (5.4.19) in the book of Hansen and McDonald (1986)).

$$c(\vec{r}) = -\frac{eV(\vec{r})}{T} + h(\vec{r}) - \log[h(\vec{r}) + 1] \qquad (2.7.11)$$

which when substituted into the Ornstein–Zernicke equation yields an integral equation for $g_i(r)$ in terms of the two-particle interaction potential,

$$\log g_i(\vec{r}) = -\frac{eV(\vec{r})}{T} + n_i \int d^3 r' [g_i(\vec{r} - \vec{r}') - 1]$$

$$\times \left(g_i(\vec{r}') - 1 - \log g(\vec{r}') - \frac{eV(\vec{r}')}{T} \right) \qquad (2.7.12)$$

If $V(r)$ is known, this equation can be used to solve for $g_i(r)$. The HNC scheme offers an extremely accurate description of the correlations in intermediate to weakly coupled ($\Gamma \leq 10$) plasmas (Ichimaru et al., 1985). For $\Gamma \leq 1$, the HNC results are virtually exact. Even for $1 < \Gamma \leq 10$, the HNC scheme reproduces the results of OCP with relative errors of less than 1%.

Many books and research papers have been published on the HNC model. These were reviewed here briefly only to the extent of our needs in the field of plasma spectroscopy. We recommend the book of Hansen and McDonald (1986) and the paper of Meeron (1960) for further reading.

Atomic Properties in Hot Plasmas

3.1 A Few Introductory Remarks

In chapter 2, we investigated the electrostatic potential generated by the plasma environment around an ion immersed in a plasma. In this chapter, we will discuss the direct effects of this potential on the properties of the ions. More explicitly, we will see how the modification of the electrostatic potential by the plasma environment affects the bound electrons' eigenvalues and wavefunctions and related parameters.

The eigenstates of the electrons of an isolated ion are determined by a Hamiltonian of the form

$$\mathcal{H}_0 \psi_{nlm}^{(0)} \equiv \left(-\frac{\hbar^2}{2m} \nabla^2 - eV_0(\vec{r}) \right) \psi_{nlm}^{(0)} = E_{nlm}^{(0)} \psi_{nlm}^{(0)} \tag{3.1.1}$$

where $V_0(\vec{r})$ is the potential generated by the nucleus and the ionic bound electrons. When this ion is immersed in a plasma, this Hamiltonian is modified by the presence of nearby ions and free electrons to

$$\mathcal{H} \psi_{nlm} \equiv \left(-\frac{\hbar^2}{2m} \nabla^2 - eV_0(\vec{r}) - eV_{pl}(\vec{r}) \right) \psi_{nlm} = E_{nlm} \psi_{nlm}. \tag{3.1.2}$$

where $V_{pl}(\vec{r})$ is the additional potential generated by nearby ions and free electrons. This plasma potential modifies the eigenvalues, E_{nlm}, and eigenfunctions, ψ_{nlm}, which now differ from their isolated case values, $E_{nlm}^{(0)}$ and $\psi_{nlm}^{(0)}$.

One can make several important predictions about the influence of this additional plasma potential, $V_{pl}(\vec{r})$, on the ionic properties even without knowing its explicit form. We know that the greatest contribution to $V_{pl}(\vec{r})$ inside the volume of the ion comes from the free electrons, whereas the neighboring ions have appreciable influence only at greater distances, close to the peripheries of the ion and beyond. It turns out that within the ionic volume $V_{pl}(\vec{r})$ is a negative potential, so that $-eV_{pl}(\vec{r})$ contributes a positive energy term to the original Hamiltonian. The plasma potential, therefore, shifts the eigenvalues upward relative to those of the isolated ion; in other words, it reduces the binding energy. This rather mathematical explanation can be given an intuitive interpretation: The free electrons distribution around the nucleus produces an effect of screening on the nuclear potential, thereby reducing the potential at every point in the field compared to the isolated ion case. The bound electrons are, therefore, attracted toward the nucleus with a slightly weaker force when the ion is immersed in a plasma than they would be without the plasma environment. This results in a repulsion of the wavefunctions away from the nucleus, or, more precisely, to their weaker attraction toward the nucleus.

To illustrate the above explanations, we show in figure 3.1 the wavefunction of an excited $3p_{1/2}$ electron of an H-like aluminum ion at a density of $2 \times 10^{23}\,\mathrm{cm}^{-3}$ (dashed line) in comparison to the same wavefunction of an isolated ion (solid

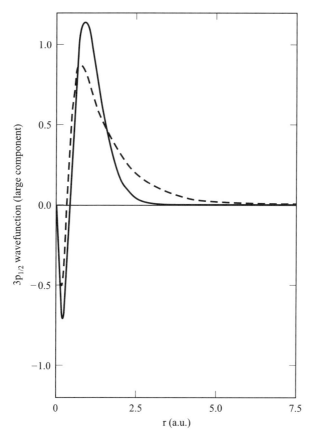

Figure 3.1 The wavefunctions of an exited $3p_{1/2}$ electron of an H-like aluminium ion at a density of $2 \times 10^{23}\,\mathrm{cm}^{-3}$ (dashed line) in comparison to the same wavefunction of an isolated ion (solid line). (From Salzmann and Szichman, 1987.)

line). The repulsion of the wavefunction from the nucleus for the high density case can be clearly seen.

Figure 3.2 shows the reduction of the bound electron binding energy of several ionic states of hydrogen-, helium-, and lithium-like aluminum ions as a function of the plasma ion density. It is interesting to note that at high enough densities the binding energy of the states may even vanish and the states cross the limit into the continuum. The phenomenon of the shift of a bound state into the continuum is termed in the literature *the pressure ionization*. It can be seen in figure 3.2 that, at densities of 10^{23}–10^{24} cm^{-3}, the $n = 3$ states of excited He-like or Li-like ions are shifted into the continuum. At densities of 10^{24}–10^{25} cm^{-3}, even the $n = 2$ states are no longer bound. For K-shell states this occurs only at densities that are beyond the scale of the figure.

3.2 Atomic Level Shifts and Continuum Lowering

The phenomenon of binding energy reduction when the plasma density increases is known as *ionization potential lowering* or *continuum lowering*, and apparently it may be regarded as the main atomic phenomenon unique to hot plasmas (More, 1982). To illustrate this phenomenon we shall first consider a simplified situation, namely, that of a homogeneous free electron distribution within the ion sphere volume and charge neutrality beyond its boundaries. For low and intermediate density plasmas this is a plausible approximation. The potential generated by such

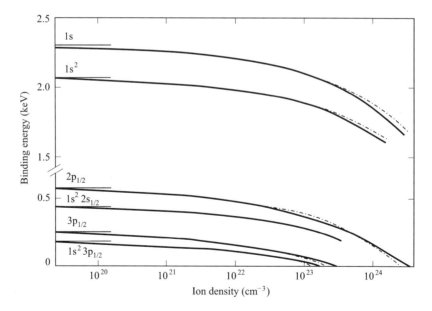

Figure 3.2 The reduction of the bound electron binding energy of several ionic states of hydrogen-, helium- and lithiumlike aluminum ions as a function of the plasma ion density. (From Salzmann and Szichman, 1987.)

a charge distribution has already been calculated in section 2.5 and is given by equation (2.5.8),

$$V_{e,pl}(r) = -\frac{3}{2}\frac{\zeta e}{R_i}\left[1 - \frac{1}{3}\left(\frac{r}{R_i}\right)^2\right] = -\frac{3}{2}\left(\frac{4\pi}{3}\right)^{1/3}\zeta e n_i^{1/3}\left[1 - \frac{1}{3}\left(\frac{r}{R_i}\right)^2\right] \quad (2.5.8)$$

where, we recall, R_i is the ion sphere radius, ζ is the number of free electrons in the ion sphere, and in the transition to the second equation we have used the explicit form of R_i in terms of the ion density, equation (1.2.7). Assume that the nucleus is located at $r = 0$. Considering only bound electrons whose orbitals are much smaller than R_i, the $(r/R_i)^2$ term is negligibly small within the ionic volume and can be omitted. In this approximation the perturbation by the plasma environment is

$$-eV_{pl}(\vec{r}) \approx -eV_{e,pl}(r = 0) = \frac{3}{2}\left(\frac{4\pi}{3}\right)^{1/3}\zeta e^2 n_i^{1/3} \quad (3.2.1)$$

where $r = 0$ is the location of the ion's nucleus. The upward shift, $\chi_{\zeta,nl}$, of the energy level $|nl\rangle$ is obtained as

$$\chi_{\zeta,nl} = E_{\zeta,nl} - E_{\zeta,nl}^{(0)} = \langle nl| - eV_{pl}|nl\rangle = \frac{3}{2}\left(\frac{4\pi}{3}\right)^{1/3}\zeta e^2 n_i^{1/3} \quad (3.2.2)$$

or, numerically,

$$\chi_{\zeta,nl} = 3.48\zeta\left(\frac{n_i}{10^{21}\text{ cm}^{-3}}\right)^{1/3} \text{ eV} \quad (3.2.3)$$

In spite of their approximate character, several interesting features can be learned from equations (3.2.2)–(3.2.3). First, all the ionic states are shifted by a constant amount, independently of their initial unperturbed energy, their quantum numbers, or any other property. This means that the energy difference between any two states is not modified (to first approximation) by the plasma environment, and this holds true also for the energy of the photons emitted during the transition between these states. This feature can be seen in figure 3.2, where all the states move approximately in parallel to each other. Second, the level shift is proportional to the 1/3 power of the ion density, indicating that large changes in the plasma density induce relatively small energy shifts. Finally, the shift is proportional to the average number of free electrons in the ion sphere volume and is therefore more important for highly charged ions.

Ionic states for which $\chi_\zeta \geq |E_{\zeta,nl}|$ are no longer bound (we have dropped the index of the ionic level). These states are shifted into the continuum and are *pressure ionized*. The explanation given above is only one of several ways of looking at this effect. In fact, there is an extensive literature on continuum lowering going back to the 1960s. A paper by Brush and Armstrong (1965) quotes almost 200 references on this subject. Early papers all agree that the general form of the continuum lowering is

$$\chi_\zeta = \frac{\zeta e^2}{R^*} \tag{3.2.4}$$

where R^* is some characteristic radius. The problem centered around the question of what value to give to R^*.

In the 1960s, generally, the Debye screening length, D, equation (2.2.5), was substituted for R^*. This, however, raised the difficulty that for low density plasmas, where $D \geq R_i$, excited states whose orbital radius is larger than the ion sphere radius overlap several neighboring ions. While in principle it is not impossible that such a state can still be regarded as a bound state, this is a highly improbable situation in real plasmas. Such a negative-energy bound state is very sensitive to the local instantaneous microfields and changes its shape rapidly as the local plasma environment fluctuates. A negative-energy electron in a highly excited state is better regarded as bound to the plasma as a whole, rather than to any particular ion. Such an electron, when moving in the plasma, will undergo collision within a short time with positive-energy electrons and will rapidly be absorbed in the Maxwellian distribution of the electrons in the plasma. The choice of $R^* = D$ seems, therefore, to be inappropriate on physical grounds.

R. M. More (1982) gave three reasons why the ion sphere radius, R_i should be taken as the characteristic radius in equation (3.2.4). He based his arguments on three different criteria. Remarkably, these criteria provide mathematically identical results.

(1) The first criterion is that pressure ionization occurs when corresponding wavefunctions of adjacent ions overlap, leading to a broadening of the atomic states into energy bands. This occurs when

$$r_n \simeq R_i \tag{3.2.5}$$

where r_n is the average radius of the bound electron orbital. Equation (3.2.5) means that the ion sphere radius is a characteristic length that determines the largest radius of the ionic orbital beyond which a bound electron is no longer localized to its original ion but to a cluster of neighboring ions. This can be regarded as a plausible criterion for the localization of an electron.

(2) A second criterion says that pressure ionization occurs when the binding energy, $E_{\zeta,nl}$, of an excited state is exceeded by the (negative) electrostatic potential generated by its nearest neighbor (which is located on the average at a distance $2R_i$),

$$E_{\zeta,nl} \simeq -\frac{\zeta e^2}{2R_i} \tag{3.2.6}$$

Equation (3.2.6) is the limit at which a bound electron in state nl is attracted to the adjacent ion with the same force as to its original binding ion. Since $E_{\zeta,nl} \simeq -\zeta e^2/2r_n$, equation (3.2.6) is equivalent mathematically to equation (3.2.5).

(3) A third criterion states that pressure ionization occurs when enough free electrons are compressed inside the bound electron orbit to screen it completely from its own nucleus. Referring to the discussion leading to equation (2.4.10),

such a complete screening occurs when $\zeta = Z - Z_b$ free electrons are located inside the electron orbital. But ζ is also the number of free electrons inside the ion sphere. If they are uniformly distributed over the ion volume, the fractional number inside an orbital of radius r_n is approximately

$$\delta\zeta = \zeta \left(\frac{r_n}{R_i} \right)^3 \tag{3.2.7}$$

Pressure ionization occurs when $\delta\zeta \simeq \zeta$, or $r_n = R_i$, so that this criterion is again identical to equation (3.2.5).

A slightly different definition of the pressure ionization was used by Stein and Salzmann (1992). Their starting point is the observation that a bound electron is localized within the volume of a given ion as long as its binding energy is well below the lowest potential barrier that separates its ion from the neighboring ones. When the energy approaches this lowest potential barrier, the electron may tunnel out from the ionic potential well. When this happens, its wavefunction overlaps two or more ions, see figure 3.3. Electronic states of increasingly higher energy overlap greater and greater "clusters" of ions. Above some (negative) critical value of the electron energy, one of the clusters *percolates*, that is, the wavefunction of an electron having this or higher energy overlaps a macroscopic portion of the plasma ions. When this occurs, the electron can be regarded as a negative-energy continuum electron. Results of computations (Stein and Salzmann, 1992) based on this type of percolation theory for a krypton plasma indicate that an additional upward shift of 10–15% of the ionic levels relative to equation (3.2.2) should be incorporated. This additional shift originates from the different definition of the continuum lowering, namely, that the electron energy has to be high enough to overlap a large number of ions, rather than just two, as in the previous formulation.

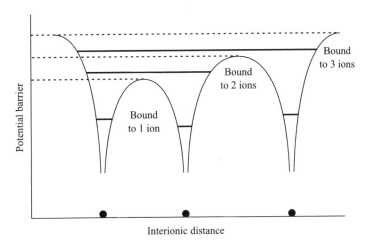

Figure 3.3 Schematic illustration of the electronic energy levels in hot and dense plasmas. (From Stein and Salzmann, 1992.)

In spite of its basic importance, experimental evidence for the continuum lowering is scarce. In fact, so far only three experiments have been reported in the literature that provide direct evidence for this effect (Bradley et al., 1987; Riley et al., 1989). The experimental difficulty arises for several reasons. First, continuum lowering in plasmas is generally not a pure effect but rather is mixed with other level shifting effects, which are not connected to the plasma potential. Second, in low density plasmas the shift is small and difficult to measure, whereas in high density plasmas the shift changes rapidly in time and its measurement requires special techniques.

S. J. Rose, D. K. Bradley, J. Kilkenny, and J. D. Hares measured the energy of the K-shell photoabsorption edge in a chlorine plasma compressed to high density by a strong laser generated shock wave (Bradley et al., 1987). The K-edge of natural uncompressed chlorine lies between 2820 and 2824 eV. They irradiated layered targets with high intensity (5×10^{14}–2×10^{15} W/cm^2) laser beams. The targets consisted of 1 μm bismuth, a variable thickness of polyester, 5 μm of KCl, and 10 μm of parylene-N, see figure 3.4. The laser beam was incident on the bismuth-coated side of the target. Under these experimental conditions, two things happen: First, the bismuth emits a bright quasicontinuum of x-rays in the spectral region between 2600 and 2900 eV, in the region around the chlorine K-edge. Second, a strong shock wave propagates into the target, compressing the KCl slab to a few times its original density. The time history of the x-ray pulse approximately follows the 0.9 ns laser pulse. Time-integrated absorption spectra of x-rays by the chlorine ions are shown in figure 3.5 for three experimental conditions. Spectrum (a) shows the absorption by a weakly compressed chlorine target; spectrum (c) corresponds to a strongly compressed case; and (b) is an intermediate case. The shift and the broadening of the edge are clearly seen. The total shift of the edge under their experimental conditions is the sum of

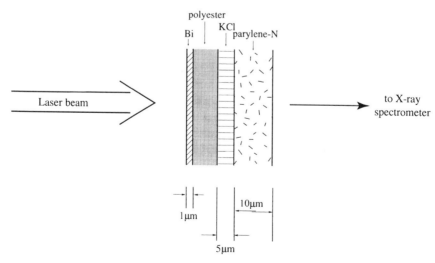

Figure 3.4 The setup in the experiment of Rose et al. (From Bradley et al., 1987.)

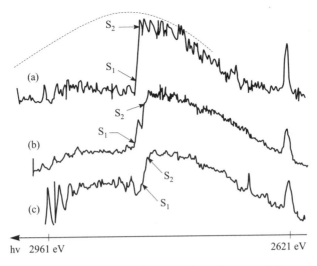

$h\nu$ 2961 eV   2621 eV

Figure 3.5 Time-integrated spectra from the rear of the
target showing the chlorine K photoabsorption edge for
increasing heating of KCl. The downward shift of the
edge is clearly seen. (From Bradley et al., 1987.)

three contributions: The first is the effect of the pressure ionization of outer shells,
which provides the K-shell electrons with a stronger binding energy and thereby
shifts the K-edge to higher energy. This effect shifts the edge by $+47.56$ eV. The
second contribution is the continuum lowering, as explained above, which in their
experiments shifts the edge downward by -45.2 eV. Finally, one must also consider the partial degeneracy of the free electron states close to the ionization
energy. This effect shifts the edge an additional -6.1 eV to lower energies. The
sum of the three effects is -3.7 eV, which is in fairly good agreement with the
experimental shift. This experimental shift could not be explained without taking
the continuum lowering into account.

A similar experiment was carried out by S. Rose, D. Riley, O. Willi and T.
Afshar-Rad (Riley et al., 1989). In this experiment, a total *blue* shift (that is,
toward higher energies) was found, because the shift due to the pressure ionization of the outer shells was larger than in the previous experiment. The contribution of the continuum lowering is, however, always a downward shift, and again,
as in the previous experiment, the total shift cannot be explained without the
presence of the continuum lowering.

A more direct measurement of the continuum lowering was made in an experiment by K. Eidmann and W. Schwanda (Schwanda and Eidmann, 1992). They
used x-rays from a laser-produced gold plasma for two purposes: first, to heat a
thin beryllium foil and, second, to absorb the x-rays incident upon this foil. An x-
ray streak camera was used to measure the time variation of the transmitted
spectrum. Their results are reproduced in figure 3.6. Figure 3.6(a) shows the
unabsorbed spectrum emitted by the gold source. Figure 3.6(b) shows the same

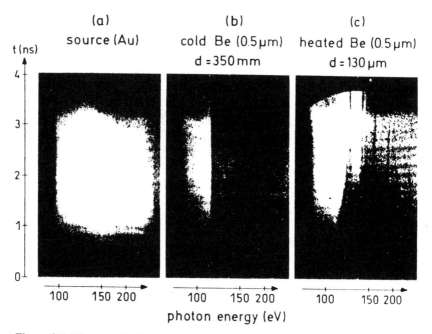

Figure 3.6 Time-resolved spectrum obtained with the source, and with a cold and x-ray-heated 0.5 μm thick beryllium target. (From Schwanda and Eidmann, 1992.)

spectrum after absorption by a cold beryllium foil. The absorption by the sharp K-edge of beryllium above $\simeq 111$ eV is clearly seen. The third spectrum, figure 3.6(c), was taken when the beryllium foil was heated by the x-rays (in practice this was done by bringing the foil to a distance of 0.013 cm from the emitting gold foil). In figure 3.6(c) one can clearly observe the shift of the K-edge from its original position at $t \simeq 0.5$ ns to about 140 eV at $t = 2$ ns. Similar experiments were carried out with aluminum and beryllium foils (Eidmann et al., 1994). The K-edge shifts obtained in these experiments are shown in figure 3.7.

3.3 Continuum Lowering in Weakly Coupled Plasmas

The difficulty with the above approaches is the fact that in a real plasma it is hard to define a single value for the continuum lowering that can be applied to all the ions In fact, every ion feels a different continuum lowering, depending on the local instantaneous microfield in its vicinity. So, rather than looking for one quantity, it seems more realistic to address a question of a more probabilistic nature: What is the *probability* that an ion in the plasma will be subject to a continuum lowering of value χ? Such a probabilistic approach seems to have more physical meaning than trying to select one single quantity to represent the entire phenomenon.

To answer this question, we shall consider only the case of weakly coupled plasmas. In such plasmas, every ion has only one closest neighbor, which makes the largest contribution to the local plasma potential. Obviously, the probability

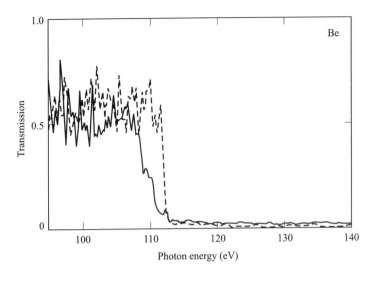

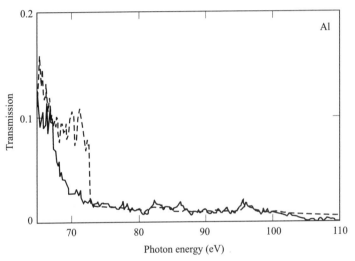

Figure 3.7 Edgelike structure measured early in time (at ≈ 1 ns) during heating (solid line), in comparison to the measured edge of the cold material (dashed line) for beryllium (top) and aluminum (bottom). (Reprinted from Eidmann, K., Schwanda, W., Földes, I. B., Sigel, R., and Tsakiris, G. D., 1994, *J. Quant. Spectrosc. Rad. Trans.*, **51**, 77, with permission from Elsevier Science Ltd.)

for having a continuum lowering between χ and $\chi + d\chi$ is closely related to the probability that the nearest neighbor will be at some given distance. The probability density to have the nearest neighbor at a distance R_n was calculated in section 2.7 for weakly coupled plasmas, see equation (2.7.1). Assuming two identical ions, each of charge ζ, at a distance R_n from each other, the continuum lowering (see figure 3.8) is

$$\chi_\zeta = -2 \frac{\zeta e^2}{R_n/2} \tag{3.3.1}$$

For each interionic distance, R_n, there corresponds a different continuum lowering. The probability $P_\chi(\chi)d\chi$, having a continuum lowering between χ and $\chi + d\chi$ is connected to the probability, $P_R(R_n)\,dR_n$, of having the nearest neighbor at a distance between R_n and $R_n + dR_n$ by (for clarity, the index ζ is dropped from χ)

$$P_\chi(\chi)\,d\chi = P_R(R_n) \frac{dR_n}{d\chi}\,d\chi \tag{3.3.2}$$

Substituting equation (2.7.1) for $P_R(R_n)$ and calculating the derivative by using (3.3.1), one obtains the probability density function for the continuum lowering:

$$P_\chi(\chi)d\chi = -3 \left(\frac{\epsilon_0}{\chi}\right)^4 \frac{d\chi}{\epsilon_0} \exp\left[-\left(\frac{\epsilon_0}{\chi}\right)^3\right] \tag{3.3.3}$$

where

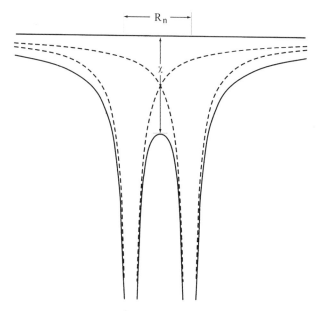

Figure 3.8 Definition of the quantities in equation (3.3.1).

$$\epsilon_0 = -\frac{4\zeta e^2}{R_i} \qquad (3.3.4)$$

is the continuum lowering when the two ions are at a distance of the ion sphere radius from each other. (Note that ϵ_0 is a negative energy, so that $P_\chi(\chi)$ is always positive.)

The probability function, $P_\chi(\chi)$, is shown in figure 3.9. It goes steeply to zero for small values of χ, reaches its maximum around $\chi/\epsilon_0 = 0.90$, with the values of the half-maximum at $\chi/\epsilon_0 = 0.68$ and $\chi/\epsilon_0 = 1.37$. It turns out that χ has a rather limited range of variation around ϵ_0, which is, of course, a consequence of the limited variations in the distances to the nearest neighbor, R_n.

The moments of the continuum lowering are readily calculated from equation (3.3.3),

$$\langle \chi^k \rangle = \int_{-\infty}^{0} \chi^k P_\chi(\chi)\, d\chi \qquad (3.3.5)$$

Substituting equation (3.3.3) into equation (3.3.5) and using the new integration variables, $y = (\epsilon_0/\chi)^3$, $\chi = \epsilon_0 y^{-1/3}$, $d\chi = -(1/3)\epsilon_0 y^{-4/3}\, dy$, one obtains

$$\langle \chi^k \rangle = \epsilon_0^k \Gamma\left(1 - \frac{k}{3}\right), \qquad k < 3 \qquad (3.3.6)$$

where $\Gamma(x)$ is the Gamma function. The average continuum lowering is obtained by setting $k = 1$,

$$\langle \chi \rangle = \epsilon_0 \Gamma\left(\frac{2}{3}\right) = -\frac{4\zeta e^2}{R_i}\Gamma\left(\frac{2}{3}\right)$$

$$= -4\left(\frac{4\pi}{3}\right)^{1/3}\zeta e^2 n_i^{1/3}\Gamma\left(\frac{2}{3}\right) = -12.57\,\text{eV}\,\zeta\left(\frac{n_i}{10^{21}\,\text{cm}^{-3}}\right)^{1/3} \qquad (3.3.7)$$

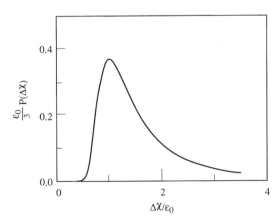

Figure 3.9 Probability density function of the continuum lowering.

This continuum lowering has the same dependence on n_i and ζ as in equation (3.2.3). The difference in the numerical coefficient stems from the different definitions used to derive these quantities.

The standard deviation due to the fluctuations of the continuum lowering is

$$\sigma_\chi = \sqrt{\langle \chi^2 \rangle - \langle \chi \rangle^2} = |\epsilon_0| \sqrt{\Gamma\left(\frac{1}{3}\right) - \Gamma^2\left(\frac{2}{3}\right)} = 0.919 |\epsilon_0| \qquad (3.3.8)$$

and the relative amplitude of the fluctuations is

$$\left| \frac{\sigma_\chi}{\langle \chi \rangle} \right| = 0.679 \qquad (3.3.9)$$

This last result indicates the strongly fluctuating character of the continuum lowering in weakly coupled plasmas.

From equation (3.3.3) one can derive another result, namely, the probability function that a given ionic bound state with energy E ($E < 0$) is bound to only one ion. This is exactly the probability that the energy of the state, E, is lower than the local value of the continuum lowering, χ,

$$P_E\{E \text{ bound}\} = P_E(E \le \chi) = \int_E^0 P_\chi(\chi)\, d\chi \qquad (3.3.10)$$

Substituting equation (3.3.3) into (3.3.10), one obtains

$$P_E\{E \text{ bound}\} = -3 \int_E^0 \left(\frac{\epsilon_0}{\chi}\right)^4 \frac{dx}{\epsilon_0} \exp\left[-\left(\frac{\epsilon_0}{\chi}\right)^3\right] = \exp\left[-\left(\frac{\epsilon_0}{E}\right)^3\right] \qquad (3.3.11)$$

This probability equals 1 for deeply bound states, well below ϵ_0, and tends to 0 for highly excited states, when $E \to -0$. Denoting by $E_{1/2}$ the energy of the state that has a probability of $P_E(E_{1/2}) = 1/2$ of being bound, one finds

$$E_{1/2} = \frac{\epsilon_0}{(\log 2)^{1/3}} = 1.3\epsilon_0 = -10.5\,\text{eV}\,\zeta\left(\frac{n_i}{10^{21}\,\text{cm}^{-3}}\right)^{1/3} \qquad (3.3.12)$$

The states that have probability of $P_E(E_{0.9}) = 0.9$ and $P_E(E_{0.1}) = 0.1$ of being bound have binding energies of $E_{0.9} = 2.12\epsilon_0$ and $E_{0.1} = 0.75\epsilon_0$, respectively.

Further simplifications occur in plasmas consisting of hydrogenlike ions only. The energy levels of such ions are given by $E_{n_p} = -Z^2 e^2 / 2a_0 n_p^2$, where a_0 is the Bohr radius, n_p is the principal quantum number, and $\zeta = Z - 1 (Z \ne 1)$. Comparing $E_{1/2}$ with E_{n_p}, one finds the principal quantum number $n_{p,1/2}$ corresponding to the state that has a probability of $1/2$ of being bound,

$$n_{p,1/2} = \left(-\frac{Z^2 E_H}{\epsilon_0}(\log 2)^{1/3}\right)^{1/2} = 1.44 \frac{Z}{\sqrt{Z-1}}\left(\frac{n_i}{10^{21}\,\text{cm}^{-3}}\right)^{-1/6} \qquad (3.3.13)$$

The states that are bound with probabilities of $P_E(E_{0.9}) = 0.9$ and $P_E(E_{0.1}) = 0.1$ have principal quantum numbers of

$$n_{p,0.9} = 0.728 n_{p,1/2}, \qquad n_{p,0.1} = 1.22 n_{p,1/2} \qquad (3.3.14)$$

respectively. $n_{p,1/2}$ is a very slowly varying function of the ion density, decreasing, for example, by only a factor of 3 when the plasma density increases by 3 orders of magnitude. It also varies rather weakly with the atomic number Z of the plasma material.

A numerical example may help to illustrate these results. Assume a neon plasma at ion density of $10^{17}\,\text{cm}^{-3}$ and electron temperature of $100\,\text{eV}$. Under these conditions H-like ions are the most abundant species in the plasma. At this density the ion sphere radius is $3.5 \times 10^{-6}\,\text{cm}$. The ion–ion coupling constant is $\Gamma = Z^2 e^2 / R_i T = 0.11$, which is low enough for the plasma to be regarded as weakly coupled. The radius of a ground state ion is $a = a_0/Z = 5.3 \times 10^{-10}\,\text{cm}$, very much smaller than the ion sphere radius. The average distance to the nearest neighbor and the corresponding fluctuations are obtained from equations (2.7.5) and (2.7.7): $R_n = 0.893 R_i \pm 36\% = (3.1 \pm 1.1) \times 10^{-6}\,\text{cm}$. The average continuum lowering and its fluctuations are calculated from equations (3.3.7)–(3.3.8): $\chi = -5.8\,\text{eV} \pm 68\% = -(5.8 \pm 3.9)\,\text{eV}$. The ionic states that have probabilities of $P_E = 0.9, 0.5$ and 0.1 of being bound have energies of $E_{0.9} = -9.1\,\text{eV}$, $E_{0.5} = -4.9\,\text{eV}$, and $E_{0.1} = -3.2\,\text{eV}$, corresponding to $n_p = 13, 18$, and 22, respectively. Figure 3.10 describes the probability function for this example.

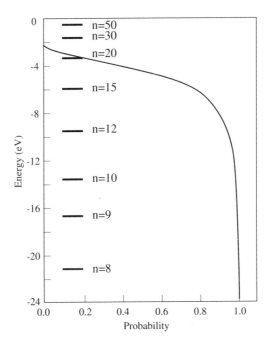

Figure 3.10 Probability density function for an ionic level to be bound. The figure represents the probability for neon plasma at ion density of $10^{17}\,\text{cm}^{-3}$ and electron temperature of $100\,\text{eV}$.

At a higher ion density and temperature, $n_i = 10^{21}$ cm^{-3} and $T = 200$ eV, equation (3.3.14) predicts that only the four innermost states ($n_p \leq 4$) are bound with probability higher than 90%. Under these conditions, however, $\Gamma \simeq 0.5$ so that the assumption of weak coupling is fulfilled only marginally.

3.4 The Partition Function

The excitational part of the partition function of the ions, equations (1.3.3) and (1.3.5), is, perhaps, the parameter that is most profoundly affected by the exact value of the continuum lowering. The formal definition of the excitational partition function of an ion of charge ζ is

$$Z_\zeta(T) = \sum_{n_p=1}^{\infty} g_{\zeta,n_p} e^{-E_{\zeta,n_p}/T} \qquad (3.4.1)$$

where n_p is the principal quantum number, and g_{ζ,n_p} and E_{ζ,n_p} are the statistical weight and the energy of the atomic shells, see equation (1.3.3). In equation (3.4.1), E_{ζ,n_p} is measured relative to the ground state energy of the ion. The partition function is a basic thermodynamic quantity, from knowledge of which one can deduce other thermodynamic parameters, such as the Gibbs free energy, the plasma pressure, internal energy, entropy, and other related quantities.

Mathematically, $Z_\zeta(T)$ diverges for an isolated ion. This can be understood from the fact that for highly excited states E_{ζ,n_p} converges to an almost constant value, namely, the ionization energy, whereas g_{ζ,n_p} grows as $2n_p^2$, so that the sum in equation (3.4.1) is infinite. Moreover, $\partial Z_\zeta(T)/\partial T$ diverges as well, which means that every tiny change in the temperature generates an infinite change in Z_ζ, which is, of course, in contradiction to all the laws of physics. The obvious flaw in this argument is that the partition function, being a statistical thermodynamic quantity, is meaningless for one isolated ion. On the other hand, an ensemble of ions in a plasma is subject to their collective continuum lowering, which excludes the possibility of the existence of an infinite number of bound states. One can consider the above argument as a proof of the statement that an ion immersed in a plasma cannot have an infinite number of excited states (otherwise its partition function would diverge).

The uppermost excited state of an ion that can still be regarded as bound is determined by the requirement that the energy of the state is below the lowered continuum; therefore, the average value of the continuum lowering, $\langle \chi_\zeta \rangle$, can be taken as a natural limit to truncate the sum in equation (3.4.1). Such a truncation immediately yields a finite result for $Z_\zeta(T)$. Such a procedure is, however, not very satisfactory, because it is sensitive to the exact value of the truncation point, and the result of the summation may alter significantly with moderate changes in the number of terms in the sum. Moreover, in a real plasma every ion is subject to a different continuum lowering, in accordance with the local instantaneous microfield in its vicinity, so that, to better describe the conditions in such a plasma, the sum should be truncated at a different value for every single ion.

The influence of the continuum lowering fluctuations on the partition function was considered by Zimmerman and More (1980). They included these fluctuations through a correction term to the statistical weight, g_{ζ,n_p}, which depends on the ratio of the radius of the ionic orbital to the ion sphere radius. This is, in fact, the continuation of the idea leading to the criterion expressed by equation (3.2.5). They proposed to use in equation (3.4.1) a corrected statistical weight, g^*_{ζ,n_p}, that has the form

$$g^*_{\zeta,n_p} = g_{\zeta,n_p} \frac{1}{1 + (\alpha r_{n_p}/R_i)^{\beta}} \tag{3.4.2}$$

Here, r_{n_p} is the radius of the atomic orbital, and $\alpha \cong 3$, $\beta \cong 4$ are numerical constants, determined by some semiempirical method. The meaning of equation (3.4.2) is that the statistical weights of ionic states with radii close to the ion sphere radius are diminishing continuously. Such a continuous decrease makes the summation much less sensitive to the number of terms than truncating the sum abruptly at a single cutoff value.

In a slightly different approach, one may try to use the energy, E_{n_p}, of the ionic state, rather than its radius, as the characteristic parameter to define a criterion for the correction of the statistical weight. In this case, the probability function $P_E\{E \text{ bound}\}$, equation (3.3.10), can be taken as a natural correction factor to be incorporated in equation (3.4.2). For weakly coupled plasmas, one obtains from equation (3.3.11)

$$g^*_{\zeta,n_p} = g_{\zeta,n_p} \exp\left[-\left(\frac{\epsilon_0}{E_{\zeta,n_p}}\right)^3\right] \tag{3.4.3}$$

with ϵ_0 given in equation (3.3.4). This approach seems to have a better theoretical foundation than equation (3.4.2). Inserting equation (3.4.3) into (3.4.1), we find for the partition function,

$$Z_\zeta(T) = \sum_{n_p} g^*_{\zeta,n_p} \exp\left(-\frac{E_{\zeta,n_p}}{T}\right)$$

$$= \sum_{n_p} g_{\zeta,n_p} \exp\left[-\left(\frac{\epsilon_0}{E_{\zeta,n_p}}\right)^3\right] \exp\left(-\frac{E_{\zeta,n_p}}{T}\right) \tag{3.4.4}$$

The terms in the sum vanish rapidly for $\epsilon_0 < E_{\zeta,n_p} < 0$, thereby reducing its sensitivity to the details of the model.

For hydrogenlike plasmas, this formula can be reduced to a simpler form. The term $(\epsilon_0/E_{\zeta,n_p})^3$ in equation (3.4.4) is proportional to n_p^6, whereas the term $E_{\zeta,n_p}/T$ decreases as n_p^{-2}. Assume that n_0 is the smallest principal quantum number for which the first term in the exponent becomes larger than the second one. Straightforward calculation shows that this occurs for

$$E^4_{\zeta,n_0} \leq |\epsilon_0|^3 T \tag{3.4.5}$$

Inserting the level energies of hydrogenlike ions, $E_{\zeta,n_0} = -Z^2 E_H/n_0^2$, one obtains

$$n_0 \geq \left[\frac{Z^2 E_H}{|\epsilon_o|}\left(\frac{|\epsilon_o|}{T}\right)^{1/4}\right]^{1/2} = n_{p,1/2}\frac{1}{(\log 2)^{1/6}}\left(\frac{|\epsilon_o|}{T}\right)^{1/8}$$

$$= 1.15Z\left[\frac{1}{(Z-1)^3}\frac{E_H}{T}\left(\frac{n_i}{10^{21}\,\text{cm}^{-3}}\right)^{-1}\right]^{1/8} \tag{3.4.6}$$

where $E_H = 13.6\,\text{eV}$ is the neutral hydrogen ground state binding energy. An estimate of the partition function is obtained if the sum can be carried out accurately up to n_0, and the contribution of all the higher closely spaced states is estimated by an integral.

$$Z_\zeta(T) \approx \sum_{n=1}^{n_0} g_n \exp\left[-\left(\frac{\epsilon_0}{E_n}\right)^3\right]\exp\left(-\frac{E_n}{T}\right)$$

$$+ \int_{n_0}^{\infty} 2n^2 \exp\left[-\left(\frac{\epsilon_0}{(-Z^2 E_H)}\right)^3 n^6\right] dn \tag{3.4.7}$$

The integration can be carried out in terms of tabulated special functions. The result is

$$Z_\zeta(T) \approx \sum_{n=1}^{n_0} g_n \exp\left[-\left(\frac{\epsilon_0}{E_n}\right)^3\right]\exp\left(-\frac{E_n}{T}\right) + \frac{1}{3}\left(\frac{-Z^2 E_H}{\epsilon_0}\right)^{3/2}\Gamma\left[\frac{1}{2}; \left(\frac{\epsilon_0}{E_{n_0}}\right)^3\right] \tag{3.4.8}$$

where $\Gamma(q;x) = \int_x^\infty t^{q-1}e^{-t}\,dt$ is the incomplete Gamma function; (see equation 6.5.3 in Abramowitz and Stegun, 1965). The algorithm in equation (3.4.8) is not very sensitive either to the exact value of n_0 or to the details of the model.

Practically, the exact shape of the correction factor to g_{ζ,n_p} in equations (3.4.2–3) is of no importance. Any function that has the value of 0.5 in the vicinity of the average continuum lowering, and decreases from 0.9 to 0.1 within a few neighboring levels, will give acceptable results. Since any such function will provide the correct values of 1 and 0 for low and high lying levels, respectively, only for states near the average continuum limit would one expect moderate inaccuracies.

3.5 Line Shift in Plasmas

The shift of spectral lines in hot dense plasmas, called the *polarization shift* in some papers, is a genuine plasma effect generated by the modification of the ionic potential by the free electron environment. Two effects can cause the shift in the positions of spectral lines: The first is electron–ion or ion–ion collisions, the second is generated by the spatial variation of the electric potential produced by the free electrons. The first of these effects, termed *collisional dynamic shift*, will be discussed in chapter 7. The second effect, which is caused by the static component of the electrostatic potential, is the topic of the present section.

In section 3.2, we saw that to first order all the ionic levels move upward by a constant amount, so that the energy difference between the levels does not change. In this approximation, therefore, one would not expect line shifts. As a second order approximation, however, the polarization of the free electron distribution around the ionic nucleus influences the various ionic states energies slightly differently, thereby modifying the energy differences and the corresponding energy of a photon emitted during a transition between two levels. An experimental identification of this effect is important, because it provides a very sensitive mapping of the free electron distribution within the ionic volume.

The shift of a spectral line, $\Delta h\nu$, is the difference between the energy of the photon, $h\nu$, emitted in a transition from an initial state, $|i\rangle$, to a final state, $|f\rangle$, when the ion is immersed in a dense plasma, and the transition energy, $h\nu_0$, between the same two states when the ion is isolated,

$$
\begin{aligned}
\Delta h\nu &= h\nu - h\nu_0 \\
&= [\langle i|\mathcal{H}_0 - eV_{pl}(r)|i\rangle - \langle f|\mathcal{H}_0 - eV_{pl}(r)|f\rangle] - [\langle i|\mathcal{H}_0|i\rangle - \langle f|\mathcal{H}_0|f\rangle] \\
&= \langle i| - eV_{pl}(r)|i\rangle - \langle f| - eV_{pl}(r)|f\rangle
\end{aligned} \tag{3.5.1}
$$

where $\mathcal{H}_0$ is the Hamiltonian of the isolated ion, and $V_{pl}(r)$ is the electrostatic potential generated by the free electrons, equations (3.1.1–2). As already mentioned in section 3.4, $V_{pl}(r)$ is always negative, therefore the perturbation to the Hamiltonian, $-eV_{pl}(r) = e|V_{pl}(r)|$, is always positive. Moreover, an electron distribution polarized around the nucleus generates a potential that is a monotonically decreasing function of the distance r. It turns out that the expectation value of the perturbation $e|V_e(r)|$ at the initial excited state is always smaller than at the final state, implying that the line shift $\Delta h\nu$, is negative, that is, the shift is toward lower energy and higher wavelength—it is a *red shift*.

To get some intuitive feeling about this effect, we again take the paradigm of a homogeneous free electron distribution ,which is a rather good approximation for weakly and intermediately coupled plasmas, and which we have already used to estimate the continuum lowering (equation (2.5.8), quoted also in section 3.4),

$$
V_{e,pl}(r) = -\frac{3}{2}\frac{\zeta e}{R_i}\left[1 - \frac{1}{3}\left(\frac{r}{R_i}\right)^2\right] = -\frac{3}{2}\left(\frac{4\pi}{3}\right)^{1/3}\zeta e n_i^{1/3}\left[1 - \frac{1}{3}\left(\frac{r}{R_i}\right)^2\right] \tag{2.5.8}
$$

Inserting this into equation (3.5.1), one obtains

$$
\begin{aligned}
\Delta h\nu &= h\nu - h\nu_0 \\
&= \left\langle i\left|\frac{3}{2}\frac{\zeta e^2}{R_i}\left[1 - \frac{1}{3}\left(\frac{r}{R_i}\right)^2\right]\right|i\right\rangle - \left\langle f\left|\frac{3}{2}\frac{\zeta e^2}{R_i}\left[1 - \frac{1}{3}\left(\frac{r}{R_i}\right)^2\right]\right|f\right\rangle \\
&= \frac{1}{2}\frac{\zeta e^2}{R_i^3}(r_f^2 - r_i^2) = \frac{2\pi}{3}\zeta e^2 n_i(r_f^2 - r_i^2)
\end{aligned} \tag{3.5.2}
$$

where $r_i^2 = \langle i|r^2|i\rangle$ is the average value of r^2 at state i, and likewise for r_f^2.

For hydrogenic Rydberg states (Condon and Shortley 1987),

$$r_{n_p l}^2 = \frac{a_0^2 n_p^2}{2Z^2} [5n_p^2 + 1 - 3l(l+1)] \tag{3.5.3}$$

so that, for instance, for a hydrogenic $2p$–$1s$ Lyman-α transition one obtains

$$\Delta h\nu = -18\pi \frac{e^2 a_0^2}{Z} n_i = -0.22\,\text{eV}\,\frac{1}{Z}\frac{n_i}{10^{21}\,\text{cm}^{-3}} \tag{3.5.4}$$

Measurements of the plasma line shifts started in the early 1960s and 1970s (Berg et al., 1962; Greig et al., 1970; Gabriel and Volonte, 1973; Goto and Burgess, 1974; Neiger and Griem, 1976), mainly with neutral hydrogen or singly ionized helium atoms in plasmas at densities below $10^{18}\,\text{cm}^{-3}$. The accuracy of these measurements was relatively low (Griem, 1988) and in some experiments blue shifts were found. The experimental line shifts in these measurements were, however, only a few percent of the lines' full width (Griem, 1974), and were barely larger than their error bars.

The theory of line shifts was developed in a series of papers during the 1970s and 1980s. Self-consistent calculations that take into account the mutual interactions of the free and bound electrons within the framework of a finite volume Thomas–Fermi model were carried out by Rozsnyai (1972, 1975) and Weisheit and Rozsnyai (1976). S. Skupsky (1980) suggested using the line shifts of hydrogenlike neon and argon ions in a compressed deuterium–tritium pellet as a diagnostics tool to measure densities of laser compressed plasmas. Computations of the line shifts in aluminum were carried out by J. Davis and M. Blaha (1982) using detailed continuum wavefunctions for the free electrons, and by D. Salzmann and H. Szichman (1987), who used a high accuracy Hartree–Dirac–Slater code to calculate the bound electron wavefunctions and energy levels. Nguyen Hoe et al. (1986) included static and dynamic effects in their calculations. Generally, the results of all these calculations differ only slightly from equation (3.5.3), which only means that a homogeneous free electron distribution is a plausible approximation for all practically interesting laboratory plasmas. Recently, H. Griem (1988) has carried out refined calculations of the contributions of the $\Delta n = 0$ transitions to the line widths and shifts of hydrogen and singly ionized helium in the dipole approximation.

Experiments in the 1970s and 1980s with hydrogen and He^+ (Grützmacher and Wende, 1978; Pittman and Fleurier, 1986) have confirmed that the lines are red shifted. Recently, high accuracy experiments have been carried out by H. J. Kunze and his collaborators (Bödecker et al., 1995), who have measured the position of the Ly_α line of hydrogen atoms immersed in a helium gas with electron density of 10^{18}–$10^{19}\,\text{cm}^{-3}$. Results of these experiments are in good agreement (Bödecker et al., 1995) with the detailed computations of Griem (1988).

While the theoretical and experimental results of line shifts of H and He^+ have started to converge, experiments aimed at measuring the line shifts in higher-Z material have been less successful. A measurement of the Lyman lines of Be(IV), B(V) and C(VI) in laser compressed plasmas (Goldsmith et al., 1984) resulted in questionable results, mainly because of difficulties in the wavelength calibration

(Griem, 1988). Several other experiments carried out for high-Z ions in high density laser generated plasmas did not show measurable line shifts.

To date, the only successful experiment for measuring line shifts in materials of higher Z has been reported by Jamelot et al. (1990), who used an x-ray streak camera to measure the wavelength of the Ly_α spectral line from a laser produced lithium plasma. A measurable shift was observed at early times of the plasma's evolution (see figure 3.11) when the density of the emitting region is still high. The Ly_α line ($\lambda = 135$ Å) was shifted approximately by 0.1 Å. This is about 16% of the line's FWHM, which was about 0.65 Å in their experiment.

It should be emphasized that an experimental measurement of plasma line shifts is a very difficult undertaking for several reasons:

1. The shift is rather small. For example, for $Z = 10$, the shift achieves a measurable value, say $\Delta h\nu = 0.1$ eV, only for $n_i \sim 5 \times 10^{21}$ atoms/cm^3 or higher. Such a high density plasma at a temperature of 100 eV, for which H-like ions are abundant, can be generated at present only by large lasers for a short fraction of the plasma lifetime.
2. Line shifting is a second order effect which is very sensitive to the details of the free electron distribution around the nucleus. The formulas found in the literature, as well as equation (3.5.3), may turn out to be model dependent oversimplifications, and may be altered significantly when this distribution is computed more accurately.
3. For high density plasmas the lines are not only shifted but broadened as well. As the density increases, both the shift and the broadening grow. At very high densities, where one would expect a measurable shift, the broad-

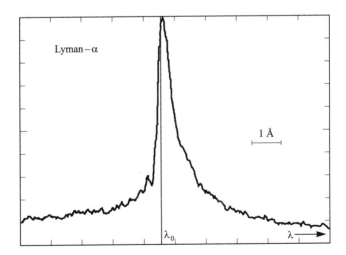

Figure 3.11 Time-resolved measurement of the Lyman-α line in a lithium plasma. The shift from the original position (λ_0) is clearly seen. (Reprinted from Jamelot, G., Jaegle, P., Lemaire, P., and Carillon, A., 1990, *J. Quant. Spectrosc. Rad. Trans.*, **44**, 71, with permission from Elsevier Science Ltd.)

ening may be so large that the shift is only a small percentage of the total width and is no longer observable. Experimentally, line shift measurements appear to be very difficult.

4. In view of the discussion above, it would seem that laser produced plasmas are natural candidates in which to look for this effect. The emission spectrum of laser produced plasmas is, however, the superposition of spectra emitted from different density and temperature regions. The high intensity unshifted spectrum emitted from the low density, high temperature portion of a plasma may overlap the shifted spectrum originating from the higher density regions.

At present, the question of line shifts is still not closed and more experimental data is required to advance this topic.

4

Atomic Processes in Hot Plasmas

4.1 Classification of the Atomic Processes

The plasma constituents, namely the electrons, ions and photons, interact through their electromagnetic fields, thereby transferring energy from one particle to the other. The interactions of the greatest interest for plasma physics are those in which an ion interacts with an electron or a photon, and which result in changes either in the state of the ionization or in the excitation of the interacting ion. The most abundant processes are listed in table 4.1. This chapter is devoted to a study of these processes, the various methods for their classification, their behaviour as a function of the plasma temperature and density, and working formulas to compute their rates.

We start this chapter by introducing the processes and the corresponding terminology. The list of the processes includes two ionic transitions in which an ion decays spontaneously from an excited state to a lower energy state, without interaction with any other particle, see table 4.1. These transitions are:

- *Spontaneous decay* in which an excited ion decays into a lower (ground or excited) state, emitting a photon whose energy equals the energy difference between the upper and lower states. A schematic description of this process is shown in figure 4.1(a).
- *Autoionization.* In this process the initial state is a doubly excited ion, in which two electrons are in excited shells, see figure 4.1(b). Such doubly excited ions are generated quite frequently by the dielectronic recombination process. It can also be produced by two successive excitations by two electrons, or even through a double excitation by a single electron. During autoionization one of the excited electrons decays to a lower state, mostly the ground state, while the second takes its energy and is emitted into the continuum. Obviously, autoionization can occur only if the sum of the energies of the two electrons in the initial states is higher than the ionization energy of the ion.

Table 4.1 Atomic Processes in Hot Plasmas

Reaction	Direct process	Rate (cm^{-3} s^{-1})	Inverse process	Rate (cm^{-3} s^{-1})
$A_{m'}^{+\zeta} \Leftrightarrow A_m^{+\zeta} + h\nu$	Spontaneous decay	$N_{\zeta,m'}A(\zeta, m' \to m)$	Resonant photoabsorption	$n_\gamma N_{\gamma,m}B(\zeta, m \to m')\Delta h\nu$
$A_m^{+\zeta} + e \Leftrightarrow A_{m'}^{+\zeta+1} + e + e$	Electron impact ionization	$n_e N_{\zeta,m}\mathcal{I}(\zeta, m \to \zeta+1)$	3-Body recombination	$n_e^2 N_{\zeta+1}\mathcal{R}^{(3)}(\zeta+1 \to \zeta, m)$
$A_m^{+\zeta} + e \Leftrightarrow A_{m'}^{+\zeta} + e$	Electron impact excitation	$n_e N_{\zeta,m}\mathcal{E}(\zeta, m \to \zeta, m')$	Electron impact deexcitation	$n_e N_{\zeta,m'}\mathcal{D}(\zeta, m' \to \zeta, m)$
$A_m^{+\zeta} + h\nu \Leftrightarrow A_{m'}^{+\zeta+1} + e$	Photoionization	$n_\gamma N_{\zeta,m}\mathcal{I}_{ph}(\zeta, m \to \zeta+1)$	Radiative recombination	$n_e N_{\zeta+1}\mathcal{R}^{(r)}(\zeta+1 \to \zeta, m)$
$A_{mm'}^{+\zeta} \Leftrightarrow A_0^{+\zeta+1} + e$	Autoionization	$N_{\zeta,mm'}\mathcal{A}(\zeta, mm' \to \zeta+1)$	Dielectronic recombination	$n_e N_{\zeta+1}\mathcal{R}^{(d)}(\zeta+1 \to \zeta, mm')$
$A_m^{+\zeta} + e \Leftrightarrow A_m^{+\zeta} + h\nu$	Bremsstrahlung	see Chapter 6	Inverse Bremsstrahlung	see Chapter 9

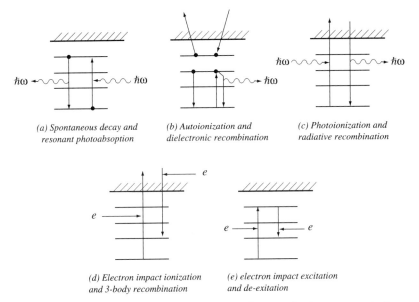

Figure 4.1 Schematic description of the most frequent ionic processes in hot plasmas.

Although not the result of two-particle interactions, these transitions change the state of ionization or excitation of an ion, and are therefore classified among the other ionic processes.

There are several ways to classify ionic processes. One of the possible methods of classification is by the interacting particles. A schematic diagram of the various processes is shown in figure 4.1. The *photon–ion* processes include the following:

- *Resonant photoabsorption*, see figure 4.1(a), in which a photon having the energy of the difference between two ionic states is absorbed from the radiation field, thereby inducing a transition of the ion from the lower to the higher state. This process is, in fact, the inverse of spontaneous decay.
- *Photoionization*, in which an absorbed photon moves a bound electron into the continuum, thereby ionizing the ion, see table 4.1 and figure 4.1(c).

The *electron–ion* interactions incorporate more processes. These are the following:

- *Electron impact ionization*, figure 4.1(d), in which a free electron hits an ion, knocking out a bound electron into the continuum.
- *Three-body recombination*, figure 4.1(d) (also called *electron impact recombination*). In this process, two free electrons enter at the same time into the volume of an ion. One of the electrons is captured into an ionic state, while the second carries away the extra energy. This process is the inverse of electron impact ionization.
- In *radiative recombination*, figure 4.1(c), an electron is captured into one of the ionic states, with the emission of a photon which takes the extra energy. This recombination is the inverse of photoionization.

- *Dielectronic recombination*, figure 4.1(b). Here a free electron is captured into one of the ionic excited states, but now the released energy is absorbed by one of the bound electrons, mostly a ground state one, as a result of which it moves into a higher excited state, resulting in a double excited ion.
- In the process of *electron impact excitation*, figure 4.1(e), a free electron that moves near an ion loses energy by inducing a transition of a bound electron from a lower into a higher state.
- *Electron impact deexcitation*, figure 4.1(e), is the inverse of electron impact excitation. In this process an electron moving near an excited ion induces a *downward* ionic transition from an upper to a lower ionic state. The electron takes the extra energy.

The *ion–ion* interactions can ionize, recombine, excite, or deexcite one or both of the interacting ions. In fact, every ion–ion interaction has a parallel electron–ion interaction from the list above, and we shall, therefore, not repeat them here separately. One process that has no parallel among the electron–ion processes is

- The process of *charge exchange* between ions, in which an electron jumps from one to the other of two ions in the encounter. This process has the effect of charge redistribution, and is of little importance in steady state plasmas.

It should be emphasized that all the ion–ion processes have much lower rates in hot plasmas than the electron–ion processes and are, in general, neglected in plasma spectroscopy calculations.

The *photon–electron* processes incorporate two processes which have relatively high rates. These are,

- *Bremsstrahlung* (which in German means "braking radiation"). In this process an electron moving close to an ion is accelerated by the ion's Coulomb field, thereby emitting a photon.
- *Inverse bremsstrahlung* is the inverse of the above process. Here an electron that moves near an ion absorbs a photon from the radiation field.

These processes will be discussed in the context of emission and absorption spectra in chapters 6 and 9.

There are, of course, many more processes in hot plasmas, mainly many-body interactions, as well as double excitations or ionizations. Such complex processes may be important under some special conditions, but in most of the present-day laboratory plasmas the above list is complete for all practical purposes.

A different method of classification is by means of the number of particles that take part initially in the interaction. Most of the processes mentioned above are *two-particle* processes. Exceptions are spontaneous decay and autoionization, which are *single-particle* processes, and three-body recombination, which is, as its name indicates, a *three-particle* process. This type of classification has importance in the context of the density dependence of the processes. For instance, the three-body recombination can occur only when two electrons enter simultaneously into the volume of an ion. This has nonzero probability only for high density plasmas. At the other extreme, the single-particle transitions are independent of the electron or photon densities, and proceed at low densities at the same rate as at high densities (the formation of the initial excited state is accomplished, however, by a two-particle process, and therefore depends on the plasma density). Their relative importance will be the highest in low density plasmas, where three-body processes have negligible rates.

The following sections will be devoted to a more detailed explanation of the behavior of these processes and their relative importance in the various temperature and density domains. An updated list of working formulas will be recommended that are in general use for the computation of the cross sections and rate coefficients.

It is not the intention of this book to repeat the methods through which the cross sections and rate coefficients are calculated from basic principles. Excellent books have been published on this subject, and a repetition of these methods here would shift our attention from the mainstream of this book to the fields of quantum mechanics and atomic physics. For readers who are interested in the derivation of the various cross sections from the basic formulas of quantum mechanics, we suggest several sources. These include the classic book of Griem (1964) in which one can find a full derivation of the formulas relevant to the radiative processes. The books of Sobelman, Vainshtein and Yukov (1981), Cowan (1981), Condon and Shortley (1987), and a recent book by Lisitsa (1994) also give information about the electron–ion interactions. References to the various research papers are given below in the corresponding sections.

During the last twenty years, an active group of data centers has been collecting and analyzing atomic physics data for hot plasmas (Post, 1995). A list of these centers is shown in table 4.2. Much of the activities of these centers has concen-

Table 4.2 Atomic and Molecular Data Centers for Fusion

Institute	Type of data
International Atomic Energy Agency, Vienna, Austria	Bibliography, cross sections, rate coefficients
National Institute of Standards and Technology, Washington D. C., USA	Bibliography, spectroscopic data, wavelengths, transition probabilities
A&M Data Center at the Kurchatov Institute, Moscow, Russia	Atomic collision data, sputtering
Max Planck Institute for Plasma Physics, Garching, Germany	Neutral reflection, sputtering
JILA, University of Colorado, Boulder, Colorado, USA	Bibliography, atomic collision data
Centre de Donnees Graphyor, Orsay, France	Bibliography, atomic collision data, energy levels
National Institute for Fusion Science, University of Nagoya, Nagoya, Japan	Bibliography, atomic collision data
Oak Ridge National Laboratory, Controlled Fusion Atomic Center Oak Ridge, Tennessee, USA	Bibliography, atomic collision data
Nuclear Data Center, JAERI, Tokai, Japan	Bibliography, atomic collision data
Atomic and Molecular Data Unit, Queen's University of Belfast, Northern Ireland	Bibliography, atomic collision data
Joint European Torus, Abingdon, UK	Atomic collision data for plasma diagnostics

trated on compiling and classifying bibliographies of atomic physics literature, but also on the collection, the evaluation, and the processing of data for atomic level energies, oscillator strengths, and collision rates. These centers also include strong atomic physics groups, which help to keep the atomic physics community aware of and working on some key problems that have importance for hot plasmas. These groups also carry out independent research on the various topics dealt with in this study, and have been partially responsible for the progress in the field. Data accumulated in these centers are accessible through international computer networks for researchers in laboratories around the world. Researchers in need of data about some atomic property or process are encouraged to access the centers.

4.2 Definitions and General Behavior

The quantity that is calculated from the basic equations of quantum mechanics is the *transition rate*, which is the number of reactions undergone by one particle per unit time. This quantity is given by *Fermi's Golden Rule*,

$$w_{i \to f} = \frac{2\pi}{\hbar} |\langle i | \mathcal{H} | f \rangle|^2 \rho_f(E) \tag{4.2.1}$$

Here $\mathcal{H}$ is the Hamiltonian of the interaction, $\langle i |$ is the initial state of the particles, $|f\rangle$ is the final state, that is, the quantum state of the outgoing particles, and $\rho_f(E)$ is the density of the final states (number of states per unit energy) of the process. The units of $w_{i \to f}$ are transitions/s.

   The *cross section* is an alternative method of giving a quantitative description of the probability of an ionic process. It describes an effective area of the interacting particles for a given process to occur. To make the discussion below more realistic, we shall illustrate its application by means of the electron impact excitation. Assume that $\sigma(\zeta, \ell \to u; v)$ is the cross section for an electron that has a velocity v to collide with an ion of charge ζ, and that as a result of this collision the ion is excited from its original lower level ℓ to an upper level u. Denote by $N_{\zeta, \ell}$ the density of the ions with charge ζ that are in the initial state ℓ (ground or excited). Take a very thin slab of material of thickness d, which is only a few times larger than the ion's diameter. The fraction of the area occupied by these ions for this specific process, as seen by an incoming electron, is $N_{\zeta, \ell} \sigma(\zeta, \ell \to u; v) d$. This fraction is the probability that an electron traversing the slab will react with one of the ions, and so represents the number of reactions of this kind generated by one electron per unit area of the slab. If the slab is thin, the number of reactions is small, much less than 1. The *mean free path* is defined as the distance λ_{mfp} along the path of the electron, in which, on the average, one reaction occurs,

$$N_{\zeta, \ell} \sigma(\zeta, \ell \to u; v) \lambda_{mfp} = 1 \tag{4.2.2}$$

An electron that has a velocity v traverses this distance within a time of λ_{mfp}/v. The average number of reactions per unit time is therefore v/λ_{mfp}. Comparing this result to equation (4.2.1), one gets the relationship between the transition rate and the cross section,

$$w_{\zeta,\ell \to \zeta,u} = \frac{v}{\lambda_{mfp}} = N_{\zeta,\ell} v \sigma(\zeta, \ell \to u; v) \qquad (4.2.3)$$

The picture underlying equation (4.2.3), as well as equation (4.2.1), is of one electron with a well-defined velocity, v, penetrating into a target whose particles are stationary. This is not the case in a plasma. In a real plasma the electrons have a distribution of velocities, which is a Maxwell–Boltzmann distribution. Each group of electrons with velocity in the range between v and $v + dv$ generates reactions at a rate given by equation (4.2.3). The density of these electrons obeys a Maxwell–Boltzmann distribution, $f_v(v)\,dv$ (see equation (1.3.13). The total rate of the reactions per unit volume is therefore

$$\frac{\text{number of reactions}}{\text{cm}^3\,\text{s}} = N_{\zeta,\ell} \int f_v(v)\,dv\, v\sigma(\zeta, \ell \to u; v) \qquad (4.2.4)$$

Substituting the explicit expression for $f_v(v)$ from equation (1.3.13) into (4.2.4), one obtains

$$\frac{\text{reactions}}{\text{cm}^3\,\text{s}} = n_e N_{\zeta,\ell} 4\pi \left(\frac{m_e}{2\pi T_e}\right)^{3/2} \int dv\, v\sigma(\zeta, \ell \to u; v) v^2\, \exp\left(-\frac{mv^2}{2T_e}\right)$$

$$= n_e N_{\zeta,\ell} \frac{2\sqrt{2}}{\sqrt{\pi}} \frac{c}{(mc^2 T^3)^{1/2}} \int_0^\infty E_e \sigma(\zeta, \ell \to u; E_e) e^{-E_e/T}\, dE_e \qquad (4.2.5)$$

In the second of these equations the cross section is expressed in terms of the incoming electron energy, rather than its velocity, and the electron energy distribution in equation (1.3.13) was used as the weighting function.

The electron impact excitation *rate coefficient* is defined by equation (4.2.5) with the electron and ion densities factored out in front of the right hand side,

$$\mathcal{E}(\zeta, \ell \to \zeta, u; T_e) = 4\pi \left(\frac{m_e}{2\pi T_e}\right)^{3/2} \int dv\, v\sigma(\zeta, \ell \to u; v) v^2\, \exp\left(-\frac{mv^2}{2T_e}\right)$$

$$= \frac{2\sqrt{2}}{\sqrt{\pi}} \frac{c}{(mc^2 T^3)^{1/2}} \int_0^\infty E_e \sigma(\zeta, \ell \to u; E_e) e^{-E_e/T}\, dE_e$$

$$(4.2.6)$$

The rather bulky notation for the rate coefficient is necessary to define accurately the initial and final states of the process. When there can be no ambiguity regarding the initial and final states, a shorter notation is possible. Equation (4.2.6) shows that the rate coefficient is the average of the product of the electron velocity with the cross section with the electron velocity distribution as the weighting function. More accurately, the weighting factor is the velocity distribution *per electron*, which is equation (1.3.13) without the factor n_e in front of the distribution function,

$$\mathcal{E}(\zeta, \ell \to \zeta, u; T_e) = \langle v\sigma(\zeta, \ell \to u; v) \rangle \qquad (4.2.7)$$

The rate coefficients for all the other ionic reactions are defined in a similar manner. The units of the rate coefficients are cm^3/s. The number of reactions per unit volume per unit time is obtained by multiplying the rate coefficient by the densities of the reacting particles,

$$\frac{\text{number of reactions}}{\text{cm}^3 \text{ s}} = n_e N_{\zeta,\ell} \mathcal{E}(\zeta, \ell \rightarrow \zeta, u; T_e) \tag{4.2.8}$$

The rate coefficient, $\mathcal{E}(\zeta, \ell \rightarrow \zeta, u; T_e)$, no longer depends on the densities of the interacting particles but depends on the electron temperature only. The study of the rate coefficients of the various atomic processes as a function of the electron temperature will be the subject of the coming sections.

The special form of equation (4.2.8) corresponds to two-particle reactions only. The units of the rate coefficient for single-particle transitions, namely, spontaneous decay and autoionization, is s^{-1}, and of course there is no factor n_e in the calculation of the reaction rate. In contrast, the rate coefficient for the three-body recombination has the units of cm^6 s^{-1}, and the density factor is $n_e^2 N_{\zeta,\ell}$, which contains n_e^2 corresponding to the two electrons taking part in the process.

4.3 The Detailed Balance Principle

As shown in table 4.1, every process has its inverse, such that the initial state of the interacting particles is the final state of the inverse process and vice versa. To find a relationship between a process and its inverse, one writes an expression similar to equation (4.2.1) for the inverse process, but with the indices i and f having reversed roles,

$$w_{f \rightarrow i} = \frac{2\pi}{\hbar} |\langle f | \mathcal{H} | i \rangle|^2 \rho_i(E) \tag{4.3.1}$$

Since the Hamiltonian is a Hermitian operator, the matrix element for a process and its inverse are complex conjugates, therefore

$$|\langle i | \mathcal{H} | f \rangle|^2 = |\langle f | H | i \rangle|^2 \tag{4.3.2}$$

It turns out that by dividing the transition rates of the two processes, equation (4.2.1) by (4.3.1), the matrix elements cancel out, and one finds the simple result,

$$\frac{w_{i \rightarrow f}}{w_{f \rightarrow i}} = \frac{\rho_f(E)}{\rho_i(E)} \tag{4.3.3}$$

which is independent of the matrix element, and thereby of the details of the interaction. This equation is known as the *detailed balance principle*. Verbally, this principle states that the ratio between the rates of the direct and the inverse processes is proportional to the ratio of the final state densities of the two processes only, and not on any other parameter of the interaction.

The detailed balance principle is of great help in the computations of the rate coefficients of the various ionic processes. Knowing the rate coefficient of one of the two processes, the application of this principle immediately enables the calculation of the rate coefficient of its inverse. This method is widely used in the computation of the various rate coefficients.

4.4 Atomic Energy Levels

Before discussing the various processes in table 4.1 in detail, we first look for the availability of data on two ionic properties, namely, the atomic binding energies and the transition probabilities. Although these two quantities are listed generally in the category of ionic properties, we will see later that they are basic ingredients of the rate coefficients of all the ionic processes.

We first discuss the sources for the computation of the ionic energy levels. The best way to get this information is just to compute it from basic principles. Today excellent computer codes are available for carrying out such computations. Among the codes in the public domain, the MCDF is recommended as a high accuracy multiconfiguration Dirac–Fock code (Fischer, 1991). There are several other codes in public use, known under names including HEX and CATS, which can provide spectroscopic accuracy of the ionic energy levels. These codes are very efficient and a computation of all the ionic energy levels (up to some highest level) of all the charge states can be carried out within reasonable time limits on present-day computers. The calculated levels are then stored on the computer hard disk to be used by successive codes. Thus, the question of the computation of the ionic energy levels, which presented a serious mathematical challenge even a decade ago, is no longer a problem.

If a computer code is not available, one can use the older but still valid tabulations of the atomic and ionic energy levels of the US-NBS (Moore, 1949; Wiese et al., 1996, 1969). These excellent tables have been widely used for many years in all laboratories around the world.

The Screened Atom Model

A low accuracy but very fast algorithm for the computation of the ionic energy levels was developed by R. M. More (1982), based on the screened atom model of Mayer (see reference in More, 1982). In this model, the ionic levels are approximated by a formula of the form

$$E_{n_p} = E^0_{n_p} - \frac{Q^2_{n_p} e^2}{2a_0 n^2_p} \tag{4.4.1}$$

where n_p is the principal quantum number, Q_{n_p} is the nuclear charge screened by the electrons inner to the n_p state, and $E^0_{n_p}$ is the energy associated with electrons outside of the n_p orbital. As a simple approximation, the dependence of the screened charge on the number of inner electrons is a linear one, and one can write

$$Q_{n_p} = Z - \sum_{m<n_p} \sigma(n_p, m) P_m - \frac{1}{2}\sigma(n_p, n_p) P_{n_p} \tag{4.4.2}$$

Here, P_m, $m = 1, 2, \ldots, n_p$, are the populations of the occupied ionic shells, and the coefficients, $\sigma(n, m)$, describe the screening of the nth shell by the mth. Spin–orbit interactions and other refinements of this basic electrostatic interaction are neglected. With this simple formula for the energy levels, the orbit radius is $r_{n_p} = a_0 n_p^2 / Q_{n_p}$. An ionic state with principal quantum number n_p is screened mainly by electrons in states inner to state n_p, and to a lesser extent also by states outer to n_p because the wavefunction of an electron in an outer state may penetrate into the volume of the n_p state. The outer screening is approximated, within the accuracy of this model, by

$$E_{n_p}^0 = \frac{1}{2} \frac{e^2}{r_{n_p}} \sigma(n_p, n_p) P_{n_p} + \sum_{m > n_p} \frac{e^2}{r_m} \sigma(m, n_p) P_m \qquad (4.4.3)$$

A list of the screening coefficients was calculated by More (1982) from a database of 800 ionization potentials for 30 elements, which was itself developed by J. Scofield using the Hartree–Fock–Slater theory (quoted in More, 1982) with a least squares procedure. His results are reproduced in table 4.3. The use of equations (4.4.1–3) gives an rms deviation of 25% from the ionization potentials in the database. An accuracy of this magnitude is acceptable for calculations of ionization potentials, but must be refined for calculations of emission spectra.

The screened atom model has been found useful for calculations of equation of state of hot dense matter and plasma opacities (Rickert and Meyer-ter-Vehn, 1990). Perrot (1989) included ℓ-splitting of the major shells to improve the accuracy of the method.

Faussurier et al. (1997) have recently recomputed the screening coefficients by using a data base of 15,636 atomic data. Comparison between the energy levels computed using their screening coefficients and the values in the data base yields a mean and root mean square errors better than 5% and 16%, respectively, substantially better than those of More.

Table 4.3 The Screening Constants $\sigma(N, M)$

M	\multicolumn{10}{c}{N}									
	1	2	3	4	5	6	7	8	9	10
1	0.3125	0.9380	0.9840	0.9954	0.9970	0.9980	0.9990	0.9999	0.9999	0.9999
2	0.2345	0.6038	0.9040	0.9722	0.9779	0.9880	0.9900	0.9990	0.9990	0.9999
3	0.1093	0.4018	0.6800	0.9155	0.9796	0.9820	0.9860	0.9900	0.9920	0.9999
4	0.0622	0.2430	0.5150	0.7100	0.9200	0.9600	0.9750	0.9830	0.9860	0.9900
5	0.0399	0.1597	0.3527	0.5888	0.7320	0.8300	0.9000	0.9500	0.9700	0.9800
6	0.0277	0.1098	0.2455	0.4267	0.5764	0.7248	0.8300	0.9000	0.9500	0.9700
7	0.0204	0.0808	0.1811	0.3184	0.4592	0.6098	0.7374	0.8300	0.9000	0.9500
8	0.0156	0.0624	0.1392	0.2457	0.3711	0.5062	0.6355	0.7441	0.8300	0.9000
9	0.0123	0.0493	0.1102	0.1948	0.2994	0.4222	0.5444	0.6558	0.7553	0.8300
10	0.0100	0.0400	0.0900	0.1584	0.2450	0.3492	0.4655	0.5760	0.623	0.7612

The Model of Pankratov and Meyer-ter-Vehn

The screening coefficients of table 4.3 were used by Pankratov and Meyer-ter-Vehn (1992a) to obtain level energies of even higher accuracy, without substantial increase of the computation time. They used a combination of More's screened atom model with the WKB approximation to improve the accuracy of the level energies. Their starting point is the Bohr–Sommerfeld quantization condition,

$$\int_a^b P_{nl}(r)\, dr = \pi(n - l - \tfrac{1}{2}) \tag{4.4.4}$$

where P_{nl} is the momentum of an electron in state nl. (In this subsection we use atomic units: $e^2 = m = \hbar = 1$. To transform back to conventional units, the energy of the nl state, which results from the computation below, must be multiplied by $e^2/a_0 = 27.21\,\mathrm{eV}$.) The turning points a and b are determined by $P_{nl}(a) = P_{nl}(b) = 0$, and the momentum is given by

$$P_{nl}(r) = \left(2E_{nl} + 2V(r) - \frac{(l+\tfrac{1}{2})^2}{r^2}\right)^{1/2} \tag{4.4.5}$$

The energy, E_{nl}, that simultaneously satisfies these two equations provides an approximation to the energy of the level. In their paper, Pankratov and Meyer-ter-Vehn show that within the WKB approximation the electrostatic potential to be inserted into equation (4.4.5) is

$$V_{nl}^{el}(r) = \frac{1}{r}\left(Z - \frac{1}{\pi}\sum_{n',l'} \frac{N_{n'l'}^{nl}}{n'^2}\Omega_{n'l'}(r)\right) \tag{4.4.6}$$

where $N_{n'l'}^{nl}$ is the number of electrons in each $n'l'$ shell, except for the shell nl, and,

$$\Omega_{n'l'}(r) = \begin{cases} \pi n'^2 & \text{for } \Theta_{n'l'} < -1 \\[2mm] -n'rP_{n'l'}^{(0)} + (rQ_{n'} - n'^2)\arcsin[\Theta_{n'l'}(r)] \\ +\dfrac{\pi}{2}(rQ_{n'} + n'^2) & \text{for } |\Theta_{n'l'}| \leq 1 \\[2mm] \pi r Q_{n'l'} & \text{for } \Theta_{n'l'} > 1 \end{cases} \tag{4.4.7}$$

Here

$$\Theta_{n'l'}(r) = \left(1 - \frac{rQ_{n'}}{n'^2}\right) \Bigg/ \left(1 - \frac{(l'+\tfrac{1}{2})^2}{n'^2}\right)^{1/2} \tag{4.4.8}$$

$$P_{n'l'}^{(0)} = \left(-\frac{Q_{n'}}{n'^2} + \frac{2Q_{n'}}{r} - \frac{(l'+\tfrac{1}{2})^2}{r^2}\right)^{1/2} \tag{4.4.9}$$

and $Q_{n'}$ is the screened charge according to equation (4.4.2).

To get even better results, the electrostatic potential in equation (4.4.6) should be corrected for the exchange potential, and for high-Z material relativistic effects must also be included. The original paper indicates how these corrections should be carried out. When these corrections are done, the results from this algorithm compare favorably, to within a few percent or less, with computations carried out with full Hartree–Fock type computer codes. These deviations from the Hartree–Fock results are generally no larger than the differences among the various self-consistent-field methods themselves.

4.5 Atomic Transition Probabilities

Definitions

The atomic transition probability is the rate at which an ion in an upper excited state u decays to a lower (excited or ground) state ℓ. The transition probability is related to the dipole matrix element which connects the upper level u and the lower state ℓ. There are several quantities related to the transition probability, and we start the discussion with their definitions and the relationships between them.

The basic quantity is the *line strength* defined as the absolute value of the corresponding dipole matrix element squared,

$$S_{\zeta,u\to\zeta,\ell} = \sum_{m,m'} |\langle \zeta, u|e\vec{r}|\zeta, \ell\rangle|^2 = |\langle \zeta, u\|e\vec{r}\|\zeta, \ell\rangle|^2 \tag{4.5.1}$$

The summation goes over all unobserved quantum numbers; for example, in the case of the absence of magnetic fields, over the magnetic quantum numbers of the initial and final states, m and m'. $S_{\zeta,u\to\zeta,\ell}$ is obviously symmetric in the initial and the final states. The units of the line strength are $eV\,cm^3$. In the following we suppress the notation of the ionic charge ζ.

The *oscillator strength* is connected to the line strength by a set of atomic constants, plus an averaging over the initial (upper) state,

$$
\begin{aligned}
f_{u\to\ell} &= \frac{2}{3}\frac{mc^2}{e^2(\hbar c)^2}\Delta E_{u,\ell}\frac{1}{2J_u+1}\sum_{m,m'}|\langle u|e\vec{r}|\ell\rangle|^2 \\
&= \frac{2}{3}\frac{mc^2}{e^2(\hbar c)^2}\Delta E_{u,\ell}\frac{1}{2J_u+1}|\langle u\|e\vec{r}\|\ell\rangle|^2 = \frac{2}{3}\frac{mc^2}{e^2(\hbar c)^2}\Delta E_{u,\ell}\frac{1}{2J_u+1}S_{u\to\ell}
\end{aligned}
$$

$$\tag{4.5.2}$$

Here, J_u is the quantum number of the total angular momentum of the upper state, and $\Delta E_{u,\ell} = E_u - E_\ell$ is the energy difference between the initial and final states. The oscillator strength is a dimensionless quantity. The corresponding excitational oscillator strength, $f_{\ell\to u} = -g_u f_{u\to\ell}/g_\ell$ is only rarely used in the literature.

The Einstein A-coefficient is a measure of the number of spontaneous transitions (decays) undergone by one atom in unit time from the upper to the lower state. It has the form

$$A_{u \to \ell} = \frac{4}{3} c \frac{(\Delta E_{u,\ell})^3}{(\hbar c)^4} \frac{1}{2J_u + 1} S_{u \to \ell} \qquad (4.5.3)$$

The relations between $A_{u \to \ell}$ and $f_{u \to \ell}$ are easily inferred from equation (4.5.2) and (4.5.3),

$$A_{u \to \ell} = 2c \frac{e^2}{(\hbar c)^2 mc^2} (\Delta E_{u,\ell})^2 f_{u \to \ell} \qquad (4.5.4)$$

The units of $A_{u \to \ell}$ are transitions/s. The number of transitions from the upper to the lower state per second per cm^3 in a plasma is

$$\frac{\text{transitions}}{\text{cm}^3 \text{ s}} = -\frac{dN_{\zeta,u}}{dt} = N_{\zeta,u} A_{u \to \ell} \qquad (4.5.5)$$

where $N_{\zeta,u}$ is the density of ions of charge ζ excited to the upper state u.

The Einstein A-coefficient is closely connected to two other Einstein coefficients, namely, the *resonant photoabsorption*, denoted by $B_{\ell \to u}$, and the *induced emission probability*, denoted by $B_{u \to \ell}$. The first of these describes the direct absorption of the radiation by ions that are initially in a state ℓ and are excited by means of the absorption to an upper state u. For a transition of this kind to happen, a photon of energy $\hbar \omega_{u,\ell} = E_u - E_\ell = \Delta E_{u,\ell}$ has to be absorbed from the radiation field. To give a more realistic description of this process, one must take into account the fact that the absorption line has a finite width, $\Delta \hbar \omega$. A full discussion of line profiles is given in chapter 7. For our purposes here, the exact shape of the line is of no interest and it suffices to assume that the line width, $\Delta \hbar \omega$, is much smaller than the transition energy, $\hbar \omega_{u,\ell}$.

Let $I(\vec{e}, \hbar \omega)$ be the *specific intensity* of the radiation field in some point in the plasma, see chapter 9. More accurately, $I(\vec{e}, \hbar \omega) \, dS \, d\Omega \, d(\hbar \omega)$ is the amount of radiant energy in the spectral region $[\hbar \omega, \hbar \omega + d\hbar \omega]$, which propagates into a solid angle $d\Omega$ around direction $\vec{e}$, crossing an area dS perpendicular to the direction of propagation per unit time. Let $J(\hbar \omega) \, d(\hbar \omega)$ be the *average radiant intensity* in the same spectral region, averaged over all directions of propagation, $J(\hbar \omega) = \int I(\vec{e}, \hbar \omega) \, d\Omega / 4\pi$. The units of $J(\hbar \omega)$ are erg/(cm^2 s eV), where we have made a distinction between the energy carried by the radiation field (ergs) and the spectral energy range measured in eV. The Einstein $B_{\ell \to u}$ coefficient is defined such that the number of upward transitions per unit time per unit volume in the plasma is

$$\frac{\text{transitions}}{\text{cm}^3 \text{ s}} = -\frac{dN_{\zeta,\ell}}{dt} = N_{\zeta,\ell} h J(\hbar \omega_{u,\ell}) B_{\zeta,\ell \to \zeta,u} \qquad (4.5.6)$$

where $N_{\zeta,\ell}$ is the density of ions of charge ζ in state ℓ (ground or excited). The explicit appearance of the Planck constant h in equation (4.5.6) is due to the fact that in the original formulation of the theory the spectral region is defined in terms of the frequency, ν, rather than the energy, $\hbar \omega$.

The coefficient of the *stimulated emission probability*, $B_{\zeta,u \to \zeta,\ell}$, describes transitions by the radiation field. The presence of this process was suggested by Einstein in 1917, and its basis is the phenomenon that photons with energy $\hbar \omega_{u,\ell}$ in the vicinity of an excited ion in state u induce transitions from u to a lower state ℓ, thereby enhancing the downward transition rate relative to the rate of the spontaneous decays. The induced emission is particularly important in plasmas that have high radiation density. In fact, this process is the basis for the lasing action. $B_{\zeta,u \to \zeta,\ell}$ is defined in a manner similar to equation (4.5.6), that is, the rate of the induced transitions per unit volume per unit time is

$$\frac{\text{transitions}}{\text{cm}^3 \text{ s}} = -\frac{dN_{\zeta,u}}{dt} = N_{\zeta,u} hJ(\hbar\omega_{u,\ell}) B_{\zeta,u \to \zeta,\ell} \tag{4.5.7}$$

The units of the two B coefficients are $\text{cm}^2/(\text{erg s})$.

Consider a system in thermodynamic equilibrium. In such a system, the total rate of absorption necessarily equals the total rate of emission, therefore

$$N_{\zeta,\ell} hJ(\hbar\omega_{u,\ell}) B_{\zeta,\ell \to \zeta,u} = N_{\zeta,u} A_{\zeta,u \to \zeta,\ell} + N_{\zeta,u} hJ(\hbar\omega_{u,\ell}) B_{\zeta,u \to \zeta,\ell} \tag{4.5.8}$$

In systems in equilibrium, the level populations satisfy a Boltzmann distribution

$$\frac{N_{\zeta,u}}{N_{\zeta,\ell}} = \frac{g_{\zeta,u}}{g_{\zeta,\ell}} \exp\left(-\frac{E_u - E_\ell}{T}\right) = \frac{g_{\zeta,u}}{g_{\zeta,\ell}} \exp\left(-\frac{\hbar\omega_{u,\ell}}{T}\right) \tag{4.5.9}$$

where the gs are the statistical weights of the states. Substituting equations (4.5.9) into (4.5.8), and reordering the terms, we find

$$hJ(\hbar\omega_{u,\ell}) = \frac{A_{\zeta,u \to \zeta,\ell}}{(g_{\zeta,\ell}/g_{\zeta,u}) \exp(\hbar\omega_{u,\ell}/T) B_{\zeta,\ell \to \zeta,u} - B_{\zeta,u \to \zeta,\ell}} \tag{4.5.10}$$

The left hand side of this equation, the average radiant intensity, in thermodynamic equilibrium is given by the well-known Planck formula, see chapter 9,

$$hJ(\hbar\omega) = \frac{2h\nu^3}{c^2} \frac{1}{\exp(\hbar\omega/T) - 1} \tag{4.5.11}$$

Comparing equations (4.5.10) with (4.5.11), it follows that the two B-coefficients have to satisfy

$$g_{\zeta,u} B_{u \to \ell} = g_{\zeta,\ell} B_{\ell \to u} \tag{4.5.12}$$

and the relation between the B and the A coefficients is given by

$$A_{u \to \ell} = \frac{2h\nu_{u,\ell}^3}{c^2} B_{u \to \ell} \tag{4.5.13}$$

where $\nu_{u,\ell} = \omega_{u,\ell}/2\pi$ is the frequency of the absorbed photon. Although these relations were derived by assuming a condition of thermodynamic equilibrium, the A- and B-coefficients are parameters that depend solely on the ionic properties and are independent of whether the ions are in the thermodynamic equilibrium. These relations, therefore, always hold true.

As a final remark we list here the scaling properties with the nuclear charge of the A, f, and S coefficients for hydrogenlike species:

$$A_{u \to \ell}(Z) = Z^4 A_{u \to \ell}(Z = 1)$$
$$f_{u \to \ell}(Z) = f_{u \to \ell}(Z = 1) \qquad (4.5.14)$$
$$S_{u \to \ell}(Z) = \frac{1}{Z^2} S_{u \to \ell}(Z = 1)$$

Oscillator Strength of Hydrogen

There are no analytical formulas of general validity for the oscillator strength and its computations is best carried out numerically. Only for hydrogen can one find analytical expressions, but even these are not very useful practically and are used mainly as benchmark points for computer codes.

The derivation of the formulas for hydrogen are too complicated to be reproduced here. We give only the results. A lengthy calculation shows that the oscillator strength of hydrogen between an upper state of principal quantum number n_u and angular quantum number l_u, and a lower state $n_\ell l_\ell$ is given by (see Mihalas, 1970),

$$f(n_u l_u \to n_\ell l_\ell) = \frac{1}{3} \left(\frac{1}{n_u^2} - \frac{1}{n_\ell^2} \right) \frac{\max(l_u, l_\ell)}{(2l_u + 1)} \left(\int_0^\infty P_{n_u l_u}(r) r P_{n_\ell l_\ell}(r)\, dr \right)^2 \qquad (4.5.15)$$

where $P_{n_u l_u}(r)$ and $P_{n_\ell l_\ell}(r)$ are the radial parts of the wavefunctions of the upper and lower states, respectively. For hydrogen, these radial parts have known analytical form in terms of the Laguerre polynomials. The integrals in (4.5.15) were first calculated by Gordon (1929),

$$\left(\int_0^\infty P_{n_u l_u}(r) r P_{n_\ell l_\ell}(r)\, dr \right)^2 = \left\{ \frac{(-1)^{n_1 - l_\ell}}{4(2l_\ell - 1)!} \left[\frac{(n_2 + l_\ell)!(n_1 + l_\ell - 1)!}{(n_2 - l_\ell - 1)!(n_1 - l_u)!} \right]^{1/2} \right.$$

$$\times \frac{(4n_2 n_1)^{l_\ell + 1}(n_2 - n_1)^{n_2 + n_1 - 2l_\ell - 2}}{(n_2 + n_1)^{n_2 + n_1}}$$

$$\times \left[F\left(-n_2 + l_\ell + 1, -n_1 + l_\ell; 2l_\ell; \frac{4n_2 n_1}{(n_2 - n_1)^2} \right) - \left(\frac{n_2 - n_1}{n_2 + n_1} \right)^2 \right.$$

$$\left. \left. \times F\left(-n_2 + l_\ell - 1, -n_1 + l_\ell; 2l_\ell; \frac{-4n_2 n_1}{(n_2 - n_1)^2} \right) \right]^2 \right\} \qquad (4.5.16)$$

In equation (4.5.16), n_2 is the principal quantum number of the state with the larger l-value, and n_1 is the quantum number of the state with the smaller l-value. $F(a, b; c; x)$ is the hypergeometric function (Abramowitz and Stegun, 1965),

$$F(a, b; c; x) = 1 + \frac{ab}{c} x + \frac{a(a+1)b(b+1)}{2!c(c+1)} x^2 + \dots \qquad (4.5.17)$$

Since in hydrogen atoms the l-states are normally degenerate, we often desire f-values for the entire transition $n_u \rightarrow n_\ell$. A summation of equation (4.5.16) over all initial and final angular momentum quantum numbers, l_u and l_ℓ was carried out by Menzel and Pekeris. Their result is

$$
f(n_u \rightarrow n_\ell) = \frac{32}{3} n_\ell^4 n_u \frac{(n_\ell - n_u)^{2n_\ell + 2n_u - 4}}{(n_\ell + n_u)^{2n_\ell + 2n_u + 3}}
$$

$$
\times \left\{ \left[F\left(-n_u, -n_\ell + 1; 1; -\frac{4n_u n_\ell}{(n_u - n_\ell)^2} \right) \right]^2 \right.
$$

$$
\left. - \left[F\left(-n_u + 1, -n_\ell; 1; -\frac{4n_u n_\ell}{(n_u - n_\ell)^2} \right) \right]^2 \right\} \tag{4.5.18}
$$

The factor in the braces is customarily abbreviated as $\Delta(n_u, n_\ell)$. A convenient factorization of equation (4.5.18) was derived by Kramers,

$$
f(n_u, n_\ell) = f_K(n_u, n_\ell) G(n_u, n_\ell) \tag{4.5.19}
$$

where

$$
f_K(n_u, n_\ell) = \frac{32}{3\pi\sqrt{3}} \left(\frac{1}{n_\ell^3} - \frac{1}{n_u^2} \right)^{-3} \frac{1}{n_\ell^3 n_u^5} \tag{4.5.20}
$$

is the result of a semiclassical analysis of the oscillator strength, and

$$
G(n_u, n_\ell) = \pi\sqrt{3} \frac{n_u n_\ell}{(n_\ell - n_u)} \left(\frac{n_\ell - n_u}{n_\ell + n_u} \right)^{2n_\ell + 2n_u} \Delta(n_u, n_\ell) \tag{4.5.21}
$$

is the *Gaunt factor*, which takes into account the quantum mechanical effects on the oscillator strength. The introduction of the Gaunt factor is a convenient formalism for treating transitions between bound ionic states, and is also used for bound–continuum transitions.

It can be seen from equations (4.5.16–20) that the oscillator strength is a rapidly decreasing function of the principal quantum numbers of both the lower and the upper states. An ion in a low lying excited state decays rapidly to a lower state, or the ground state, whereas the transitions from highly excited states are much slower. This has great influence on the population distribution of highly excited states relative to low lying ones.

The oscillator strengths of charged hydrogenlike ions are inferred from those of neutral hydrogen by means of the scaling laws, equation (4.5.14).

To simplify the above rather bulky formulas, an analytical approximation for the oscillator strength of hydrogenlike ions was suggested by E. Minguez, who also included relativistic corrections, which may be important for high-Z ions (Minguez, 1990). His result for a transition from an upper state $\gamma_u = n_u l_u j_u$ (j is the total angular quantum number) to a lower state $\gamma_\ell = n_\ell l_\ell j_\ell$ is

$$
f_{\gamma_u \rightarrow \gamma_\ell} = K_{\gamma_u \rightarrow \gamma_\ell} \Delta E_{\gamma_u \rightarrow \gamma_\ell} |R_{\gamma_u \rightarrow \gamma_\ell}|^2 \tag{4.5.22}
$$

where $\Delta E_{\gamma_u \to \gamma_\ell}$ is the energy difference between the states,

$$K_{\gamma_u \to \gamma_\ell} = \frac{2}{3} \max(l_u, l_\ell)(2j_u + 1) \left\{ \begin{matrix} l_\ell & j_\ell & \frac{1}{2} \\ j_\ell & l_u & 1 \end{matrix} \right\}^2 \tag{4.5.23}$$

In this last equation $\{\ldots\}$ is the $6j$ symbol (Cowan, 1981). Minguez has fitted the R matrix to the following very simple forms,

$$R_{\gamma_u \to \gamma_\ell} = \begin{cases} [1.290/(n-1)^{1.32}](a_0/Z) & 1s-np, \ n \le 5 \\[2mm] [2.17/n^{3/2}](a_0/Z) & 1s-np, \ n > 5 \\[2mm] [3.065/(n-2)^{1.257}](a_0/Z^x) & 2s-np \\[2mm] [4.748/(n-2)^{1.43}](a_0/Z) & 2p-nd \\[2mm] [0.938/(n-2)^{1.3}](a_0/Z) & 2p-ns \\[2mm] [R_{H,n_u l_u n_\ell l_\ell}/(n_u - n_\ell)^{1.3}](a_0/Z^x) & n_u l_u - n_\ell l_\ell, n_u, n_\ell \geqslant 3 \end{cases} \tag{4.5.24}$$

where $x = 1$ for elements with $Z < 40$, and $x = 1.02$ for higher Z material, and $R_{H,n_u l_u n_\ell l_\ell}$ is the R-matrix for hydrogen. The accuracy claimed is better than 10% for all the transitions in hydrogenlike ions, for all Z (Minguez, 1990).

Where To Get Data about Transition Probabilities?

The best data about the transition probabilities in the literature are still the NBS compilations (Moore 1949; Wiese et al., 1966, 1969). More updated data can be found in the compilations of *Atomic Data and Nuclear Data Tables*.

   In general, the computer codes that generate the wavefunctions and energy levels of the various ionic states can be modified to compute the dipole matrix elements as well, and therefrom the line strength, the oscillator strength, and the Einstein coefficients. This may be the only method to get information about these parameters for transitions that are not tabulated in the literature. A word of warning, however, should be included: If the number of relevant ionic states is M, then the number of the oscillator strengths that have to be computed and stored is M^2. Such a computation, while certainly possible, requires much more computational resources than the calculation of just the energy levels.

The Pankratov–Meyer-ter-Vehn Algorithm

As a continuation of their method for the ionic energy levels, Pankratov and Meyer-ter-Vehn (1992b) developed a very simple algorithm for the dipole matrix element. We give here only their final result. The interested reader is encouraged to read the original paper for the derivation.

   Our interest is the computation of the dipole matrix element of an ion of charge ζ, between an upper state u, having energy $E_{n_u l_u}$, and a lower state ℓ with energy $E_{n_\ell l_\ell}$. In the following we use atomic units, $e^2 = m = \hbar = 1$. The final result, equation (4.5.27) has to be multiplied by ea_0 to transform it back into standard units. Denote by ν and ν' effective quantum numbers of the lower and upper states, defined by

$$E_{n_\ell \ell_\ell} = -\frac{\bar{\zeta}}{2\nu^2}, \qquad E_{n_u l_u} = -\frac{\bar{\zeta}}{2\nu'^2} \qquad (4.5.25)$$

where $\bar{\zeta} = \zeta - 1$ is the charge of the ion core, without the active electron. Denote by γ and ν_c the following quantities,

$$\gamma = \frac{\omega \nu_c^3}{Z^2} = \left\langle \frac{\omega Z}{|E_{n_\ell \ell_\ell} + E_{n_u l_u}|^{3/2}} - \Delta \nu \right\rangle + \Delta \nu \qquad (4.5.26)$$

where $\langle \cdots \rangle$ denotes the closest integer. The approximate dipole matrix element suggested by Pankratov and Meyer-ter-Vehn is given by

$$|\langle \zeta, u \| e\vec{r} \| \zeta, \ell \rangle| = \frac{\bar{\zeta} \nu_c^2}{\omega (\nu \nu')^{3/2}} [U_\gamma(\epsilon \gamma) \cos(\pi \Delta \nu) - V_\gamma(\epsilon \gamma) \sin(\pi \Delta \nu)] \qquad (4.5.27)$$

where

$$\epsilon = \sqrt{1 - \left(\frac{l_u + l_\ell + 1}{2\nu_c}\right)^2}, \qquad \omega = E_u - E_\ell, \qquad \Delta \nu = \nu' - \nu \qquad (4.5.28)$$

The functions U_γ and V_γ are defined by

$$U_\gamma(y) = \frac{1}{2}[J_{\gamma-1}(y) - J_{\gamma+1}(y)] + \Delta \lambda \frac{(1 - \epsilon)^{1/2}}{\epsilon} J_\gamma(y)$$

$$V_\gamma(y) = \frac{1}{2}[E_{\gamma-1}(y) - E_{\gamma+1}(y)] + \Delta \lambda \frac{(1 - \epsilon)^{1/2}}{\epsilon}\left(E_\gamma(y) - \frac{1}{\pi\gamma}\right) + \frac{(1 - \epsilon)}{\pi} \qquad (4.5.29)$$

where $\Delta \lambda = l_u - l_\ell = \pm 1$, and $J_\gamma(y)$ and $E_\gamma(y)$ are the Anger and Weber functions. These are given by

$$J_\gamma(y) = \frac{\sin \gamma \pi}{\gamma \pi} f_\gamma(y) + \frac{\sin \gamma \pi}{\pi} g_\gamma(y),$$

$$E_\gamma(y) = \frac{1 - \cos \gamma \pi}{\gamma \pi} f_\gamma(y) - \frac{1 + \cos \gamma \pi}{\pi} g_\gamma(y) \qquad (4.5.30)$$

with the expansions

$$f_\gamma(y) = 1 - \frac{y^2}{2^2 - \gamma^2} + \frac{y^4}{(2^2 - \gamma^2)(4^2 - \gamma^2)} + \cdots$$

$$g_\gamma(y) = \frac{y}{1^2 - \gamma^2} - \frac{y^3}{(1^2 - \gamma^2)(3^2 - \gamma^2)} + \cdots \qquad (4.5.31)$$

The line strength, the oscillator strength, and the Einstein coefficients then follow by equations (4.5.1–3). Pankratov and Meyer-ter-Vehn showed that the oscillator strength derived from this algorithm tends to the accurate hydrogenic values, equation (4.5.16), in the limit $n_u \gg n_\ell \gg 1$ and $n_u \gg l_\ell$.

Comparison of results from this algorithm with accurate dipole matrix elements of hydrogen shows that the differences are less than 15% for transitions to the $1s$ state. The accuracy greatly improves for transitions into higher states; it is $\sim 3\%$ for transitions to the $2p$ state, and less than 1% for transitions to the $n = 10$ states. A comparison of experimental and computational results for sodiumlike ions also shows good agreement.

4.6 Electron Impact Excitation and Deexcitation

In the process of electron impact excitation, an electron encounters an ion of charge ζ, which is in a low energy (ground or excited) state ℓ. As a result of the interaction, the ion is excited to an upper excited state, u, whereas the electron loses energy $\Delta E = E_{\zeta,u} - E_{\zeta,\ell}$. Electron impact deexcitation is the inverse process, in which an excited ion and an electron collide. In the final state, the ion is deexcited from an upper state, u, to one of its lower (ground or low lying excited) states, ℓ, and the electron takes away the extra energy. Neither process changes the charge of the ion but only its state of excitation,

$$A_\ell^{+\zeta} + e \Leftrightarrow A_u^{+\zeta} + e \tag{4.6.1}$$

We shall denote the rate coefficients by $\mathcal{E}(\zeta, \ell \to \zeta, u)$ and $\mathcal{D}(\zeta, u \to \zeta, \ell)$, respectively. These two rate coefficients are connected through the detailed balance principle,

$$\frac{\mathcal{D}(\zeta, u \to \zeta, \ell; T_e)}{\mathcal{E}(\zeta, \ell \to \zeta, u; T_e)} = \frac{g_{\zeta,\ell}}{g_{\zeta,u}} \exp\left(-\frac{\Delta E}{T_e}\right) \tag{4.6.2}$$

The gs in equation (4.6.2) denote the statistical weights of the two states.

The computations of the electron impact excitation cross sections have reached a very high level of sophistication. In figure 4.2 a comparison is shown between experimental and theoretical cross sections of the electron impact excitation on atomic hydrogen (Callaway, 1982; Williams, 1988). The agreement is quite impressive. These high accuracy results were, however, obtained with lengthy computer codes that cannot be incorporated directly into plasma simulation codes. Simpler but more efficient computable formulas are required.

The parameter that is calculated, in general, is the *collision strength*, $\Omega(\zeta, \ell \to \zeta, u; E_e)$, a dimensionless quantity that is connected to the cross section, $\sigma(\zeta, \ell \to \zeta, u; E_e)$, through the expression

$$\sigma(\zeta, \ell \to \zeta, u; E_e) = \pi a_0^2 \frac{E_H}{g_{\zeta,\ell}E_e} \Omega(\zeta, \ell \to \zeta, u; E_e) \tag{4.6.3}$$

where E_e is the energy of the incoming electron, and we recall that a_0 and E_H are the Bohr radius and the ground state energy of the hydrogen atom, respectively. The advantage of using the collision strength, rather than the cross section, is that $\Omega(\zeta, \ell \to \zeta, u; E_e)$ is symmetrical with respect to the direct and the inverse processes,

$$\Omega(\zeta, \ell \to \zeta, u; E_e) = \Omega(\zeta, u \to \zeta, \ell; E_e) \tag{4.6.4}$$

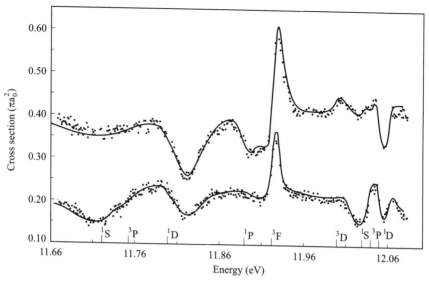

Figure 4.2 Integrated cross sections for inelastic electron scattering by hydrogen just below the threshold for $n = 3$ excitations of the hydrogen atom (12.09 eV). (From Callaway, 1982.)

In the following we give a short list of published formulas for the calculation of the electron impact excitation rate coefficient.

Van Regemorter's Rate Coefficient

The most widely used rate coefficient for the electron impact excitation is still van Regemorter's thirty-plus years old formula (van Regemorter, 1962),

$$\mathcal{E}(\zeta, \ell \to \zeta, u; T_e) = 3.2 \times 10^{-7} \, \text{cm}^3 \, \text{s} \left(\frac{E_H}{T_e} \right)^{3/2} f_{u \to \ell} \, \frac{\exp(-y)}{y} \, G(y) \qquad (4.6.5)$$

where

$$y = \frac{E_{\zeta,u} - E_{\zeta,\ell}}{T_e} \qquad (4.6.6)$$

is the energy difference between the upper and lower states, in units of the electron temperature. This formula is the result of a comparison between the first order Born approximation and experimental results. In equation (4.6.5), $f_{u \to \ell}$ is the oscillator strength and $G(y)$ is the Gaunt factor, which accounts for quantum mechanical corrections. Values of $G(y)$ are tabulated in table 4.4. Its asymptotic behavior for small and large arguments is

$$G(y) \simeq \begin{cases} \sqrt{3} E_1(y)/2\pi & y \to 0 \\ 0.200 & y > 1 \end{cases} \qquad (4.6.7)$$

where $E_1(y) = \int_y^\infty dt\, e^{-t}/t$ is the exponential integral of first order (Abramowitz and Stegun, 1965).

Sampson and Zhang (1992) have carried out a comparison between results obtained from van Regemorter's formula with those of more accurate computations. Their conclusion is that equation (4.6.5) is reasonably correct only for transitions within the same shell $n_u = n_\ell$. Even in this case, a limiting value of $G(y > 1) = 0.8$ gives better results. For $\Delta n \geq 1$ transitions, errors may be as large as an order of magnitude in each direction.

The Rate Coefficient of Sampson et al.

D. Sampson, in collaboration with L. B. Golden, R. E. H. Clark, C. J. Fontes, S. J. Goett, H. L. Zhang and G. V. Petrou, has carried out a comprehensive study of the behavior of the collision strength for ions with 2 to 5 bound electrons for all values of Z. They have used a Coulomb–Born exchange method, which is equivalent to the use of first order time-dependent perturbation theory in which the perturbation consists of the entire electrostatic interaction of the electrons plus all relativistic corrections. They have studied cases in which the ion is initially in its ground state, and also those of excited ions whose final state may be singly or doubly excited. In reference for Sampson (1985) we include a partial list of their publications.

Sampson and his collaborators have suggested a parametrization of the form

$$\Omega(\varsigma, \ell \rightarrow \varsigma, u; E_e) = c_0 + \frac{c_r}{(a+\epsilon)^r} + \frac{c_{r+1}}{(a+\epsilon)^{r+1}} + \frac{4}{3} Z^2 S \log \epsilon \qquad (4.6.8)$$

where S is the line strength of the transition,

$$\epsilon = \frac{E_e}{\Delta E} \qquad (4.6.9)$$

is the energy of the incoming electron divided by the energy of the transition, $r = 1$ for allowed transitions, and $r = 2$ when $Z^2 S = 0$. a, c_0, c_r and c_{r+1} are fitting parameters whose values for the various transitions are tabulated in the papers of the group.

Table 4.4 The Gaunt Factor for the Electron Impact Excitation Rate Coefficient

y	Atoms	Positive ions	y	Atoms	Positive ions
≤ 0.005	$(\sqrt{3}/2\pi)E_1(y)$	$(\sqrt{3}/2\pi)E_1(y)$	0.4	0.209	0.290
0.01	1.160	1.160	1	0.100	0.214
0.02	0.0956	0.977	2	0.063	0.201
0.04	0.758	0.788	4	0.040	0.200
0.1	0.493	0.554	10	0.023	0.200
0.2	0.331	0.403	≥ 10	$0.066/\sqrt{y}$	0.200

The product of the electron velocity with the cross section can be readily averaged over a Maxwellian energy distribution to find the rate coefficient for the process of electron impact excitation. The result is

$$\mathcal{E}(\zeta, \ell \to \zeta, y) = \frac{\pi a_0^2}{g(\zeta, \ell) Z_{eff}^2} \left(\frac{8T_e}{\pi m}\right)^{1/2} \left(\frac{E_H}{T_e}\right)$$

$$\times \left[c_0 e^{-y} + \frac{4}{3} Z^2 SE_1(y) + y e^{ay}\right.$$

$$\times \left.\left(\frac{c_r}{(a+1)^{r-1}} E_r(ay + y) + \frac{c_{r+1}}{(a+1)^r} E_{r+1}(ay + y)\right)\right] \qquad (4.6.10)$$

Here $Z_{eff} = Z - \sigma$ is the effective nuclear charge, corrected for the screening of the inner electrons. A tentative list of screening constants are given in Sampson (1985(iii)), but one may use values taken from other sources as well. In equation (4.6.10), $E_r(x) = \int_x^\infty dt \, (e^{-t}/t^r)$ are the exponential integrals, and y has the same meaning as in equation (4.6.6). The parameters a, c_0, c_r, and c_{r+1} for the various isoelectronic sequences and ionic transitions are tabulated in several of the papers of the group. The accuracy of equation (4.6.10) was compared with other computational and experimental results. These comparisons have indicated that this equation, with a few exceptions, is accurate to within 30% when Z is in the range $2Z_b \le Z \le 74$, where Z_b is the number of bound electrons. This accuracy is much better than van Regemorter's formula, although more parameters are required for its application.

Other Proposals

Many other attempts have been made in the literature to parametrize the electron impact excitation collisions strengths and rate coefficients. We mention here only a few. For hydrogenlike ions, Poquerusse (1993) proposed the following rate coefficient:

$$\mathcal{E}(\zeta, \ell \to \zeta, u; y) = 1.68 \times 10^{-7} \frac{n_u/n_\ell g_{u\ell}'}{(n_u/n_\ell - n_\ell/n_u)^4} \exp\left(\frac{n_u^{-2} - n_\ell^{-2}}{U}\right)$$

$$\times \frac{\log V + D_{u\ell}}{Z^3 \sqrt{U}} \, cm^3 \ s \qquad (4.6.11)$$

where

$$U = \frac{T}{Z^2 E_H}$$

$$V = n_i^2 U + \left(\frac{1.87}{n_\ell} - 2.84 + \frac{3.7 n_\ell - 1.9}{n_u}\right) n_\ell \sqrt{U} + 3.7 \frac{n_u - n_\ell}{n_u} + \frac{2.3}{n_u - n_\ell}$$
$$+ \frac{3 - 6 n_\ell}{n_u^2} - \frac{1.59}{n_\ell}$$

$$D_{u\ell} = n_\ell \left(1 + \frac{1.2 - n_\ell^2}{n_u^2}\right)\left(2.48 - \frac{1.18 n_\ell + 2.6}{n_u^2}\right) - 1.85 + \frac{0.49}{n_\ell}$$

$$g'_{u\ell} = 1 - 0.202 \left(\frac{n_\ell/n_u + n_u/n_\ell - 0.8}{n_u - n_\ell + 0.05}\right)^{0.7}$$

$$(4.6.12)$$

The author claims an accuracy of a few percent. Of the other proposals we mention here a very accurate fitting formula for *helium atoms* which was proposed by T. Kato and R. K. Janev (1992). U. I. Safronova et al. (1992) have proposed a fitting formula for *berylliumlike* ions that gives somewhat better results for these species than Sampson's formula.

The Behavior of the Electron Impact Excitation

Figure 4.3 shows a representative behavior of the electron impact rate coefficient as a function of the electron temperature. The figure corresponds to the Lyman-α transition, $1s$–$2p$, in hydrogenlike neon. Three rate coefficients are compared: those of van Regemorter, Poquerusse, and Sampson and Zhang. The energy difference between the initial and final states is $\Delta E = 1020$ keV. The following features are noteworthy: (i) The rate coefficient is low when the temperature is much lower than the transition energy, $T \ll \Delta E$. It increases rather sharply to a maximum around $T \approx \Delta E$, and decreases slowly at higher temperatures. It must be recalled, however, that at temperatures higher than a few times the transition energy, the energy of most of the electrons is too high for excitation into a discrete state, and the ionization of the electron into the continuum becomes the more probable process. (ii) The excitation rate coefficient is proportional to the oscillator strength, which attains its highest values between two low lying states. Between two highly excited states both the oscillator strength and the excitation rate coefficient decrease rapidly.

Multielectron Impact Excitation

The case depicted in all the above proposals is of a single electron hitting an ion and, as a result of their mutual interaction, the ion being excited to a higher quantum state. This picture is correct only in low density plasmas in which the

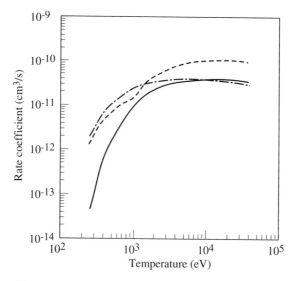

Figure 4.3 Comparison between the rate coefficients of the electron impact excitation of the $1s \rightarrow 2p$ transition in hydrogenlike neon, as computed by the formulas of Poquerusse (solid line); van Regemorter (dashed line); and Sampson and Zhang (dash-dot line).

duration of the interaction between the ion and the electron is much shorter than the time between two successive collisions of the ion with the plasma electrons. In higher density plasmas, the ion is interacting simultaneously with several electrons and the above picture has to be corrected to account for multielectron or *collective* excitation. The interaction time of the electron with the ion is approximately $t_{int} = D/v$, where D is the ion diameter and v is the electron velocity. The average time between two collisions is of the order of

$$t_{coll} = \frac{1}{n_e \mathcal{E}(i \rightarrow f)} = \frac{1}{n_e \langle v\sigma \rangle}$$

Comparing these two times, one finds that single electron impact excitation, as described above, is valid as long as $t_{int} \ll t_{coll}$, or $n_e \leq 1/D\sigma$. Inserting representative values for D and σ, one finds that the single electron impact picture is valid up to rather high electron densities, of the order of $n_e < 10^{22}-10^{23}$ electrons/cm³. In plasmas around these and higher densities, the above picture breaks down and multielectron effects must be accounted for.

The interaction between the ion and the plasma electrons proceeds by means of the fluctuation of the local Coulomb field. In fact, the largest contribution to the excitation is caused by the Fourier component of the field whose frequency equals the transition frequency between the two states. The higher the amplitude of this component the larger is the transition probability between the two states. If multielectron collective excitation is to be calculated, one has first to give an estimate of this frequency component of the field. At the rather high densities at which the effect dominates, the motions of the electrons in the plasma are strongly correlated, and this gives additional complexity to the problem.

An interesting attempt was made by J. Weisheit to include the many-electron or collective effects into the impact excitation rate coefficients (Weisheit, 1988). His calculation is based on previous work by Vinogradov and Shevelko (1976), who were the first to derive the rate of atomic excitation in a random field. The transition probability, when the density is not too high, is given by (Weisheit, 1988)

$$\omega_{\ell \to u} = 16\pi(\alpha a_0 c)^2 \left(\frac{E_H}{\Delta E}\right) n_e \int_0^\infty dk\, f_{\ell \to u}(k) \left(\frac{S_{ee}^{(0)}(k, \Delta\omega)}{|\varepsilon_e^{(0)}(k, 0)|^2}\right) \qquad (4.6.13)$$

where

$$S_{ee}^{(0)}(k, \Delta\omega) = \frac{1}{2\pi n_e} \int_{-\infty}^{\infty} dt\, e^{i\omega t} \int d\vec{r}\, e^{-i\vec{k}\cdot\vec{r}} \langle n_e(\vec{r}, t) n_e(0, 0)\rangle \qquad (4.6.14)$$

is the structure factor which is related to the electron–electron spatial and temporal autocorrelation function, $\varepsilon_e^{(0)}(k, 0)$ is the plasma dielectric response function, and

$$f_{\ell \to u}(k) = \frac{2mc^2\, \Delta E}{E_e^2(2J_u + 1)} \sum_{m_u m_\ell} |\langle \ell | e^{i\vec{k}\cdot\vec{r}} | u\rangle|^2 \qquad (4.6.15)$$

is the *generalized oscillator strength*, which differs from equation (4.5.2) mainly by the appearance of the exponential factor, rather than the electric dipole moment, in the matrix element.

Weisheit has shown that for uncorrelated plasmas equation (4.6.14) tends to the plane wave Born approximation of the electron impact excitation. This work has pioneering importance in the sense that it shows the direction in which the atomic rate coefficients need to be modified to account for multiparticle interactions in high density plasmas. It seems, however, that more effort must be invested to bring this work into a directly applicable form for inclusion in high density plasma codes.

4.7 Electron Impact Ionization and Three-Body Recombination

Electron impact ionization is probably the most important process in hot plasmas, particularly in optically thin ones. Its inverse process, three-body recombination, is important in high density plasmas only. The process of impact ionization occurs when an electron hits an ion and knocks out one bound electron into the continuum. Three-body recombination occurs when two free electrons enter at the same time into the volume of one ion and one of the two is captured in a bound state, whereas the other takes up the extra energy.

$$A_m^{+\zeta} + e \Leftrightarrow A_{m'}^{+\zeta+1} + e + e$$

Since three-body recombination requires the presence of two electrons inside the ionic volume, its rate is higher in high electron density plasmas.

We denote by $\mathcal{I}(\zeta, m \rightarrow \zeta + 1, m'; T_e)$ the rate coefficient for electron impact ionization, where ζ is the ionic charge and $m = 0, 1, 2, \ldots$ is the state of excitation in ascending order of energy before the interaction, with $m = 0$ denoting the ground state, $\zeta + 1$ and m' denote the same properties after the interaction. In most cases the knocked-out electron is initially in an *optical* state, namely, in the outermost populated ground state shell. In these cases the ionized ion is left in its ground state, $m' = 0$. However, ionizations of inner shell electrons may also occur, as well as ionizations with simultaneous excitation of a second electron, in which case the ion is left in an excited state. The total number of ionizations of charge state ζ per unit time per unit volume is given by $N_\zeta n_e \mathcal{I}(\zeta, m \rightarrow \zeta + 1, m'; T_e)$.

The rate coefficient of three-body recombination will be denoted by $\mathcal{R}^{(3)}(\zeta + 1, m' \rightarrow \zeta, m; T_e)$, with a notation similar to that used for the ionization rate coefficient. The total number of recombinations of this kind per unit time per unit volume is proportional to the square of the electron density, reflecting the number of electrons taking part in the process: $N_\zeta n_e^2 \mathcal{R}^{(3)}(\zeta + 1, m' \rightarrow \zeta, m; T_e)$.

The ratio between the rate coefficient of these two processes is determined by the detailed balance principle,

$$\frac{\mathcal{I}(\zeta, m \rightarrow \zeta + 1, m'; T_e)}{\mathcal{R}^{(3)}(\zeta + 1, m' \rightarrow \zeta, m; T_e)} = 2\left(\frac{mc^2 T_e}{2\pi(\hbar c)^2}\right)^{3/2} \frac{g(\zeta + 1, m')}{g(\zeta, m)} \exp\left(-\frac{E_{\zeta+1,m'} - E_{\zeta,m}}{T_e}\right)$$

(4.7.1)

where the first factor on the right-hand side is the final state density of the outgoing free electron, the factor 2 accounts for its two possible spin states, and the last factor, including the ratio of the statistical factors, is the final state density of the ion.

The Electron Impact Ionization Rate Coefficient

There exists a long list of rate coefficients proposed for the electron impact ionization. Most of them apply to the case in which the final state of the ionized ion is a ground state, $m' = 0$. In the following, therefore, we drop the notation of the final state and simply write $\zeta + 1$ instead of the full notation $\zeta + 1, 0$.

Of the earlier proposals, one should mention those suggested by McWhirter (1965), Drawin (1968), and Landshoff and Perez (1976). All these authors agree that the electron ionization rate coefficient has the general form

$$\mathcal{I}(\zeta, nl \rightarrow \zeta + 1; T_e) = C\xi_{\zeta,nl} T_e^{-3/2} \frac{e^{-y}}{y^q} F(y)$$

(4.7.2)

where

$$y = \frac{|E_{\zeta,nl}|}{T_e}$$

(4.7.3)

is the ionization energy of an electron in state nl in units of the electron temperature and $\xi_{\zeta,nl}$ is the number of electrons in the ionizing shell. If the continuum lowering is to be taken into account, then y should be replaced by

$$y' = \frac{|E_{\zeta,nl} - \chi_\zeta|}{T_e} \qquad (4.7.4)$$

The original formulas differ only in the values of the constants C, q and the function $F(y)$. Table 4.5 gives a comparison between the values of these constants according to the various models. Figure 4.4 shows the behavior of the electron impact ionization of hydrogenlike neon according to the formula of Lotz (see below).

A simple parametrization was proposed by Seaton (1964),

$$\mathcal{I}(\zeta, nl \to \zeta + 1; T_e) = 2.2 \times 10^{-6} \frac{1}{T_e^{3/2}} \frac{e^{-y}}{y} \quad \text{cm}^3\,\text{s} \qquad (4.7.5)$$

Although this formula has low accuracy, it has some "historical" importance, because several authors still use it and express their results in terms of corrections to this formula.

Lotz's Formula

A widely used formula for the electron ionization rate coefficient is due to Lotz (1968),

$$\mathcal{I}(\zeta, nl \to \zeta + 1; T_e) = 3 \times 10^{-6} \left(\frac{\text{cm}^3}{\text{eV}^{3/2}\,\text{s}} \right) \xi_{\zeta,nl} T^{-3/2} \frac{1}{y} E_1(y) \qquad (4.7.6)$$

where $E_1(y)$ is the exponential integral. In spite of its simplicity, the Lotz formula gives reasonably accurate results for the ionization rate coefficient, see below.

The Sampson and Golden Formula

Sampson and Golden used the fact that in the limit $Z = \infty$ one can express the cross section $\sigma(\zeta, nl \to \zeta + 1; E_e)$ for ionization from sublevel nl of a complex ion

Table 4.5 Proposed Values of C, q, and $F(u)$

	C $(\text{cm}^3/\text{eV}^{3/2}\,\text{s})$	q	$F(u)$
Lotz	3×10^{-6}	1	$Ei(u)\,\exp(u)$, ($Ei(u)$ is the exponential integral)
Seaton	2.2×10^{-6}	1	1
Drawin	2.9×10^{-6}	1	$\psi(u)\,\exp(u)$, ($\psi(u)$ is defined in Drawin, 1968)
McWhirter	2.34×10^{-7}	7/2	1
Landshoff–Perez	1.24×10^{-6}	2	$0.915(1 + 0.64/u)^{-2} + 0.42(1 + 0.5/u)^{-2}$

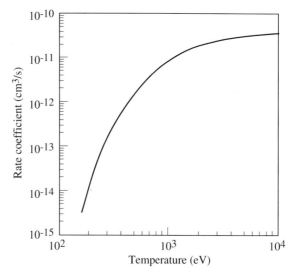

Figure 4.4 Rate coefficient of the electron impact ionization of hydrogenlike neon, according to the formula of Lotz, equation (4.7.6).

in terms of the reduced hydrogenic ion cross section $\sigma^H(0, nl \rightarrow 1; E_e)$ using the relation (Golden and Sampson, 1978)

$$\sigma(\zeta, nl \rightarrow \zeta + 1; E_e) = \pi a_0^2 \left(\frac{n}{Z_{eff}(nl)}\right)^4 [\xi_{nl}\sigma^H(0, nl \rightarrow 1; E_e)]_{Z=\infty} \qquad (4.7.7)$$

This relationship is valid within the nonrelativistic Coulomb–Born–Oppenheimer approximation. Z_{eff} is the ion's effective charge, whose values are tabulated in Golden and Sampson (1977), and $[\xi_{nl}\sigma^H(0, nl \rightarrow 1; E_e)]_{Z=\infty}$ is the ionization cross section of a hydrogenlike ion with a very large nuclear charge. They parametrized the ionization cross section of hydrogenlike ions to the form.

$$\sigma^H(\zeta = 0, nl \rightarrow \zeta = 1; u) = \pi a_0^2 \frac{1}{u}\left[A(nl) \ln(u) + D(nl)\left(1 - \frac{1}{u}\right)^2\right.$$

$$\left. + \left(\frac{c(nl)}{u} + \frac{d(nl)}{u^2}\right)\left(1 - \frac{1}{u}\right)\right] \qquad (4.7.8)$$

In equation (4.7.8), u is the ratio between the energy of the incident electron and the binding energy of the ionizing state,

$$u = \frac{E_e}{|E_{\zeta,nl}|} \qquad (4.7.9)$$

A, D, c, and d are constants that depend on the ionizing state but are independent of Z; their values for $1s$, $2s$ and $2p$ states are tabulated in several papers of the group (Golden and Sampson, 1978, 1980; Moores et al., 1980). At very large incident electron energies this formula reduces to Bethe's formula,

$$\sigma_{Bethe}^{H}(\zeta = 0, nl \to \zeta = 1; E_e) = \pi a_0^2 \frac{1}{u} \left(\frac{A'(nl)}{n} \ln(u) + D'(nl) \right) \qquad (4.7.10)$$

The coefficients A' and D' are connected to A and D by simple relationships. Combining equations (4.7.7) and (4.7.8), and accounting for the continuum lowering, their formula for the cross section of the electron impact ionization becomes

$$\sigma(\zeta, nl \to \zeta + 1; u') = \pi a_0^2 \left(\frac{n}{Z_{eff}(nl)S_0} \right)^4 \frac{\xi_{nl}}{u'} \left[A(nl) \ln(u') + S_0^2 D(nl) \left(1 - \frac{1}{u'} \right)^2 \right.$$

$$\left. + S_0 \left(\frac{c(nl)}{u'} + \frac{d(nl)}{u'^2} \right) \left(1 - \frac{1}{u'} \right) \right] \qquad (4.7.11)$$

where

$$S_0 = \frac{E_{\zeta,nl} - \chi_\zeta}{E_{\zeta,nl}} \qquad (4.7.12)$$

is the fraction of the continuum lowered binding energy relative to the unlowered one, χ_ζ is the average continuum lowering, and the other parameters have the same meanings as in equation (4.7.8). The electron impact rate coefficient obtained from this formula is

$$\mathcal{I}(\zeta, nl \to \zeta + 1; T_e) = \pi a_0^2 \left(\frac{8T_e}{\pi m} \right)^{1/2} \left(\frac{n}{Z_{eff}(nl)S_0} \right)^4 \xi_{nl} y'$$

$$\{ S_0^2 D(nl) e^{-y'}$$

$$- S_0 y' d(nl) E_3(y') + [A(nl) + S_0 y'(c(nl) - 2S_0 D(nl))] E_1(y')$$

$$+ S_0 y' [S_0 D(nl) + d(nl) - c(nl)] E_2(y') \} \qquad (4.7.13)$$

Younger's Formula

In a series of papers, S. Younger published results for the ionization rate coefficient of several isoelectronic sequences. The reference list gives his papers on this subject. He has carried out computations of the rate coefficients for hydrogenlike and lithiumlike (1980a), heliumlike (1980b), berylliumlike (1981a), neonlike (1981b), sodiumlike (1981c), chlorinelike (1982a), argonlike (1982b), xenonlike (1988) and palladiumlike (1986) isoelectronic sequences, as well as some other ions that are of interest for experimental purposes.

Younger's calculations were carried out by means of a Coulomb–Born distorted wave approximation and are expected to have very good accuracy. He found that a parametrization of the cross sections to the form

$$y|E_{\zeta,nl}|^2 \sigma(\zeta, nl \to \zeta + 1; E_e) = A\left(1 - \frac{1}{u}\right) + B\left(1 - \frac{1}{u}\right)^2 + C \ln u + D\frac{\ln u}{u}$$

$$(4.7.14)$$

gives a very good approximation to the results obtained by his more elaborate calculations. In equation (4.7.14), $u = E_e/|E_{\zeta,nl}|$ is again the incident electron energy in units of the ionization threshold energy, see equation (4.7.9), and A, B, C, and D are constants computed in his papers.

To allow isoelectronic interpolation of the cross section, the parameters A, B, C, and D were fitted by the inverse power series

$$A = \sum_{i=0}^{3} \frac{a_i}{(Z - Z_b + 1)^i}, \qquad B = \sum_{i=0}^{3} \frac{b_i}{(Z - Z_b + 1)^i} \qquad (4.7.15)$$

etc., where Z is the nuclear charge and Z_b is the number of bound electrons. The constants a_i and b_i are tabulated in Younger's papers. These constants are of the order of 10^{-14} cm^2 eV^2.

The product of the cross section in equation (4.7.14) with the electron velocity can readily be averaged with a Maxwellian velocity distribution. Younger expressed the resulting rate coefficient in the form

$$\mathcal{I}(\zeta, nl \to \zeta + 1; T_e) = F(\chi) \mathcal{I}_{Seaton}(\zeta, nl \to \zeta + 1; T_e) \qquad (4.7.16)$$

where $\mathcal{I}_{Seaton}$ is given by equation (4.7.5), $\chi = 1/y$ (see equations 4.7.3–4) and

$$F(\chi) = \frac{3.0 \times 10^{13}}{\chi}\left(A + B\left(1 + \frac{1}{\chi}\right)\right.$$

$$+ \left\{C - \frac{1}{\chi}\left[A + B\left(2 + \frac{1}{\chi}\right)\right]\right\}\alpha(\chi) + \left.\frac{D}{\chi}\beta(\chi)\right) \qquad (4.7.17)$$

$$\alpha(\chi) = \frac{0.001\,93 + 0.9764\chi + 0.6604\chi^2 + 0.025\,90\chi^3}{1.0 + 1.488\chi + 0.2972\chi^2 + 0.004\,925\chi^2} \qquad (4.7.18a)$$

$$\beta(\chi) = \frac{-0.000\,5725 + 0.013\,45\chi + 0.8691\chi^2 + 0.034\,04\chi^3}{1.0 + 2.197\chi + 0.2475\chi^2 + 0.002\,053\chi^3} \qquad (4.7.18b)$$

The Tawara–Kato Compilation

A comprehensive compilation of the ionization cross sections available in the literature, experimental as well as computational, was carried out by H. Tawara and T. Kato (1987). The graphs in their paper illustrate the behavior of the ionization cross sections as a function of the nuclear charge Z and the quantum numbers of the ionization state. The authors compared these cross sections to the Lotz's formula (4.7.6) for purposes of reference. A representative graph is shown in figure 4.5. The points are compiled from experiments available in the literature.

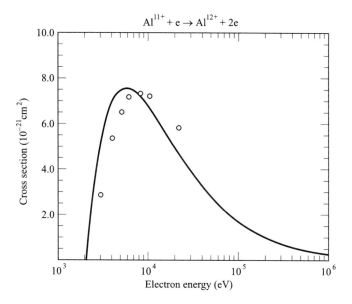

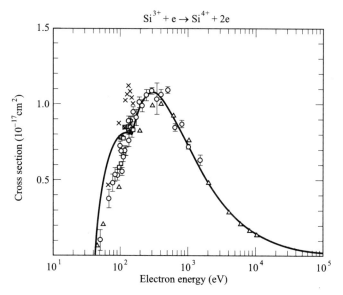

Figure 4.5 Ionization cross section of heliumlike aluminium and sodiumlike silicon, as function of the electron impact energy. The points are from experiments. (From Tawara and Kato, 1987.)

From their graphs it can be seen that the Lotz formula gives reasonably good results for low-Z material. For high-Z ions the Lotz formula is less accurate.

The Fitting Formula of Voronov

In a recent paper Voronov (1997) suggested a new type of rate coefficient for the electron impact ionization. His formula is

$$\mathcal{I}(\zeta, nl \to \zeta + 1; T_e) = A \frac{1 + P y^{1/2}}{X + y} y^k e^{-y} \qquad (4.7.19)$$

In equation (4.7.19), A, k, and X are adjustable parameters. The parameter P can acquire two values only: $P = 0$ or $P = 1$. This formula was used to fit recommended results from the Belfast library (table 4.2). More than 400 cases were tested from H to Ni^{27+}. The fitting parameters for $Z = 1$–28 and all ions are listed in Voronov (1997). The paper claims that the deviation of the fitted values from the values recommended by the library, less than 10% in most cases, is far less than the estimated accuracy of the recommended data, which is 40–60%.

Behavior of the Ionization Rate Coefficient

Figure 4.4 shows that the behavior of the ionization rate coefficient is very similar to the behavior of the excitation rate coefficient, figure 4.3. In both of these processes the rate coefficient increases rather sharply at low temperatures, and attains its maximum when the temperature is a few times the binding energy. At higher temperatures the rate coefficient decreases slowly. Such behavior is common to all the ionizing states, ground or excited, of every charged ion.

At a given temperature, the highly excited states are ionized with greater ease than the low lying ones. For example, according to the Lotz formula, the ionization rate coefficient for hydrogenlike ions increases as $n_p^2 \log n_p$, where n_p is the principal quantum number of the ionizing shell. Three-body recombination also proceeds with the highest probability into the uppermost states. Highly excited states, therefore, undergo rapid ionizing and recombining processes that make the population of these states rather unstable. This special behavior of these two processes has an important effect on the population probabilities of highly excited states.

4.8 Photoionization and Radiative Recombination

In the process of photoionization, a photon is absorbed by a bound electron, which is ejected into the continuum. The inverse process is radiative recombination, in which an incident electron is captured in one of the ionic bound states, emitting a photon that takes up the excess energy and momentum:

$$A_m^{+\zeta} + e \Leftrightarrow A_{m'}^{+\zeta+1} + \hbar\omega$$

The cross sections of these two processes are connected by the detailed balance principle, in this case called *the Milne relations*, after the physicist who first wrote them down,

$$\frac{\sigma_{phi}(\zeta, nl \to \zeta + 1; \hbar\omega)}{\sigma_{rr}(\zeta + 1 \to \zeta, nl; E_e)} = \frac{2mc^2 E_e}{(\hbar\omega)^2} \frac{g_{\zeta+1}}{g_{\zeta, nl}} \tag{4.8.1}$$

where σ_{phi} and σ_{rr} are the cross sections for photoionization and radiative recombination, respectively, (ζ, nl) is the state with the electron bound and $(\zeta + 1)$ is the ionized state. In equation (4.8.1), $E_e = \hbar\omega - |E_{\zeta, nl}|$ is the energy of the outgoing electron.

Photoionization is an important process in hot plasmas only if the local radiation field density, that is, the photon density, is high enough to induce a sufficiently large rate of photoionization relative to electron impact ionization. This occurs only in *optically thick* plasmas, in which the mean free path of the photons, $\lambda_{mfp} = 1/n_i \sigma_{phi}(\hbar\omega)$, is shorter than the plasma dimensions and the probability for photon reabsorption is reasonably large. In *optically thin* plasmas, whose dimensions are smaller than the mean free path, photoionization has a negligibly small influence.

The simplest approximation for the photoionization cross section of K-shell electrons is due to Heitler, whose used the nonrelativistic high energy ($\hbar\omega \gg |E_\zeta|$) Born approximation. His expression for K electrons is

$$\sigma_K(\hbar\omega) = \begin{cases} \sigma_{Th}\alpha^4 Z^5 2^{5/2}(mc^2/\hbar\omega)^{7/2} & \hbar\omega \geq |E_{\zeta, K}| \\ 0 & \hbar\omega < |E_{\zeta, K}| \end{cases} \tag{4.8.2}$$

where $\alpha = 1/137$ is the fine structure constant, $|E_{\zeta, K}|$ is the binding energy, and

$$\sigma_{Th} = \frac{8\pi}{3} r_0^2 = 6.653 \times 10^{-25}\, \text{cm}^2, \qquad r_0 = \frac{e^2}{mc^2} \tag{4.8.3}$$

is the classical cross section for Thomson scattering. There are several important things that one can learn from this approximate formula. First of all, the cross section has a threshold at $\hbar\omega = |E_\zeta|$, which simply means that photons whose energy is lower than the electron binding energy cannot ionize the ion. Second, note the very strong dependence on the atomic number, Z^5, which indicates that this effect is important for high-Z materials and less important for low-Z ones. Third, the cross section attains its highest value at the threshold; from there on it decreases as the 7/2 power of the photon energy. More accurate computations suggest a power law between 2.5 and 3, but the decrease is always a power law, that is, a decreasing straight line on a log–log plot, except, perhaps, near the threshold.

Calculations of the photoionization cross section have reached a very high level of sophistication. F. Sauter (1931), M. Gavrila (1959), and R. H. Pratt (1960) have suggested a series of corrections to equation (4.8.2). Pratt obtained the following formula for the high energy limit of the K-shell cross section for all values of αZ,

$$\sigma_K(\hbar\omega) = \frac{3}{2}\,\sigma_{Th}\alpha^4 Z^5 \left(\frac{mc^2}{\hbar\omega}\right)^5 (\alpha Z)^{2\xi}\,\frac{\beta^3}{(1-\beta^2)^{3/2}}\,M(\beta)$$

$$\times \exp[-2(\alpha Z/\beta)\arccos(Z)]\left(1 + \pi\alpha Z\,\frac{N(\beta)}{M(\beta)} + R(\alpha Z)\right) \qquad (4.8.4)$$

with

$$\xi = \sqrt{1 - (\alpha Z)^2} - 1$$

$$M(\beta) = \frac{4}{3} + \frac{1 - 3(1-\beta^2)^{1/2} + 2(1-\beta^2)}{\beta^2(1-\beta^2)^{1/2}}\left(1 + \frac{(1-\beta^2)}{2\beta}\log\frac{1-\beta}{1+\beta}\right)$$

$$N(\beta) = \frac{1}{\beta^3}\left[-\frac{4}{15}\frac{1}{(1-\beta^2)^{1/2}} + \frac{34}{15} - \frac{63}{15}(1-\beta^2)^{1/2} + \frac{25}{15}(1-\beta^2)\right.$$

$$\left. + \frac{8}{15}(1-\beta^2)^{3/2} + (1-\beta^2)^{1/2}\left(\frac{1 - 3(1-\beta^2)^{1/2} + 2(1-\beta^2)}{2\beta}\right)\log\frac{1-\beta}{1+\beta}\right]$$

$$(4.8.5)$$

In equation (4.8.4) $R(\alpha Z)$ is a correction function which is small for all values of αZ, and $\beta = v/c$ where v is the velocity of the ejected electron.

A comprehensive description of the theory, computational results, and comparison with experiments, for photon energy above 10 keV, is given in the excellent review paper of R. H. Pratt, A. Ron and H. K. Tseng (1973). Recent updated results for $10\,\text{eV} \leq \hbar\omega \leq 100\,\text{GeV}$ were published by a group from Livermore Lawrence National Laboratory (Cullen et al., 1989).

Hydrogenlike Ions

Formulas (4.8.2–5) are valid only for the K-shell of neutral atoms. For hydrogenlike ions one can use the explicit hydrogenic wavefunctions to calculate the transition from a bound state to the continuum by absorbing a photon,

$$\sigma_{phi}^{(H\text{-}like)}(Z-1, n_p \rightarrow Z; \hbar\omega) = \frac{64\pi\alpha a_0 n_p}{3^{3/2}Z^2}\left(\frac{|E_{Z-1,n_p}|}{\hbar\omega}\right)^3 \qquad (4.8.6)$$

which predicts a $1/(\hbar\omega)^3$ behavior of the photoionization cross section above threshold as a function of the energy of the incident photon. Moreover, since the binding energy $|E_{Z-1,n_p}|$ of an electron in shell n_p is proportional to Z^2/n_p^2, equation (4.8.6) implies that the photoionization cross section is proportional to Z^4, which means that high-Z ions absorb photons much more efficiently than low-Z ones. The cross section is also proportional to $1/n_p^5$, where n_p is the principal quantum number. This indicates that only the innermost electrons are absorbing photons, while outer ones are affected much less by the radiation field. This behavior is opposite to electron impact ionization, which preferentially ionizes electrons from the outermost ionic states.

Photoionization of Inner Shell Electrons in Complex Ions

An interesting attempt was made by R. H. Pratt, D. J. Botto, J. McEnnan and S. D. Oh (1978; see Pratt et al., 1973) to derive analytical expressions for the photoionization cross sections for positively charged ions. They expanded the ionic potential and the bound electron wavefunctions into power series in the ion's interior and found self-consistent screened solutions for both. Using these solutions, they have found approximate analytical expressions for the photoio-nization cross sections. It would be too lengthy to repeat their results here, but it is not too difficult to insert them into computer codes. One of the interesting results of their studies is a scaling law they have found for the photoionization cross sections of the K-shell and L-shell electrons as a function of the charge state of iron ions. Their results are reproduced in figure 4.6 and 4.7. From the

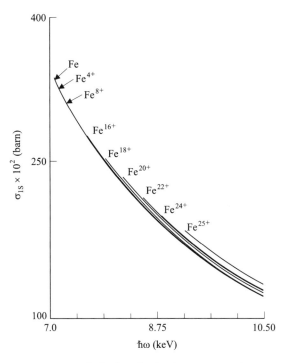

Figure 4.6 *K*-shell photoionization cross sections for various ions of iron. The cross section is in barn, and the photon energy in keV. (From Botto, 1978.)

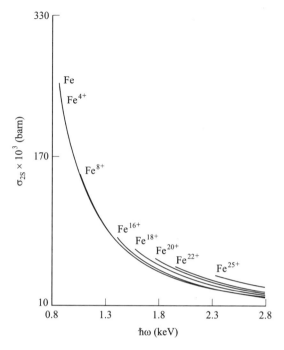

Figure 4.7 Same as figure 4.6 for the *L*-shell photoionization cross section.

figures it can be seen that, apart from a shift of the energy threshold, the cross
section for the photoionization of an electron in a given shell is changed only
moderately even for large changes in the ionic charge. This fact can be of great
help in building an algorithm for an approximate photoionization cross section
for all the electrons in a complex ion. One assumes cross sections of the form of
equation (4.8.2) for all the ionic states by using the photoionization cross sections of the neutral atom for all the bound electrons, and shifts the threshold to
the correct ionization energy.

Radiative Recombination

The rate coefficient for radiative recombination will be denoted here by
$R^{(r)}(\zeta + 1, m' \to \zeta, m; T_e)$, where we use the same convention to denote the initial
and final states as in the three-body recombination case. We shall drop the notation m' if the recombining ion is initially in its ground state, and denote the final
state m by its quantum numbers nl. The radiative recombination is a two-body
recombination process and has its largest influence in low density plasmas. The
total number of recombinations of this kind is given by $N_{\zeta+1} n_e R^{(r)}(\zeta + 1 \to \zeta,$
$n_p l; T_e)$, where n_p is the principal quantum number.

The Radiative Recombination Rate Coefficient

The calculation of the radiative recombination rate coefficient starts from equation (4.8.6); then one uses the Milne relations to obtain the cross section, $\sigma_{rr}(\zeta + 1 \rightarrow \zeta, n_p l; E_e)$, as a function of the incident electron energy. The product $v\sigma_{rr}(\zeta + 1 \rightarrow \zeta, n_p l; v)$ is then averaged with a Maxwellian velocity distribution to obtain the rate coefficient. The calculation is straightforward and we quote only the result, which is known as *Kramer's formula*,

$$R^{(r)}(\zeta + 1 \rightarrow \zeta, n_p; T_e) = \frac{16\sqrt{\pi}}{3\sqrt{3}} \frac{\alpha^3 \hbar c a_0}{mc^2} (\zeta + 1) y^{3/2} e^y E_1(y)$$

$$= 5.20 \times 10^{-14} n_p \sqrt{\frac{|E_{\zeta,n_p}|}{E_H}} y^{3/2} e^y E_1(y) \ \text{cm}^3/\text{s} \qquad (4.8.7)$$

where $y = |E_{\zeta,n_p l}|/T_e$ is the ratio between the ionization energy of state $(\zeta, n_p l)$ and the electron temperature, and $E_1(y)$ is the exponential integral. Since for hydrogenlike ions $|E_{Z-1,n_p}| \propto Z^2/n_p^2$, this formula predicts that for these species the recombination goes preferentially into the lowest ionic states, and with much lower probability into higher states. This conclusion is correct for other ionic species as well. This behavior of the radiative recombinations is opposite to that of the three-body recombination process, which goes with the highest probability into the uppermost bound states.

In a real plasma the ionization energy of the ground state of the most abundant charge state, $|E_{\zeta,n}|$, is several times larger than the electron temperature, that is, $y > 1$. The asymptotic behavior of the exponential integral for large y has the form (Abramowitz and Stegun, 1965) $ye^y E_1(y) = 1 + 1/y + \dots$. Keeping only the first term, the radiative recombination into the ground state, which is the most probable process, is written as

$$R^{(r)}(\zeta + 1 \rightarrow \zeta, g.s.; T_e) = 1.00 \times 10^{-14} \left(\frac{\text{cm}^3}{\text{eV}^{1/2} \text{s}}\right) n_p y \sqrt{T_e} \qquad (4.8.8)$$

Apart from a factor in the numerical coefficient, this formula is known as Seaton's formula (McWhirter, 1965).

A. Costescu, F. D. Aaron and C. Dinu have published exact analytical formulas for the radiative recombination cross sections (Aaron et al., 1993). Their final result would be too lengthy to be reproduced here, and the interested reader is encouraged to read the original paper. Their results can be coded, however, without too much difficulty, and then integrated numerically to find the corresponding rate coefficient. These algorithms are recommended when high accuracy results are required.

4.9 Autoionization and Dielectronic Recombination

Autoionization and dielectronic recombination are two processes in which doubly excited states are involved. A doubly excited ion $A^{+\zeta}_{n'l',n''l''}$ (n' and n'' are the principal quantum numbers, l' and l'' are the orbital quantum numbers), in which one of the electrons is excited to state $n'l'$ and the second to state $n''l''$, can ionize spontaneously if the energy of the lower excited electron above ground state is larger than the binding energy of that in the higher energy state. During ionization, one of the excited electrons decays to the ground state, while the other takes the extra energy and is emitted into the continuum.

Dielectronic recombinations is not exactly the inverse process to autoionization. The term *dielectronic recombination* refers to a two-step process. In the first step a continuum electron recombines into an excited state of the ion, and the extra energy is taken up by one of the bound electrons, which is then excited to an upper state. The result of this process is a doubly excited state, $n'l', n''l''$. Recombination up to this point is the inverse of autoionization. The doubly excited state has two channels of decay: either by autoionization or through radiative decay of one of the excited electrons into a lower state, nl. If autoionization follows the recombination, the overall process is equivalent to a resonant scattering of a free electron, a case that does not make any change in the state of excitation or ionization of the ion, and therefore has only marginal importance in hot plasmas. The process is called dielectronic recombination only in the second case, namely, when *radiative stabilization* of the ion follows the recombination. Schematically these processes are the following:

$$A_0^{+(\zeta+1)} + e \Leftrightarrow A_{n'l',n''l''}^{+\zeta}$$

$$A_{n'l',n''l''}^{+\zeta} \rightarrow A_{nl,n''l''}^{+\zeta} + \hbar\omega \qquad (4.9.1)$$

A schematic description of both processes is shown in figure 4.8. In figure 4.8(a) the process of autoionization is shown. This is a relatively simple process.

The various steps of the dielectronic recombination are shown in figure 4.8(b 1–3). Figure 4.8(b1) shows the initial state of a free electron and a ground state ion (of course, an excited ion is also possible). The next step is the recombination of the free electron and the simultaneous excitation of a ground state bound electron, figure 4.8(b2), thereby forming a doubly excited ion. Such an ion can be produced only if

$$E_e = E_{n'l',n''l''} - E_0 \pm \delta E \qquad (4.9.2)$$

where E_e is the energy of the incident electron, $E_{n'l',n''l''} - E_0$ is the energy of the doubly excited state above the ground state, and δE is the excited state width. Finally, radiative stabilization occurs when the lower excited electron decays to the ground state, figure 4.8(b3), with a spectator electron remaining in an upper state. The radiative decay goes, in most cases, from the lower state, because this state has the larger oscillator strength.

It should be noted that radiative stabilization occurs in an electric potential field that is slightly different from the potential of an ion without the spectator electron, therefore the energy of the emitted photon is also slightly different from the main transition energy. For example, the energy of the photon emitted during

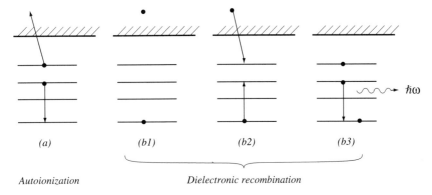

(a)   (b1)   (b2)   (b3)

Autoionization   Dielectronic recombination

Figure 4.8 Schematic description of the processes of (a) autoionization, and (b) dielectronic recombination.

the radiative stabilization of a doubly excited state in heliumlike ions, such as $2pn''l'' \rightarrow 1sn''l''$, is close to but does not exactly equal the corresponding Lyman-α line $2p \rightarrow 1s$. The line emitted from the heliumlike species is called a *satellite line*, and it is shifted to the lower energy, longer wavelength direction. Moreover, for every upper state $n''l''$ of the spectator electron one gets different emission lines, all of which are near the Ly$_\alpha$ line. There is, therefore, a whole series of satellite lines near the main transition line, the Ly$_\alpha$ in our example. The intensity of these satellite lines is not small, and they are clearly seen in every emission spectrum from plasmas on the low energy side of every strong emission line. These satellite lines in the emission spectrum are, in fact, the best experimental evidence for the existence of dielectronic recombination.

Dielectronic recombination, being a two-particle interaction, is rather frequent in hot and low density plasmas. Its rate coefficient will be denoted below by $R^{(d)}(\zeta + 1 \rightarrow \zeta, nl, n''l''; T_e)$, and the number of recombinations per unit time and unit volume is $N_{\zeta+1}n_e R^{(d)}(\zeta + 1 \rightarrow \zeta, nl, n''l''; T_e)$. In this notation $n''l''$ is the state of the spectator electron, and nl is the final state of the electron after radiative decay. Under the name *two-body recombinations* we include the radiative and the dielectronic recombinations. Whenever we shall be interested only in the sum of these two-body recombinations, we shall use a "2" as the superscript,

$$R^{(2)}(\zeta + 1 \rightarrow \zeta, nl; T_e) = R^{(r)}(\zeta + 1 \rightarrow \zeta, nl; T_e) + R^{(d)}(\zeta + 1 \rightarrow \zeta, nl, n''l''; T_e)$$
$$(4.9.3)$$

Dielectronic recombination depends strongly on the plasma density. In fact, at high plasma densities the continuum lowering cuts out the highly excited ionic levels and leaves less phase space available for the dielectronic recombination. The influence of this ionic process decreases, therefore, in high density plasmas, but is significant in low density ones.

The Detailed Balance Principle and Radiative Stabilization

The detailed balance principle in this case holds true only between autoionization and dielectronic recombination before stabilization. It has the form (Cowan, 1981)

$$R^{(d)*}(\zeta + 1 \rightarrow \zeta, n'l', n''l''; T_e)$$

$$= \frac{1}{2} \frac{g_{\zeta,n'l',n''l''}}{g_{\zeta+1}} \left(\frac{2\pi(\hbar c)^2}{mc^2 T_e} \right)^{3/2} Au(\zeta, n'l', n''l'' \rightarrow \zeta + 1) \exp(-E_{n'l',n''l''} T_e)$$

(4.9.4)

where $n'l'$, $n''l''$ is the doubly excited state before relative stabilization, $Au(\ldots)$ is the rate coefficient for autoionization, and $R^{(d)*}(\ldots)$ is the rate coefficient for recombination into the doubly excited state. The dielectronic recombination rate coefficient is that fraction of $R^{(d)*}(\ldots)$ which ends with a radiative decay,

$$R^{(d)}(\zeta + 1 \rightarrow \zeta, nl, n''l''; T_e) = R^{(d)*}(\zeta + 1 \rightarrow \zeta, n'l', n''l''; T_e)$$

$$\times \frac{A_{\zeta,n'l' \rightarrow nl}}{Au(\zeta, n'l', n''l'' \rightarrow \zeta + 1) + \sum_{n_\ell \leq n'} A_{\zeta,n'l' \rightarrow n_\ell \ell_\ell}}$$

(4.9.5)

where $A_{\zeta,n'l' \rightarrow nl}$ is the Einstein coefficient for the radiative transition from state $n'l'$ to nl. Substituting equations (4.9.5) into (4.9.4), one gets

$$R^{(d)}(\zeta + 1 \rightarrow \zeta, nl, n''l''; T_e) = \frac{1}{2} \frac{g_{\zeta,n'l',n''l''}}{g_{\zeta+1}} \left(\frac{2\pi(\hbar c)^2}{mc^2 T_e} \right)^{3/2}$$

$$\times \frac{A_{\zeta,n'l' \rightarrow nl} Au(\zeta, n'l', n''l'' \rightarrow \zeta + 1)}{Au(\zeta, n'l', n''l'' \rightarrow \zeta + 1) + \sum_{n_\ell \leq n'} A_{\zeta,n'l' \rightarrow n_\ell \ell_\ell}}$$

$$\times \exp(-E_{n'l',n''l''}/T_e)$$

(4.9.6)

Although this formula has a rather simple structure, its computation requires a knowledge of the rate coefficient for autoionization, for which there is no simple analytical formula.

Review Articles and Experiments

Dielectronic recombination and its influence on the charge state distributions in hot plasmas have been studied by several research groups. Most of these studies are theoretical and computational, but there have also been a few experiments. Here we mention only works connected directly to the cross section and rate coefficients. The applications of dielectronic recombination for the interpretation of the satellite lines in the emission spectra of hot plasmas (Gabriel, 1972; Gabriel and Jordan, 1972; Bely-Dubau et al., 1979) will be given in chapter 6.

A series of studies has been carried out by I. L. Beigman, L. A. Vainshtein, B. N. Chichkov and R. A. Sunyaev (Beigman and Chichkov 1980; Beigman et al., 1968, 1981), over almost all aspects of dielectronic recombination. This group has suggested an analytical formula for the rate coefficient of the effect, and their papers include a comprehensive discussion of the behavior of the effect as a function of the temperature, the ionic charge state, as well as the electronic configuration of the initial ion. These authors made comparisons between the influences of the dielectronic recombination and the other recombination processes in the plasma, and investigated the temperature and density domains where each of these recombination process prevails (Beigman et al., 1968). In Beigman et al., (1981) there are suggestions for correction formulas that account for the suppression of the process at high ion densities due to the shift of the upper excited states into the continuum. This group has also examined the possibility that dielectronic recombination proceeds through forbidden levels, and has found that such an effect is of the same order of magnitude as radiative recombination, and may influence the charge state distributions in low ion density, low electron temperature plasmas (Beigman and Chichkov, 1980).

A series of computations of dielectronic recombination was carried out by V. L. Jacobs, who used several models for the calculations. An example of his studies is the computation of the dielectronic recombination rate coefficients for sodium as a function of temperature (Jacobs, 1985). In other papers he attempted, with several collaborators, to use a multichannel scattering operator method (Jacobs et al., 1987) to obtain a unified description of the radiative and dielectronic recombinations. In a separate paper (Haan and Jacobs, 1989) he used a projector operator approach for the same purpose.

Most of the experimental measurements of the dielectronic recombination cross sections were carried out by measuring the intensity of the dielectronic satellites relative to the main transition line; see, for instance, a paper by Chichkov et al. (1981). Of the other methods, we mention here an experiment by H.-J. Kunze and his group (Meng et al., 1985) who measured the intensity of the emission lines of Ar^{7+} to Ar^{11+} and compared it to computed intensities obtained by including or excluding the dielectronic recombination rate coefficients. Similar measurements were carried out with the various charge states of molybdenum and titanium by Griem, Brooks, Datla and Krumbein (Brooks et al., 1980). A more direct method of measurement used crossed beams of electrons and singly ionized magnesium, Mg^{+}. The emerging neutral magnesium atoms were measured in coincidence with the emitted satellite lines (Belic et al., 1983). The experimental cross section was larger by about a factor of 5 than the computed values.

The Burgess–Merts Dielectronic Recombination Rate Coefficient

As yet there is still no universal formula for the rate coefficient of the dielectronic recombination. The main problem in obtaining a reliable formula is the difficulty in the summation of the separate rate coefficients over all the possible states of the spectator electron. A formula that is widely used is the scaled Burgess–Merts formula (Burgess, 1965; Merts et al., 1976)

$$R^{(d)}(\zeta+1\to\zeta,nl;T_e) = \sum_{n''l''} R^{(d)}(\zeta+1\to\zeta,nl,n''l'';T_e)$$

$$= 4.8\times10^{-11}(\mathrm{cm}^3/\mathrm{s})\left(\frac{E_H}{T}\right)^{3/2} B(z)\sum_{n''l''}\bar{f}_{n'l'\to nl}A(x)\exp\left(\frac{-E_s^\infty}{aT}\right)$$

$$\tag{4.9.7}$$

where

$$z = Z - Z_b + 1 \qquad x = \frac{E_s^\infty}{z+1}$$

$$B(z) = \sqrt{\frac{z(z+1)^5}{z^2+13.4}}, \qquad z \le 20$$

$$\tag{4.9.8}$$

$$A(x) = \frac{x^{1/2}}{1+0.105x+0.015x^2}, \qquad > 0.05$$

$$a = 1.0 + 0.015\frac{z^3}{(z+1)^2}$$

In this formula, Z_b is the number of bound electrons before recombination, $\bar{f}_{n'l'\to nl}$ is the average oscillator strength for the radiative part of the transition, and E_s^∞ is the mean energy of the doubly excited state.

The Burgess formula gives a reasonably good approximation for the dielectronic recombination when the radiative transition occurs within the same shell, $\Delta n = 0$. For $\Delta n > 0$ transitions it was suggested (Merts et al., 1976) that better results are obtained if the function $A(x)$ is modified to

$$A(x) = \frac{x^{1/2}}{2(1+0.210x+0.030x^2)} \tag{4.9.9}$$

The Hahn Rate Coefficient

Y. Hahn has compiled all the available data on dielectronic recombination rate coefficients, and has carried out a critical assessment of these data (Hahn, 1993). Most of the information is obtained from theoretical computations, but there are also a few experiments which provided test cases for comparison. He restricted his studies to cases with atomic number $Z < 50$, and number of bound electrons $Z_b < 13$.

Hahn's work is distinct from others in that he has fitted the available results according to the modes of the intermediate excitation. In other words, during the initial formation of the resonant states by the capture + excitation process, the target electrons are raised to upper states by the following excitation modes: (i) excitation of $1s$ electrons to upper levels, $\Delta n > 0$; (ii) excitation of $2s$ electrons to $2p$, $\Delta n = 0$; (i) excitation of $2s$ or $2p$ electrons to upper levels, $\Delta n > 0$; (iv) excitation of $3s$ or $3p$ electrons to $3p$ or $3d$ levels, $\Delta n = 0$; (v) excitation of $3s$ or $3p$ electrons to higher levels, $\Delta n > 0$.

We denote by

$$\alpha(nl) = R^{(d)}(\zeta + 1, nl \to \zeta, nl; T_e) = \sum_{n'l', n''l''} R^{(d)}(\zeta + 1, nl \to \zeta, n'l', n''l''; T_e)$$

(4.9.10)

the rate coefficient for the recombination of an electron with an ion charge $\zeta + 1$ (initially in its ground state); as a result of the recombination an electron from state nl is excited to a doubly excited state $n'l', n''l''$, from which the $n'l'$ decays back to the ground state of ion ζ, while the spectator electron is in any of the possible excited state $n''l''$.

Hahn proposed five different formulas for the rate coefficients according to the above five cases:

$$\alpha(1s) = \frac{A_1}{T^{3/2}} \exp\left(-\frac{A_2}{T}\right) \left(\frac{6}{Z_b + 4}\right)^{0.9} \exp[-A_3(Z_b - 2)^2]$$

$$A_1 = \frac{1230}{Z^{0.14}} \exp\left(-\frac{44}{Z + 2.86}\right), \quad A_2 = 0.0075\left(Z + \frac{1}{Z_b}\right)^2, \quad A_3 = 0.0222Z$$

(4.9.11)

In these equations Z_b is the number of bound electrons, T is the electron temperature in keV, and α is in units of 10^{-13} cm^3/s. The other rate coefficients are

$$\alpha(2s) = \frac{B_1}{T^{3/2}} \exp\left(-\frac{B_2}{T}\right) \frac{(Z_b - 2)(10 - Z_b)(1 + 0.3/T^{0.21})}{(Z_b + B_3)^{2.5}}$$

$$B_1 = 52\left(\frac{Z}{10 + 0.011Z^2}\right)^{0.65} \exp\left(-\frac{18}{Z_1 + 1}\right), \quad B_2 = 0.0023\frac{Z_1}{1 + 0.0015Z_1^2},$$

$$B_3 = 0.8, \quad Z_1 = Z - 2$$

(4.9.12)

$$\alpha(2p) = \frac{C_1}{T^{3/2}} \exp\left(-\frac{C_2}{T}\right) F_c \exp\left(-\frac{C_2}{T} \frac{0.0001}{Z_b + 1}\right)$$

$$F_c = \left(\frac{10}{Z_b + 1}\right)^{1/2} \exp(-C_3|Z_b - 9.6|)$$

$$C_1 = 2.15Z_1^{1.8} \exp(-0.004(Z_1 - 35)^2), \quad C_2 = 0.001\,15Z_1^2(1 - 0.003Z_1),$$

$$C_3 = 0.17$$

(4.9.13)

here Z_1 is the same as in equation (4.9.12).

$$\alpha(3s) = \frac{D_1}{T^{3/2}} \exp\left(-\frac{D_2}{T}\right) \exp\left(\frac{D_2}{T} \frac{0.15}{(Z_b - 10)^{1.5}}\right) \frac{(Z_b - 10)(Z - Z_b)}{Z_b - D_3}$$

$$D_1 = 0.16 Z_2^2 \exp(-0.11 Z_2), \qquad D_2 = 0.0024 Z_2 (1 - 0.01 Z_2),$$

$$(4.9.14)$$

$$D_3 = 6, \qquad Z_2 = Z - 10$$

$$\alpha(3p) = \frac{E_1}{T^{3/2}} \exp\left(-\frac{E_2}{T}\right) \exp[-E_3(Z_b - 12)^2] \frac{(Z_b - 10)(Z - 10)}{8}$$

$$E_1 = \frac{0.45}{Z_2} \exp\left(\frac{Z_2}{4 + 0.02 Z_2}\right), \qquad E_2 = 0.0003 Z_3^2 (1 - 0.003 Z_3), \qquad (4.9.15)$$

$$E_3 = 0.02, \qquad Z_3 = Z - 7 = Z_2 + 3$$

(Incidentally, we have taken the opportunity of correcting several misprints in the original paper). The total dielectronic recombination rate coefficient is, of course, the sum of the partial rate coefficients over all the electrons in ion $\zeta + 1$. The accuracy of the total rate coefficient derived from these formulas is estimated as 30–50% which is, perhaps, the best accuracy available at present. Although these equations do not have a clear internal structure, they suggest that the rate coefficient has the general form

$$\alpha = \frac{\text{constant}}{T^{3/2}} \exp\left(-\frac{\text{av.exc.energy}}{T}\right) F(Z, Z_b; T) \qquad (4.9.16)$$

where $F(Z, Z_b; T)$ is a function that varies very slowly, or is even independent of the temperature. In figure 4.9 are shown the rate coefficients for dielectronic recombination for the various charge states of magnesium. Although the rate coefficients are plotted for a wide range of temperatures, they are meaningful only in the range where a specific charge state has appreciable abundance in the plasma. From figure 4.9 it can be seen that the general behavior of this rate coefficient is similar to those of the electron impact ionization and excitation, namely, a sharp increase at low temperatures, a maximum around a characteristic excitation energy, and a slow decrease in the high energy portion of the graph.

High Density Corrections

It was noted by Beigman et al. (1968) that the rate coefficients for the dielectronic recombination, as calculated for low density plasmas, need additional corrections as pressure ionization of an ion in a highly excited state may compete with the effect of dielectronic recombination. This causes a reduction of the effect, particularly at high densities where the number of bound states is

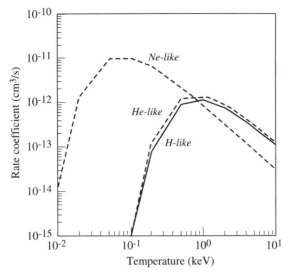

Figure 4.9 Rate coefficient of dielectronic recombination for H-, He-, and Li-like magnesium, as a function of the temperature.

rather small. It was shown by Salzmann and Krumbein (1978) that, when a correction factor suggested by Beigman et al. is taken into account, the dielectronic recombination has an appreciable effect on the charge populations in an aluminium plasma up to electron densities of about 10^{18} cm^{-3} only; its effect is around 10% at densities of 10^{20} cm^{-3}, and becomes completely negligible at densities of 10^{22} cm^{-3} and above.

Population Distributions

5.1 General Description

The rapid ionization–recombination processes in the plasma determine the relative abundance of the various charge states. For example, in plasma in which the rate of the ionization is larger than the rate of the recombinations, more and more electrons are stripped off the ions so that the ions' charge is increasing continually. Such a plasma is called an *ionizing plasma*. On the other hand, in a plasma in which the rate of the recombinations is larger than the rate of the ionizations, the ions are filling up with electrons, and the plasma is called *a recombining plasma*. A *steady state* is attained when the rate of ionizations equals the rate of recombinations. In this case the charge state distribution is constant, independent of time. The excited states behave in a similar manner. The rapid excitation and deexcitation (including spontaneous decay) processes determine the average excitation state of the ions and, again, a steady state is achieved when the rate of excitations equals the rate of deexcitations + decays. It turns out that the average charge of the ions, $\bar{Z}$ (equation 1.2.3), in a steady state depends on the relative values of the rate coefficients for the ionization and recombination processes, and so, implicitly on the electron temperature and ion density. In the following sections we shall investigate the conditions under which steady state is attained and what happens under conditions that are changing with time.

5.2 Local Thermodynamic Equilibrium

Among the various steady states in a plasma, the local thermodynamic equilibrium (LTE) has a special status. In fact, this is the plasma state that is closest to a full thermodynamic equilibrium, as described by the methods of statistical mechanics. In complete thermodynamic equilibrium (TE) all three types of particles—the ions, the electrons, and the photons—are in equilibrium, which means that the rate of each process equals exactly the rate of its inverse. In such a plasma, the electron and the ion velocity distributions follow a Maxwell–Boltzmann distribution function (equation 1.3.13), the distribution of the excited states is of a Boltzmann type (equation 1.3.5), and the photons have a Planck energy distribution function (equation (4.5.11).

Local thermodynamic equilibrium occurs in plasmas whose dimensions are significantly smaller than the mean free path (m.f.p.) of the photons emitted from the plasma, but are much longer than the collision length of the electrons and the ions. The m.f.p. of photons in a $T = 10$–$1000\,eV$ hydrogen plasma is longer by a factor of 10^4 or more than the m.f.p. of the electrons. In an aluminum plasma between the same temperatures, the photon m.f.p. is longer by a factor of at least 300 than those of the electrons. In LTE the photons move a relatively large distance from the location of their emission, and they may either escape from the plasma or be reabsorbed in some other part which may have a different temperature or density. Photons are, therefore, not necessarily in equilibrium with the material particles. The electrons and the ions are, however, colliding at a high rate and their distributions of velocities and excited states are the same as for equilibrium conditions.

In earlier publications, LTE was defined slightly differently (see for instance Zeldovitch and Raizer, 1966), by the condition that the mean free path of the photons is not much longer than the relative temperature gradients, that is, $\lambda_{m.f.p.}$ is of the same order of magnitude as $L_{plasma} = T/|\text{grad }T|$. In view of this definition, the photons arriving in a given region in the plasma have been emitted from regions having other temperatures or densities that are not much different from the conditions at the point of absorption. The idea common to both definitions is that in LTE the electrons and ions are in equilibrium among themselves, whereas the photons are not.

The charge state distribution in LTE plasmas can be calculated from basic principles. Consider a plasma consisting of ions having charges from 0 (neutral atoms) to Z (fully ionized species), and electrons. Denote by $N_0, N_1, \ldots, N_Z$, respectively, the number densities (ions/cm^3) of the various charge states, and as usual, by n_e the electron density. The derivation of the charge state distribution in LTE plasmas is carried out by minimizing the Helmholtz free energy, F, which is defined in equation (1.3.8),

$$F = -T \log[Z(T)] \qquad (1.3.8)$$

where $Z(T)$ is the partition function of the whole system. As shown in the discussion following equation (1.3.5), $Z(T)$ equals the product of the partition functions of the individual species,

$$Z(T) = Z_e(T) \prod_{\zeta=0}^{Z} Z_\zeta(T) \qquad (5.2.1)$$

LTE conditions develop when the rates of the collisional processes, namely, electron impact ionization and its inverse, three-body recombination are equal. However, as explained in chapter 4, this recombination prevails only in rather high density plasmas. It turns out, therefore, that LTE conditions can be attained only at high ion densities. More accurate criteria for the validity domain of LTE will be given later. For a reaction of the type

$$A_{\zeta-1} + e \Leftrightarrow A_\zeta + e + e$$

conservation of charge gives

$$-\delta(M_{\zeta-1}) = \delta(M_\zeta) = \delta(M_e) \tag{5.2.2}$$

where M_e, $M_{\zeta-1}$, and M_ζ are the total number of the electrons, ions of charge $\zeta - 1$, and ions of charge ζ, respectively. The variation of the free energy with respect to the number of particles gives

$$\delta F = \frac{\partial F}{\partial M_{\zeta-1}} \delta M_{\zeta-1} + \frac{\partial F}{\partial M_\zeta} \delta M_\zeta + \frac{\partial F}{\partial M_e} \delta M_e \tag{5.2.3}$$

Equilibrium conditions are obtained by minimizing the free energy, $\delta F = 0$. Substituting equations (5.2.2) into (5.2.3) the condition for equilibrium becomes

$$-\frac{\partial F}{\partial M_{\zeta-1}} + \frac{\partial F}{\partial M_\zeta} + \frac{\partial F}{\partial M_e} = 0 \tag{5.2.4}$$

The partial derivatives of the free energy with respect to the number of particles were calculated in chapter 1 (equation 1.3.10). Inserting these into equation (5.2.4), one gets

$$-T\left[-\log\left(\frac{z_{\zeta-1}}{M_{\zeta-1}}\right) + \log\left(\frac{z_\zeta}{M_\zeta}\right) + \log\left(\frac{z_e}{M_e}\right) \right] = 0 \tag{5.2.5}$$

where z_e, $z_{\zeta-1}$, and z_ζ are the partition functions of the individual particles (see explanation in section 1.3). Equation (5.2.5) is rewritten as

$$\frac{M_\zeta M_e}{M_{\zeta-1}} = \frac{z_\zeta(T) z_e(T)}{z_{\zeta-1}(T)} \tag{5.2.6}$$

The values of z_e, $z_{\zeta-1}$, and z_ζ were calculated in equation (1.3.4) for the electrons and in equation (1.3.5) for the ions. The kinetic parts of the partition functions of the ions of charge ζ and $\zeta - 1$ are equal and drop out from the equation. Some care must be taken in the calculation of the ratios of the two internal partition functions, because each employs its own ground state energy as the reference point. These two reference points differ by $E_{\zeta-1} - \Delta E_{\zeta-1}$, which is the ground state ionization energy of the charge state $\zeta - 1$ corrected for the local continuum lowering, $\Delta E_{\zeta-1}$. Inserting these values into equation (5.2.6), and replacing the number of particles by the corresponding densities, $M_e = n_e V$, $M_{\zeta-1} = N_{\zeta-1} V$, and $M_\zeta = N_\zeta V$, one obtains

$$\frac{N_\zeta n_e}{N_{\zeta-1}} = 2\left(\frac{mc^2 T_e}{2\pi(\hbar c)^2}\right)^{3/2} \frac{z_\zeta(T_e)}{z_{\zeta-1}(T_e)} \exp\left(-\frac{E_{\zeta-1} - \Delta E_{\zeta-1}}{T_e}\right) \tag{5.2.7}$$

which is the well known *Saha equation*. This equation gives the ratio of the density of charge state ζ to that of $\zeta - 1$ in terms of the electron density, the temperature, and $E_{\zeta-1}$, the ionization energy of charge state $\zeta - 1$. Writing down similar equations for all the charge states one gets a set of homogeneous linear equations. Two complementary equations which have to be satisfied together with equation

(5.2.7) are, first, the requirement that the sum of all the partial densities equals the total ion density (equation 1.2.1),

$$\sum_{\zeta=0}^{Z} N_\zeta = n_i \tag{1.2.1}$$

and, second, the charge neutrality condition in the plasma, which is given by equation (1.2.2),

$$n_e = \sum_{\zeta=0}^{Z} \zeta N_\zeta \tag{1.2.2}$$

The interactions in the plasma are, in fact, taken into account in equation (5.2.7) by the term of the continuum lowering, and the introduction of the truncated partition functions, equations (3.4.4) or (3.4.8), into (5.2.7). These functions incorporate the ionization potential reduction, and thereby, the average interaction of the plasma ions with their environment.

The charge state distribution for an aluminum plasma of ion density $n_i = 1 \times 10^{22}\,\mathrm{cm}^{-3}$ as a function of temperature is shown in figure 5.1. At such a high density the plasma is in LTE conditions up to temperatures of $\sim 300\,\mathrm{eV}$. The figure indicates that as the plasma temperature increases the dominant charge states are the more highly ionized species. Above $\sim 1000\,\mathrm{eV}$ the plasma becomes fully ionized. It is interesting to note the plateau between 100 and 400 eV, in which region the heliumlike ions are the most abundant species. These ions have a high ionization potential and therefore have a very stable configuration. This is why between 200 and 400 eV heliumlike ions are practically the only species in the

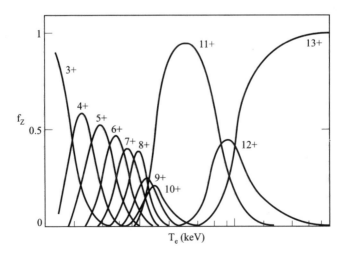

Figure 5.1 Charge state distributions for an aluminum plasma of ion density of $n_i = 1 \times 10^{22}\,\mathrm{cm}^{-3}$ as a function of temperature, calculated assuming LTE.

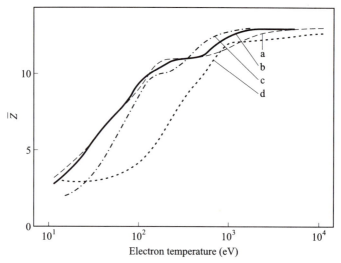

Figure 5.2 Behavior of the average ionization, $\bar{Z}$, as function of the electron temperature, for several ion densities: (a) 10^{18}, (b) 10^{20}, (c) 10^{22}, and (d) $10^{24}\,\mathrm{cm}^{-3}$. (Reprinted with permission from Salzmann, D. and Krumbein, A., 1978, *J. Appl. Phys.*, **49**, 3229. Copyright 1978 American Institute of Physics.)

plasma, their fractional density being close to 100%. The behavior of the average charge state, $\bar{Z}$, is plotted in figure 5.2 for the same density as a function of the temperature.

The partial ion densities, partition functions, and several other properties of LTE plasmas have been tabulated by Drawin and Felenbok (1965) for many elements.

Next we look into the problems of the population distribution of the electrons in the various excited states. First, we order the excited states according to their ascending energy, and denote them by a second index m, $m = 0, 1, 2, \ldots, M_\zeta$, where $m = 0$ corresponds to the ground state and $m = M_\zeta$ is the uppermost bound ionic state of charge state ζ. As discussed in chapter 3, M_ζ is finite. The density of ions having charge ζ that are excited to the mth state will be denoted by two indices, $N_{\zeta,m}$, to distinguish it from that of the whole charge state, which is denoted by N_ζ.

$$N_\zeta = \sum_{m=0}^{M_\zeta} N_{\zeta,m} \tag{5.2.8}$$

For convenience, we define *the population probability* of an ionic state (ground or excited), $P_{\zeta,m}$, as the density of this state relative to the density of the whole charge state,

$$P_{\zeta,m} = \frac{N_{\zeta,m}}{N_\zeta} \tag{5.2.9}$$

so that

$$\sum_{m=0}^{M_\zeta} P_{\zeta,m} = 1 \qquad (5.2.10)$$

Since in a plasma that fulfills the conditions of LTE the collisional processes are much more frequent than the radiative processes, the distribution of the ionic states follows a Boltzmann-type distribution, equation (1.3.5),

$$N_{\zeta,m} = N_\zeta \frac{g_{\zeta,m} \exp(-E_{\zeta,m}/T)}{\sum_{m'} g_{\zeta,m'} \exp(-E_{\zeta,m'}/T)} = N_\zeta \frac{g_{\zeta,m} \exp(-E_{\zeta,m}/T)}{Z_\zeta(T)}$$

$$\qquad (5.2.11)$$

$$P_{\zeta,m} = \frac{g_{\zeta,m} \exp(-E_{\zeta,m}/T)}{Z_\zeta(T)}$$

where $E_{\zeta,m}$ is measured with respect to the ground state energy.

5.3 Corona Equilibrium

Next we turn to the other extreme, namely, to very low density optically thin plasmas. Such plasmas occur frequently under both astrophysical and laboratory conditions. The steady state conditions in such plasmas are called *corona equilibrium* (CE). Examples in astrophysics are the interstellar nebulae or the sun's corona, from which it gained its name. In laboratory experiments one meets corona equilibrium in tokamaks and other low density plasma machines.

While the derivation of the basic equations of LTE was based heavily on the methods of statistical physics, at the low ion density extreme one can get a great deal of information from intuitive considerations. In particular, one can give a rather good estimate for the population probabilities of the ground and excited states without the need to write down complicated formulas. In fact, in low ion/electron density plasmas the upward excitation rate by collisions is so low relative to the spontaneous decay that one can safely assume that an electron excited to an upper level will most likely decay to the ground state before experiencing a second excitation. Most of the ions are, therefore, in their ground state,

$$P_{\zeta,m} = \begin{cases} 1 & \text{ground state} \\ 0 & \text{excited states} \end{cases} \qquad (5.3.1)$$

Moreover, in low density optically thin plasmas the photoionization and photoexcitation processes have very low rates, and the same is true for the three-body recombination. In fact, in such plasmas the dominant processes are electron impact ionization and two-body—namely, radiative + dielectronic—recombinations. The charge state distribution is calculated by equating the rates of these processes (see table 4.1),

$$n_e N_{\zeta-1} \mathcal{I}(\zeta - 1 \to \zeta) = n_e N_\zeta \mathcal{R}^{(2)}(\zeta \to \zeta - 1) \qquad (5.3.2)$$

wherefrom one finds

$$\frac{N_\zeta}{N_{\zeta-1}} = \frac{\mathcal{I}(\zeta - 1 \to \zeta; T_e)}{\mathcal{R}^{(2)}(\zeta \to \zeta - 1; T_e)}$$

$$= \frac{\mathcal{I}(\zeta - 1 \to \zeta; T_e)}{\mathcal{R}^{(r)}(\zeta \to \zeta - 1; T_e) + \mathcal{R}^{(d)}(\zeta \to \zeta - 1; T_e)} \qquad \zeta = 1, \ldots, Z \qquad (5.3.3)$$

By virtue of equation (5.3.1), $N_\zeta = N_{\zeta,0}$ and equation (5.3.3) is a relation between the partial densities of the ground states of two adjacent charge states. Writing equation (5.3.3) for all the charge states, one gets a set of $Z + 1$ recursive linear homogeneous equations with $Z + 1$ parameters, N_ζ ($\zeta = 0, 1, \ldots, Z$), which is complemented by the equations of the total ion density and charge neutrality, equations (1.2.1) and (1.2.2), in a manner similar to LTE.

It is interesting to note that the recursion equation (5.3.3) is independent of the electron or ion densities, which implies that the charge state distribution, too, in CE, is independent of the density. The charge state distribution of an aluminum plasma at CE is shown in figure 5.3. It is the extension of figure 5.1 to lower densities. The general features of the charge state distributions in corona equilibrium are similar to those of the higher density LTE plasmas, that is, there is an increase in the average ionization as the temperature goes up, and a plateau around the stable heliumlike species.

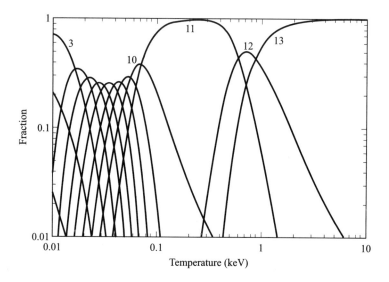

Figure 5.3 Charge state distributions for an aluminium plasma of ion density of $n_i = 1 \times 10^{18} \, \text{cm}^{-3}$ as a function of temperature, calculated assuming corona equilibrium.

An approximate rule of thumb is sometimes helpful in the context of CE. This states that the most abundant charge state in the plasma, $\zeta \approx \bar{Z}$, will be the one whose ionization energy, E_ζ, is larger by a factor of 2 to 5 than the temperature,

$$\frac{E_\zeta}{T} \simeq 2 - 5 \qquad (5.3.4)$$

In some cases, this relation helps to rapidly find an approximate value of $\bar{Z}$.

A few points should be emphasized in connection with CE. The first is the notion that in CE the rate of the collisional three-body recombinations is negligibly small relative to the radiative recombinations. In other words, this means that the energy-conserving collisional processes in a CE plasma are much less frequent than the energy-dissipating radiative processes. This brings us to a second important point, namely that, in contrast to its name, CE is not an equilibrium model at all. In fact, one photon is emitted with every radiative recombination that occurs in the plasma. This causes a slow but steady cooling of the plasma. In view of equation (5.3.3), this cooling follows a thermodynamic path that ensures that the rate of the variation of the partial densities of the various charge states will be a minimum. Alternatively, the system can be heated by some external source; for example, the sun's coronal regions are constantly heated by radiation and convection from the sun's interior. In this sense, CE is a steady state only, but not an equilibrium one.

5.4 The Collisional Radiative Steady State

The collisional radiative steady state (CRSS) is an intermediate model between LTE and CE. However, as we shall see, it is not just an interpolation between the two models but rather a generalization of both of them. In fact, the CRSS tends to the corona equilibrium in the low density limit, and to LTE for high density plasmas, thereby providing a model such that CE and LTE are special cases of CRSS at low and high densities.

Unlike LTE or CE, in CRSS the population distribution among the various states cannot be guessed from basic principles, and the density of a given excited species can be solved only from the whole set of rate equations that takes into account all the processes that increase or decrease the species' abundance.

Consider a plasma of atoms of atomic number Z, electron temperature T_e, and ion density n_i. Again, we denote the densities of the charge states by N_ζ, $\zeta = 0, 1, \ldots, Z$, and the partial density of ions of charge ζ excited to the mth excited state by $N_{\zeta,m}$, $m = 0, 1, \ldots, M_\zeta$, where $m = 0$ corresponds to the ground state and $m = M_\zeta$ to the uppermost bound state which, in general, is different for each charge state. We assume that the plasma is optically thin, and neglect the processes initiated by photon–ion interactions. With these assumptions, the rate of change of the population of a particular level, m of a given charge state, ζ, is affected by seven processes, each of which can populate or depopulate this level. As discussed in chapter 4, these processes are the following (for a reminder of the rates of these processes and the notation of the rate coefficients, see table 4.1):

- Spontaneous decay (if the state is excited)
- Electron impact ionization
- Three-body recombination
- Electron impact excitation
- Electron impact deexcitation (if the state is excited)
- Radiative recombination
- Dielectronic recombination

In a steady state, the rates of population and depopulation are equal, so that the equation that determines the partial densities of this species can be symbolically written as

$$\sum_{\substack{all \\ populating \\ processes}} n_e^k N_{\zeta',m'} \mathcal{R}(\zeta',m' \to \zeta,m) = \sum_{\substack{all \\ depopulating \\ processes}} n_e^k N_{\zeta,m} \mathcal{R}(\zeta,m \to \zeta',m')$$

$$\zeta = 0, 1, \ldots, Z, \qquad \zeta' = \zeta, \zeta \pm 1, \qquad m = 0, 1, \ldots, M_\zeta, \qquad m' = 0, 1, \ldots, M_{\zeta'}$$

$$(5.4.1)$$

where $\mathcal{R}(\zeta,m \to \zeta',m')$ represents the rate coefficient of the process that changes the state ζ, m into state ζ', m' (units cm^{3k}/s) and k is the number of electrons taking part in the process. Thus, $k = 2$ for the three-body recombination, $k = 0$ for the spontaneous decay of an excited state, and $k = 1$ for the other processes. Altogether, there are seven terms on the right hand side of equation (5.4.1), and seven terms on its left hand side. The number of equations is $M = \sum_{\zeta=0}^{Z} M_\zeta$, which may be a very high number for low density plasmas but nevertheless is not infinite. For the sake of completeness, we write equation (5.4.1) in full for a state (ζ, m),

$$N_{\zeta,m} \sum_{m'<m} A(\zeta,m \to \zeta,m') + n_e N_{\zeta,m} \left\{ \sum_{m'} I(\zeta,m \to \zeta+1,m') \right.$$

$$+ \sum_{m'} [\mathcal{R}^{(r)}(\zeta,m \to \zeta-1,m') + \mathcal{R}^{(d)}(\zeta,m \to \zeta-1,m')]$$

$$\left. + \sum_{m'>m} \mathcal{E}(\zeta,m \to \zeta,m') + \sum_{m'<m} \mathcal{D}(\zeta,m \to \zeta,m') \right\}$$

$$+ n_e^2 N_{\zeta,m} \sum_{m'} \mathcal{R}^{(3)}(\zeta,m \to \zeta-1,m')$$

$$= \sum_{m'>m} N_{\zeta,m'} A(\zeta,m' \to \zeta,m) + n_e \left\{ \sum_{m'} N_{\zeta-1,m'} I(\zeta-1,m' \to \zeta,m) \right.$$

$$+ \sum_{m'} N_{\zeta+1,m'} [\mathcal{R}^{(r)}(\zeta+1,m' \to \zeta,m) + \mathcal{R}^{(d)}(\zeta+1,m' \to \zeta,m)]$$

$$\left. + \sum_{m'<m} N_{\zeta,m'} \mathcal{E}(\zeta,m' \to \zeta,m) + \sum_{m'>m} N_{\zeta,m'} \mathcal{D}(\zeta,m' \to \zeta,m) \right\}$$

$$+ n_e^2 \sum_{m'} N_{\zeta-1,m'} \mathcal{R}^{(3)}(\zeta-1,m' \to \zeta,m)$$

$$\zeta = 0, 1, \ldots, Z, \qquad m = 0, 1, \ldots, M_\zeta \qquad (5.4.2)$$

where, we recall, $\mathcal{A}$ is the Einstein coefficient; $\mathcal{I}$, $\mathcal{E}$, and $\mathcal{D}$ are the rate coefficients of the electron impact ionization, excitation, and deexcitation, respectively; and $\mathcal{R}^{(x)}$ are the rate coefficients of the various recombination processes. The various terms in equation (5.4.2) are arranged according to ascending powers of the electron density, n_e. Equation (5.4.2) exhibits the complexity of the problem. It is one representative of a finite set of coupled nonlinear equations whose solutions yield the required densities $N_{\zeta,m}$ for $0 \leq \zeta \leq Z, 0 \leq m \leq M_\zeta$. Any attempt to solve equation (5.4.2) directly leads to great difficulties, both because of the mathematics involved and because of the large number of rate coefficients required for setting up the equations.

Things are not that bad, however, and a few plausible approximations can simplify the computations. The most important simplification is to neglect the species that presumably have very low density and solve equation (5.4.2) only for the most abundant species. We know that for any ion density, n_i, and electron temperature, T_e, there are only a few charge states participating in a plasma, see for instance figures 5.1 and 5.3, and these have only a few excited states that are important for plasma spectroscopy or other purposes. Such considerations will limit the equations to a number that can be handled by a computer code within reasonable CPU time. Nevertheless, such computations are never easy.

The solution of equation (5.4.2) becomes much simpler if only the total densities of the charge states are required, without their distribution among the various states of excitation. For this purpose one sums both sides of equation (5.4.2) over the index m. The terms of excitation, deexcitation, and spontaneous decay drop out from the equations, reflecting the fact that these terms connect states only within a given charge state and do not connect between different charge states. This can be proved rigorously by explicitly carrying out the m-sum on both sides of equation (5.4.2) and rearranging the double sums over m and m'. Summing both sides of equation (5.4.2) over m, and expressing the partial densities of the ionic states by means of the population probabilities, $P_{\zeta,m}$, equation (5.2.11), one obtains

$$n_e N_\zeta \left\{ \sum_{m,m'} P_{\zeta,m} \mathcal{I}(\zeta, m \to \zeta + 1, m') \right.$$

$$\left. + \sum_{m,m'} P_{\zeta,m} [\mathcal{R}^{(2)}(\zeta, m \to \zeta - 1, m') + n_e \mathcal{R}^{(3)}(\zeta, m \to \zeta - 1, m')] \right\}$$

$$= n_e N_{\zeta-1} \sum_{m,m'} P_{\zeta-1,m'} \mathcal{I}(\zeta - 1, m' \to \zeta, m) + n_e N_{\zeta+1} \sum_{m,m'} P_{\zeta+1,m'}$$

$$\times [\mathcal{R}^{(2)}(\zeta + 1, m' \to \zeta, m) + n_e \mathcal{R}^{(3)}(\zeta + 1, m' \to \zeta, m)] \tag{5.4.3}$$

We define the *average rate coefficients* as the average of the ordinary rate coefficients over the initial ionic state with the population probabilities, $P_{\zeta,m}$, as the weighting factors, and sum over all possible final states,

$$\mathcal{I}(\zeta \to \zeta + 1) = \sum_{\substack{0 \le m \le M_\zeta \\ 0 \le m' \le M_{\zeta+1}}} P_{\zeta,m} \mathcal{I}(\zeta, m \to \zeta + 1, m')$$

$$\mathcal{R}^{(x)}(\zeta \to \zeta - 1) = \sum_{\substack{0 \le m \le M_\zeta \\ 0 \le m' \le M_{\zeta-1}}} P_{\zeta,m} \mathcal{R}^{(x)}(\zeta, m \to \zeta - 1, m') \qquad (5.4.4)$$

with $x = r, d, 2$, or 3 for the radiative, dielectronic, 2-body (radiative + dielectronic), and 3-body recombinations, respectively. As shown in equation (5.4.4), we distinguish the average rate coefficients from the ordinary ones by suppressing the indices of the ionic states, m. Inserting the definitions in equation (5.4.4) into (5.4.3), one gets a recursive equation among the densities of the charge states,

$$N_\zeta\{\mathcal{I}(\zeta \to \zeta + 1) + \mathcal{R}^{(2)}(\zeta \to \zeta - 1) + n_e \mathcal{R}^{(3)}(\zeta \to \zeta - 1)\}$$

$$= N_{\zeta-1}\mathcal{I}(\zeta - 1 \to \zeta) + N_{\zeta+1}[\mathcal{R}^{(2)}(\zeta + 1 \to \zeta) + n_e \mathcal{R}^{(3)}(\zeta + 1 \to \zeta)] \quad (5.4.5)$$

This equation can be further simplified. Define the quantity Q_ζ by

$$Q_\zeta = N_\zeta \mathcal{I}(\zeta \to \zeta + 1) - N_{\zeta+1}[\mathcal{R}^{(2)}(\zeta + 1 \to \zeta) + n_e \mathcal{R}^{(3)}(\zeta + 1 \to \zeta)] \quad (5.4.6)$$

Q_ζ is, in fact, the difference between the rate of ionizations from charge state ζ and the rate of recombinations into this state. As seen from equation (5.4.6), an ionizing plasma is characterized by $Q_\zeta > 0$, whereas a recombining one is characterized by $Q_\zeta < 0$. In a steady state one expects $Q_\zeta = 0$. In fact, reordering the terms in equation (5.4.5) and replacing the appropriate groups using Q_ζ, equation (5.4.5) can be rewritten in a very simple form,

$$Q_\zeta = Q_{\zeta-1} \qquad (5.4.7)$$

which is solved recursively as

$$Q_\zeta = Q_{\zeta-1} = Q_{\zeta-2} = \cdots = Q_0 = 0 \qquad (5.4.8)$$

or

$$N_\zeta \mathcal{I}(\zeta \to \zeta + 1) = N_{\zeta+1}[\mathcal{R}^{(2)}(\zeta + 1 \to \zeta) + n_e \mathcal{R}^{(3)}(\zeta + 1 \to \zeta)] \qquad (5.4.9)$$

which is valid for every charge state. From equation (5.4.9) one derives the equation of the CRSS for the charge state distribution,

$$\frac{N_{\zeta+1}}{N_\zeta} = \frac{\mathcal{I}(\zeta \to \zeta + 1)}{\mathcal{R}^{(2)}(\zeta + 1 \to \zeta) + n_e \mathcal{R}^{(3)}(\zeta + 1 \to \zeta)}, \qquad \zeta = 0, 1, \ldots, Z - 1 \quad (5.4.10)$$

This set of Z equations must, again, be complemented by equations (1.2.1) and (1.2.2). A few words of caution are in order: equation (5.4.10) can be solved only if the averaged rate coefficients, equation (5.4.4), are known beforehand. These are, however, available only if the population probabilities are known, and these can be obtained only by solving the whole set of equations (5.4.2). So, apparently equation (5.4.10) is not easier to solve than the full solution of equations (5.4.2). Equation (5.4.10) can be of great help only if the population probabilities are known from other sources, such as in the case of LTE or CE.

The explicit appearance of the electron density, n_e, in the denominator of equation (5.4.10) automatically splits the results into two extreme regions:

(i) the high density region in which $n_e \mathcal{R}^{(3)}(\zeta + 1 \to \zeta) \gg \mathcal{R}^{(2)}(\zeta + 1 \to \zeta)$, and
(ii) the low density region in which $n_e \mathcal{R}^{(3)}(\zeta + 1 \to \zeta) \ll \mathcal{R}^{(2)}(\zeta + 1 \to \zeta)$.

In region (i), equation (5.4.10) reduces to

$$\frac{N_{\zeta+1}}{N_\zeta} = \frac{\mathcal{I}(\zeta \to \zeta + 1)}{n_e \mathcal{R}^{(3)}(\zeta + 1 \to \zeta)}, \qquad \zeta = 0, 1, \ldots, Z - 1 \qquad (5.4.11)$$

The two rate coefficients are connected by means of the detailed balance principle, equation (4.7.1). Inserting equation (4.7.1) into (5.4.11) it can be seen that this last is exactly the Saha equation (5.2.9)! On the other hand, in region (ii), equation (5.4.10) can be rewritten with good accuracy as

$$\frac{N_{\zeta+1}}{N_\zeta} = \frac{\mathcal{I}(\zeta \to \zeta + 1)}{\mathcal{R}^{(2)}(\zeta + 1 \to \zeta)}, \qquad \zeta = 0, 1, \ldots Z - 1 \qquad (5.4.12)$$

which is the basic equation of the corona equilibrium, equation (5.3.3). It turns out, then, that LTE and CE are special cases of CRSS at high and low plasma densities, respectively.

5.5 Low Density Plasmas

The case of low density plasmas, whose density is only slightly higher than the CE regime, is of great interest in a variety of laboratory and astrophysical plasmas. Quite surprisingly, in this region the equations of the CRSS model are solvable with several plausible approximations. To be more specific, in this section we will show how to solve equation (5.4.2) when the plasma density is so low that the population probabilities are only slightly different from the delta-function type distribution that is characteristic of CE, see equation (5.3.1). The results we will find do not have a simple structure, as in LTE or CE, but are nevertheless instructive in providing a picture of the atomic processes that influence the behavior of the population probabilities for this case.

First, we will try to get a solution in this regime for the excited states only (Salzmann, 1979). We are therefore looking for the partial density, $N_{\zeta,m}$, of the mth ionic state ($m \neq 0$) of the ζth charge state. For this purpose we make a few approximations. The first is to neglect the terms of the three-body recombination on both sides of equation (5.4.2). As the rate of this process is proportional to n_e^2, neglecting these terms is justifiable in low density plasmas. Second, we will neglect all the processes whose initial state is an excited state, because the partial density of excited states in the plasma is presumably low, albeit not zero. By virtue of this assumption we neglect the decay of higher states into the state ζ, m (the first term in the right hand side of equation (5.4.2)), and the term of electron impact deexcitation into this state (the sixth term on the right hand side). Third, we replace the partial densities by the population probabilities, $N_{\zeta,m} = N_\zeta P_{\zeta,m}$, and assume that in the density region under discussion, $P_{\zeta,m}$, is only slightly different from the values of the low density extreme, equation (5.3.1). Using these assumptions, equation (5.4.2) can be rewritten as

$$
\begin{aligned}
N_\zeta P_{\zeta,m} \sum_{m'<m} \mathcal{A}(\zeta,m \to \zeta,m') &+ n_e N_\zeta P_{\zeta,m} \left\{ \sum_{m'} \mathcal{I}(\zeta,m \to \zeta+1,m') \right. \\
&+ \sum_{m'} \mathcal{R}^{(2)}(\zeta,m \to \zeta-1,m') + \sum_{m'>m} \mathcal{E}(\zeta,m \to \zeta,m') \\
&+ \left. \sum_{m'<m} \mathcal{D}(\zeta,m \to \zeta,m') \right\} \\
&= n_e \{ N_{\zeta-1} \mathcal{I}(\zeta-1,0 \to \zeta,m) + N_{\zeta+1} \mathcal{R}^{(2)}(\zeta+1,0 \to \zeta,m) \\
&+ N_\zeta \mathcal{E}(\zeta,0 \to \zeta,m) \}
\end{aligned}
\tag{5.5.1}
$$

Within the validity range of the above set of approximations, the average rate coefficients assume the following forms:

$$
\begin{aligned}
\mathcal{I}(\zeta \to \zeta+1) &= \sum_{m'} \mathcal{I}(\zeta,0 \to \zeta+1,m') \\
\mathcal{R}^{(x)}(\zeta \to \zeta-1) &= \sum_{m'} \mathcal{R}^{(x)}(\zeta,0 \to \zeta-1,m')
\end{aligned}
\tag{5.5.2}
$$

Dividing both sides of equation (5.5.1) by N_ζ, one obtains

$$
\begin{aligned}
P_{\zeta,m} &\left\{ \sum_{m'<m} \mathcal{A}(\zeta,m \to \zeta,m') + n_e \left[\sum_{m'} \mathcal{I}(\zeta,m \to \zeta+1,m') \right. \right. \\
&+ \sum_{m'} \mathcal{R}^{(2)}(\zeta,m \to \zeta-1,m') + \sum_{m'>m} \mathcal{E}(\zeta,m \to \zeta,m') + \left. \left. \sum_{m'<m} \mathcal{D}(\zeta,m \to \zeta,m') \right] \right\} \\
&= n_e \left\{ \frac{N_{\zeta-1}}{N_\zeta} \mathcal{I}(\zeta-1,0 \to \zeta,m) + \frac{N_{\zeta+1}}{N_\zeta} \mathcal{R}^{(2)}(\zeta+1,0 \to \zeta,m) + \mathcal{E}(\zeta,0 \to \zeta,m) \right\}
\end{aligned}
\tag{5.5.3}
$$

As we are considering the low density regime, one can use the CE values as an approximation for the ratios of the densities of adjacent charge states, see equation (5.3.2). Substituting these into equation (5.5.3), and using equation (5.5.2) for the average rate coefficients, one finds that the population probabilities of the excited states in this limit are proportional to the electron density (Salzmann 1979),

$$
P_{\zeta,m} = \frac{n_e}{\nu_{\zeta,m}}
\tag{5.5.4}
$$

where $\nu_{\zeta,m}$ is a characteristic density that is a measure of the change of the population probabilities as a function of the electron density. The explicit form of $\nu_{\zeta,m}$ is

$$\nu_{\zeta,m} = \frac{A + Bn_e}{C} \tag{5.5.5}$$

where

$$A = \sum_{m'<m} \mathcal{A}(\zeta, m \to \zeta, m') \tag{5.5.6}$$

$$B = \sum_{m'} \mathcal{I}(\zeta, m \to \zeta + 1, m') + \sum_{m'} \mathcal{R}^{(2)}(\zeta, m \to \zeta - 1, m')$$

$$+ \sum_{m'>m} \mathcal{E}(\zeta, m \to \zeta, m') + \sum_{m'<m} \mathcal{D}(\zeta, m \to \zeta, m') \tag{5.5.7}$$

and

$$C = \mathcal{E}(\zeta, 0 \to \zeta, m) + \frac{\mathcal{I}(\zeta - 1, 0 \to \zeta, m)}{\sum_{m'} \mathcal{I}(\zeta - 1, 0 \to \zeta, m')} \sum_{m'} \mathcal{R}^{(2)}(\zeta, 0 \to \zeta - 1, m')$$

$$+ \frac{\mathcal{R}^{(2)}(\zeta + 1, 0 \to \zeta, m)}{\sum_{m'} \mathcal{R}^{(2)}(\zeta + 1, 0 \to \zeta, m')} \sum_{m'} \mathcal{I}(\zeta, 0 \to \zeta + 1, m') \tag{5.5.8}$$

In principle, the right hand side of (5.5.4) contains only known quantities: the electron density and the rate coefficients for the various processes, which are presumably known beforehand. Equation (5.5.4) is, therefore, an explicit expression for the direct computation of the population probabilities of the excited states. After obtaining all the population probabilities of the excited states, the population probability of the ground state is calculated from

$$P_{\zeta,0} = 1 - \sum_{1 \leq m \leq M_\zeta} P_{\zeta,m} = 1 - n_e \sum_{1 \leq m \leq M_\zeta} \frac{1}{\nu_{\zeta,m}} \tag{5.5.9}$$

The above discussion implies that the validity of equation (5.5.4) extends up to densities of the order of

$$n_e \leq 0.1 \frac{1}{\sum_{m=1}^{M_\zeta} 1/\nu_{\zeta,m}}$$

A detailed discussion of the solutions of equation (5.5.4) will be of help in understanding the behavior of the population probabilities of the excited states in this low density regime. First, we note that A in equation (5.5.6) depends only on the Einstein coefficients from state ζ, m to all the lower ionic states. As shown in chapter 4, this quantity decreases rather sharply as the principal quantum number of the initial states goes up. The second parameter that defines $\nu_{\zeta,m}$ is $n_e B$ (equation 5.5.7). This term, being proportional to n_e, is small at low electron densities. The B parameter incorporates all the rate coefficients that depopulate the state m by ionization, recombination, excitation or deexcitation. For highly excited states the dominant depopulating process among these is electron impact ionization. For low lying excited states all the terms, according to their relative importance, contribute to this parameter. Altogether, the parameters A and B in the numerator of equation (5.5.5) include all the depopulating processes of the given ionic state, where A includes the spontaneous and B includes the electron-induced processes.

The parameter C in the denominator of $\nu_{\zeta,m}$ incorporates the processes that populate the given ionic state. It consists of three components: (i) excitation from the ground state, (ii) ionization of the ground state of charge state $\zeta - 1$, and (iii) recombination into the ground state of charge state $\zeta + 1$. The first of these, the excitation from the ground state, is large only for low lying states, whereas its value decreases for states of higher principal quantum number. The second term of C, which describes the ionization of charge state $\zeta - 1$, is different from zero only for states ζ, m that can be attained by the ionization of the ground state of $\zeta - 1$. There is only a limited number of such states and they are, in general, those in which the ionized electron was in the same shell as the outermost ionic state. For example, the ionization of a $2s$ electron in a boronlike ion, $1s^2 2s^2 2p \rightarrow 1s^2 2s 2p + e$ leaves the ion in an excited state whose principal quantum number is the same as that of the ionized electron. The ionization of inner electrons (such as a $1s$ state in our example) would require more energy and has, therefore, a much lower ionization rate coefficient. For the very important cases of hydrogenlike or heliumlike ions, this term is just zero, because obviously in these species there can be no inner shell ionization. It turns out that this term is zero in most cases and its contribution can be neglected, except for some special cases in complex ions.

The last term in equation (5.5.8), the recombination term, has an appreciable contribution mainly to low lying excited states. The radiative recombination part makes a contribution to the population of the lower states, but a negligible contribution to the formation of upper ones, whereas the contribution of the dielectronic recombination to the population of all the excited states is more homogeneous, though it, too, preferentially populates the lower states. Altogether, summarizing the three terms of C, one finds that the low lying states will be populated with higher probability than the upper ones, and the relative population is smaller for states with higher principal quantum numbers.

At the low electron density end of the region under discussion, $n_e \ll A/B$, one can neglect the $n_e B$ term relative to the A term in the numerator of $\nu_{\zeta,m}$, and equation (5.5.5) further reduces to

$$\nu_{\zeta,m} \approx \frac{A}{C} \quad \text{for} \quad n_e \ll \frac{A}{B} \quad (5.5.10)$$

In this region the two main processes that determine the population probabilities of excited states are the depopulation by spontaneous decay and population by electron impact excitation from the ground state, with contributions from two-body recombinations. This means that the population of an excited state is coupled mainly to the ground state of its own charge state, and the contribution of processes that change the charge is much smaller. Adjacent charge states are coupled through their ground states and only to a small extent through their excited states. For this extreme end of the region it is, therefore, justified to compute the charge state distribution by taking into account only the ionization and recombination rate coefficients that connect the ground states of the adjacent charge states. The populations of the excited states are then computed from the population of the ground state by means of equation (5.5.10). This type of decoupling between the state of charge and the state of excitation is often used in such calculations, and the above discussion gives the limits for the validity of this approximation.

Since both the spontaneous decay and the electron excitation are proportional to the oscillator strength (see, for example, van Regemorter's formula), in this low density limit the $\nu_{\zeta,m}$ are almost constant, and depend only weakly on the principal and orbital quantum numbers, nl, of the excited state. At higher electron densities, at which the approximation in equation (5.5.10) is no longer valid, the B term must also be taken into account and the behavior of $\nu_{\zeta,m}$ as a function of the state's quantum numbers becomes more complicated.

At still higher electron densities, all the approximations used to obtain equation (5.5.4) are no longer valid, and around electron densities of the order of $n_e \sim \nu_{\zeta,m}$ the population probabilities join smoothly to their high density LTE limit (equation 5.2.13), which is a Boltzmann-type distribution.

5.6 The Average Atom Model

The *average atom model* is a model that can provide a tool for the computation of several average properties of high density plasmas, particularly those of high Z. Its basic equations were formulated by B. Rozsnyai (1972). The average atom (AA) model replaces the whole charge and excitation state distributions with one single kind of fictitious atom that represents the average charge and average excitation of all the multitude of species that are present in a real plasma.

The starting point of the AA model is the statistical Thomas–Fermi (TF) model (chapter 2.5) in conjunction with the ion sphere (IS) conditions. For the sake of completeness, we recall these equations here in brief.

Assume that a nucleus of charge Z is located at $r = 0$ and that Z electrons (bound + free) span the volume of the ion sphere. The initial equations of the AA model are identical to those of the TF model, namely, the Poisson equation for the electrostatic potential,

$$\nabla^2 V(\vec{r}) = -4\pi e[Z\delta(\vec{r}) - n_e(\vec{r})] \tag{2.4.1}$$

where $V(r) = Ze/r + V_e(r)$, $V_e(r)$ being the potential generated by the electrons,

$$V_e(r) = -4\pi e\left(\frac{1}{r}\int_0^r n_e(r')r'^2\,dr' + \int_r^{R_i} n_e(r')r'\,dr'\right) \tag{2.4.8}$$

and $R_i = (3/4\pi n_i)^{1/3}$ is the ion sphere radius, equation (1.2.7). We recall that in our notation $n_e(\vec{r})$, with explicit display of the location $\vec{r}$, is the *local electron density*, whereas the average electron density in the plasma is given by writing simply n_e. The second equation is the statistical distribution of the electrons within the ion sphere volume,

$$n_e(r) = \frac{1}{2\pi^2}\left(\frac{2mc^2 T_e}{(\hbar c)^2}\right)^{3/2} F_{1/2}\left(\frac{\mu + eV(r)}{T_e}\right) \tag{2.4.12}$$

where $F_{1/2}$ is the Fermi–Dirac integral, equation (1.3.20). The Fermi energy, μ, is determined from the charge neutrality requirement,

$$Z = \int_0^{R_i} n_e(r,\mu)\, d^3r \tag{2.4.13}$$

The free electrons are characterized by having positive energy,

$$\frac{p^2}{2m} - eV(r) \geq 0 \tag{2.4.14}$$

and their partial density, $n_{e,f}$, is calculated by

$$n_{e,f}(r) = \frac{1}{2\pi^2}\left(\frac{2mc^2 T_e}{(\hbar c)^2}\right)^{3/2} F_{1/2}\left(\frac{\mu + eV(r)}{T_e}; \left|\frac{eV(r)}{T_e}\right|\right) \tag{2.4.15}$$

where $F_{1/2}(x;\beta)$ is the *incomplete Fermi–Dirac integral*,

$$F_j(x;\beta) = \int_\beta^\infty \frac{y^j\, dy}{1 + \exp(y - x)} \tag{2.4.16}$$

The average degree of ionization, $\bar{Z}$, which is the number of free electrons per ion within the ion sphere, is given by

$$\bar{Z}(T_e, n_i) = \int_V n_{e,f}(r,\mu)\, d^3r \tag{2.4.22}$$

Of course, $\bar{Z}$ is not necessarily an integer.

Knowing $\bar{Z}$, the average number of bound electrons per ion, Z_b is calculated from

$$Z_b = Z - \bar{Z} \tag{5.6.1}$$

The fictitious ion of the AA model consists of Z_b bound electrons, which are distributed into the various atomic shells, $1s, 2s, 2p, \ldots$, according to the population probabilities determined by the Fermi–Dirac statistics,

$$P_{\bar{Z},nl} = \frac{1}{1 + \exp[(E_{nl} - \mu)/T]} \tag{5.6.2}$$

which, also, are not integers. E_{nl} are the energies of the nl states that are the eigenvalues of the Schrödinger equation,

$$\left(-\frac{\hbar^2}{2m}\nabla^2 - eV(r)\right)\psi_{nl} = E_{nl}\psi_{nl} \tag{5.6.3}$$

After having obtained the energies and the wavefunctions of all the relevant ionic states, the local bound electron density is calculated from

$$n_{e,b}(r) = \sum_{nl} P_{\bar{Z},nl} |\psi_{nl}|^2 \tag{5.6.4}$$

This, together with the free electron density, equation (2.4.15), yields the total electron density

$$n_e(r) = n_{e,b}(r) + n_{e,f}(r) \tag{5.6.5}$$

which is inserted into equation (2.4.8) to obtain the electrostatic potential. The method of calculation of the AA model is best explained by writing down the steps of its numerical algorithm:

1. Assume an average charge of the fictitious ion, $\bar{Z}$, a Fermi potential, μ, and a list of energies of the ionic states.
2. Calculate Z_b from equation (5.6.1), and the population probabilities of the ionic nl-states from equation (5.6.2).
3. Solve the Schrödinger equation (5.6.3) for all the states whose energy is below the energy of the lowered continuum, and whose fractional population probability is not negligibly small.
4. Calculate the bound electron density by equation (5.6.4.)
5. Calculate the free electron density by equation (2.4.15).
6. Calculate the total electron density by equation (5.6.5), and the potential $V_e(r)$ generated by the electrons, equation (2.4.8).
7. Find the new $\bar{Z}$ from equation (2.4.22), and new Fermi potential by equation (2.4.13).
8. Find the electrostatic potential, $V(r) = Ze/r + V_e(r)$.
9. Go to (2) to find a better approximation for all the above quantities, or stop computations if the results converge.

The final results of this computation are the properties of an ion that has a fractional charge, and its nl-states are populated by fractional numbers of electrons. Although nothing in the details of such a fictitious ion is accurate, nevertheless it can give useful results for several quantities that depend on the average properties of an ion in the plasma. For example, the AA model was very successful in computing opacities (Rozsnyai, 1972, 1991) and the equation of the state of hot and dense matter (Dhrama-wardana and Perrot, 1982).

In the formulation presented above, the AA model is applicable to high density plasmas only, because of the use of the Fermi–Dirac statistics. Its predictions are obviously much better for high-Z material that has a large number of bound electrons and a large number of free electrons. In low density plasmas in which the most abundant species have only a few bound electrons, the AA model breaks down, as was shown by a simple example given by More (1983, 1985).

5.7 Validity Conditions for LTE and CE

The use of CRSS is not always easy and convenient, and, whenever possible, one would prefer to use LTE or CE to model a plasma. These two models have, however, only a limited range of validity: LTE is applicable mainly in high density plasmas, whereas CE is expected to give accurate results in low density plasmas only. It is, therefore, important to know the temperature and density domains for which these two models are valid.

We start with LTE, because its validity domain has been investigated more extensively than that of CE. For LTE to be valid, several conditions must be fulfilled:

- The electron and ion velocity distribution is a Maxwell–Boltzmann type distribution.
- The charge state distribution of the ions follows the solutions of the Saha equation.
- The distribution of the excited states is a Boltzmann statistical distribution.

We have also assumed that

- The plasma is optically thin, and the photons are not necessarily in equilibrium with the plasma material particles. This requirement is not a necessary condition, and we will discuss the influence of radiation reabsorption in optically thick plasmas in chapter 9.

The first condition is generally fulfilled in most laboratory and astrophysical plasmas. In fact, the electron self-collision time, t_c (equation 1.3.14), is so short that in plasmas of interest in the laboratory or in astrophysics a Maxwell–Boltzmann distribution of the electrons' and ions' velocities is always a plausible approximation.

The crucial condition that underlies the validity region of LTE is that the rate of the energy-conserving collisional processes in the plasma will be much higher than the rate of energy-dissipating radiative processes. From the basic equations of CRSS in section 5.4 it was found that the Saha equation for the charge state distribution is reproduced at rather high electron densities, when

$$n_e \gg \frac{\mathcal{R}^{(2)}(\zeta+1 \to \zeta)}{\mathcal{R}^{(3)}(\zeta+1 \to \zeta)} \tag{5.7.1}$$

The ratio of the two rate coefficients is estimated in the following way. First, one uses the detailed principle to rewrite the three-body rate coefficient by means of the electron impact ionization rate coefficient, relation (4.7.1). Second, one uses the Lotz formula for the ionization rate coefficient, equation (4.7.6), and Kramer's formula for the radiative recombination rate coefficient, equation (4.8.7). The condition for the validity of the Saha equation then becomes

$$n_e \gg 1 \times 10^{13} \, \text{cm}^{-3} \text{eV}^{-3} \, T^3 \left(\frac{E_\zeta}{T}\right)^{5/2} \tag{5.7.2}$$

where E_ζ is the ionization energy of charge state ζ. Following a conjecture by Griem (1963, 1964) deviations from LTE will be less than 10% if the electron density is larger than the right hand side of equation (5.7.2) by a factor of 10,

$$n_e \geq 10 \frac{\mathcal{R}^{(2)}(\zeta+1 \rightarrow \zeta)}{\mathcal{R}^{(3)}(\zeta+1 \rightarrow \zeta)} = 1 \times 10^{14} \, \text{cm}^{-3} \, \text{eV}^{-3} \, T^3 \left(\frac{E_\zeta}{T}\right)^{5/2} \tag{5.7.3}$$

This ensures that the charge state distribution will satisfy the Saha equation to within 10%.

This is, however, not sufficient. A second condition, namely, a Boltzmann energy distribution of the excited states, has to be fulfilled as well. By a comparison between the radiative and collisional processes that populate and depopulate the ionic states, Griem (1963, 1964) proposed the following limit for deviations smaller than 10% from a Boltzmann distribution among the ionic states:

$$n_e \geq 9 \times 10^{17} \left(\frac{E_{\zeta,1} - E_{\zeta,0}}{E_H}\right)^3 \left(\frac{T}{E_H}\right)^{1/2} \, \text{cm}^{-3} \tag{5.7.4}$$

where $E_{\zeta,1} - E_{\zeta,0}$ is the energy of the first excited state above ground state. Both these conditions predict that LTE is valid at rather high electron densities. For instance, a $T = 1\,\text{eV}$ hydrogen plasma attains LTE conditions at an electron density of $n_e \geq 1 \times 10^{17}\,\text{cm}^{-3}$, while a helium plasma at a temperature of $4\,\text{eV}$ is in LTE at $n_e \geq 1 \times 10^{19}\,\text{cm}^{-3}$, and an aluminum plasma at $T = 300\,\text{eV}$ needs $n_e \geq 1.5 \times 10^{23}\,\text{cm}^{-3}$ for the Saha equation to be satisfied, according to equation (5.7.3), and $n_e \geq 6 \times 10^{24}\,\text{cm}^{-3}$ for the excited states to reach a Boltzmann distribution, according to equation (5.7.4).

The validity condition for the CE characteristic charge state distribution is derived by a reasoning similar to that used for the LTE validity conditions. The CRSS charge state distributions deviate from those predicted on the basis of the CE equation (5.3.3) by less than 10% if

$$n_e \leq 0.1 \frac{\mathcal{R}^{(2)}(\zeta+1 \rightarrow \zeta)}{\mathcal{R}^{(3)}(\zeta+1 \rightarrow \zeta)} = 1 \times 10^{12} \, \text{cm}^{-3} \, \text{eV}^{-3} \, T^3 \left(\frac{E_\zeta}{T}\right)^{5/2} \tag{5.7.5}$$

At yet there is no explicit formula in the literature that gives the limit for the excited states to satisfy the characteristic population distribution of CE, equation (5.3.1).

5.8 A Remark on the Dependence of the Sensitivity of the CRSS Calculations on the Accuracy of the Rate Coefficients

Everybody who has carried out a calculation of the charge state distributions in the framework of CRSS asks how the sensitivity of the results depend on the accuracy of the rate coefficients in the equations. As the expressions for the rate coefficients are of an approximate character, there has always been the question of how the inaccuracies in the rate coefficients of the various ionic processes affect the resulting charge state distributions. Moreover, in the derivation of the recursive equation (5.4.10), many low probability atomic processes have been neglected, and it is not clear at first sight how these approximations affect the final result. This question will be given a partial answer in this subsection. The discussions in this subsection will follow closely those in Salzmann (1980).

The recursive equation for the partial densities of the charge state, equation (5.4.10), can formally be rewritten as

$$\frac{N_\zeta}{N_{\zeta-1}} = f_\zeta \tag{5.8.1}$$

where f_ζ represents the right hand side of equation (5.4.10), and is the quantity that depends on the rate coefficients. A formal solution of equation (5.8.1) that satisfies equation (1.2.1), $\sum_{\zeta=0}^{Z} N_\zeta = n_i$, has the form

$$N_\zeta = n_i \frac{f_1 f_2 \cdots f_\zeta}{\sum_{j=0}^{Z} f_1 \cdots f_j} \tag{5.8.2}$$

Each rate coefficient contains some inaccuracy, and we assume that the combined inaccuracy of all the rate coefficients on the right hand side of equation (5.4.10) is reflected in the f_ζ as some total inaccuracy Δf_ζ. The total relative error in the partial density of charge state ζ follows from equation (5.8.2),

$$\frac{\Delta N_\zeta}{N_\zeta} = \Delta(\ln N_\zeta) = \sum_{j=0}^{Z} \frac{\partial(\ln N_\zeta)}{\partial f_j} \Delta f_j \tag{5.8.3}$$

This equation is, of course, correct only if $\Delta f_j / f_j$ is significantly smaller than 1, which we shall assume in the following. Partial derivatives are calculated explicitly from equation (5.8.2),

$$\ln N_\zeta = \ln n_i + \sum_{k=0}^{\zeta} \ln f_k - \ln\left(\sum_{j=1}^{Z} f_1 \cdots f_j\right) \tag{5.8.4}$$

wherefrom one gets by direct differentiation,

$$\frac{\partial(\ln N_\zeta)}{\partial f_j} = \begin{cases} (1/f_j)\left(1 - \sum_{k=j}^{Z} f_1 \cdots f_k / \sum_{k=1}^{Z} f_1 \cdots f_k\right) & j \le \zeta \\ -(1/f_j) \sum_{k=j}^{Z} f_1 \cdots f_k / \sum_{k=1}^{Z} f_1 \cdots f_k & j > \zeta \end{cases} \tag{5.8.5}$$

Substituting this into (5.8.3) one obtains

$$\frac{\Delta N_\zeta}{N_\zeta} = \sum_{j=0}^{\zeta} \frac{\Delta f_j}{f_j} - \sum_{j=0}^{Z} \frac{\Delta f_j}{f_j} \frac{\sum_{k=j}^{Z} f_1 \cdots f_k}{\sum_{k=1}^{Z} f_1 \cdots f_k} \tag{5.8.6}$$

This equation can be simplified as follows: Denote

$$\alpha_\zeta = \sum_{j=0}^{\zeta} \frac{\Delta f_j}{f_j} \tag{5.8.7}$$

and correspondingly the average of α_ζ by

$$\bar{\alpha} = \sum_{j=0}^{Z} \frac{\alpha_j N_j}{n_i} \tag{5.8.8}$$

After rearranging the double sum in the nominator of the second term on the right hand side, equation (5.8.6) reduces to

$$\frac{\Delta N_\zeta}{N_\zeta} = \alpha_\zeta - \bar{\alpha} \tag{5.8.9}$$

The necessary condition $\sum_\zeta \Delta N_\zeta = 0$ is automatically satisfied by equation (5.8.9).

To get some insight into the meaning of equation (5.8.9) we consider a special case, namely, that in which all the fractional inaccuracies in the f_ζ are equal, $\Delta f_\zeta / f_\zeta = p = \text{const}$. This is quite a common possibility. For this very special case one gets a simple expression for the α's, $\alpha_\zeta = p\zeta$ and $\bar{\alpha} = p\bar{Z}$. Inserting these values into equation (5.8.9), one obtains for this special case,

$$\frac{\Delta N_\zeta}{N_\zeta} = p(\zeta - \bar{Z}) \tag{5.8.10}$$

This very simple expression gives useful hints as to the influence of the inaccuracies in the rate coefficients on the charge state distributions. It tells us that the relative error is the smallest for the most abundant ion species, $\zeta \cong \bar{Z}$. As the value of $|\zeta - \bar{Z}|$ of these ions is less than 1, the relative error, $\Delta N_\zeta / N_\zeta |_{\zeta \approx \bar{Z}}$, in the calculations of the partial densities of these species is even less than the relative error in the rate coefficients, p. For less abundant ion species, the relative error increases linearly with the difference $|\zeta - \bar{Z}|$, and for very rare species whose charge is far from $\bar{Z}$, it may reach rather high values.

Although these conclusions apply rigorously only to the case of constant $\Delta f_\zeta / f_\zeta$, they are approximately correct also for cases in which the fluctuation in this ratio is not too extreme. Choosing updated rate coefficients for the various processes one can reduce significantly the relative inaccuracies in $\Delta f_\zeta / f_\zeta$ and calculate the densities of the most abundant charge states with relatively good accuracy.

The above considerations provide us with several more conclusions. Obviously, the charge states with the highest abundance, $\zeta \approx \bar{Z}$, have the greatest contribution to the plasma parameters that are dependent on the *average plasma properties*, such as the electron density, average charge $(\bar{Z})$, internal energy, pressure, optical properties, the total radiation rate, and so on. Equation (5.8.10) predicts that the results of computations of these parameters within the framework of the CRSS are only slightly affected by the inaccuracies in the rate coefficients. On the other hand, quantities that are dependent on the presence of rare species in the plasma are greatly influenced by the accuracy used for the rate coefficients. Such parameters include the intensity of the line and the bound–free spectra, opacities in spectral regions where certain rare species have their largest influence, and similar properties. Thus, it turns out that in this regard there are two kinds of plasma parameters: The first kind consists of parameters that depend on the average plasma conditions—these are only slightly influenced by the approximate nature of the rate coefficients. The second kind includes parameters that depend on the partial densities of rare species—these are critically dependent on the accuracy of the rate coefficients. There are almost no parameters in an intermediate group.

These features can be seen in figure 5.4, which shows the variations of three parameters as calculated by two model rate coefficients for the electron impact ionization. The parameters are the charge state distribution, the average charge and the intensity of the $1s2p\,{}^1P \to 1s^2$ resonant line in heliumlike aluminum. These parameters are given in figure 5.4 at an ion density of 10^{18} ions/cm^3. The

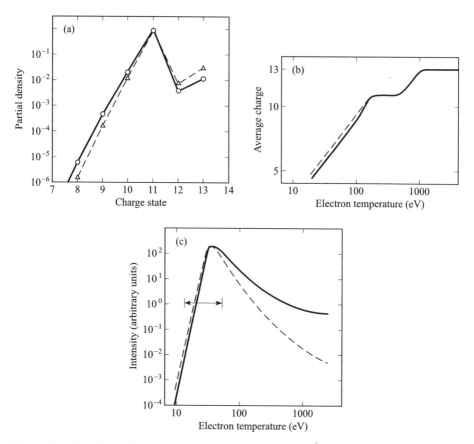

Figure 5.4 Variation of the computational results of three quantities in an aluminium plasma, when the computations are carried out by two different rate coefficients for the electron impact ionization. Solid curves were computed with the Lotz formula, equation (4.7.6) and dashed curves with the formula of Landshoff and Perez, equation (4.7.2). (a) Charge state distribution at $n_i = 10^{18}\,\mathrm{cm}^{-3}$ and $T_e = 300\,\mathrm{eV}$; (b) average charge, $\bar{Z}$, as function of temperature, for $n_i = 10^{18}\,\mathrm{cm}^{-3}$; (c) intensity of the He-like $1s2p \to 2s^2$ line as function of the temperature for $n_i = 10^{18}\,\mathrm{cm}^{-3}$. The horizontal arrow denotes the temperature region where He-like ions are the abundant species.

two electron impact ionization rate coefficients used in the solution of the charge
state distributions are those due to Landshoff and Perez (1976) and Lotz (1968),
see section 4.7, and table 4.5. It can be seen from figure 5.4 that, as explained
above, the two calculations agree on the partial density of the most abundant
species as well as on the average charge. Almost two orders of magnitude difference may occur between the two calculations for the resonant line intensity,
particularly for high temperatures where the heliumlike species is no longer the
most abundant charge state.

   S. Stone and J. Weisheit (1984) compared results from nine non-LTE codes
that had been in use in the Lawrence Livermore National Laboratory during the
early 1980s. Their results practically confirm all the above conclusions.

5.9 Time-dependent Models

For rapidly evolving plasmas the time-independent models cannot provide sufficiently accurate answers. This is the case in some high intensity laser-produced
plasmas, particularly in the recently rapidly developing field of subpicosecond
plasmas. The time-independent models can be assumed acceptably accurate if
the characteristic time for the change of the local plasma temperature and density,

$$t_{plasma} = \min\left(\frac{n_i}{\partial n_i/\partial t}, \frac{T}{\partial T/\partial t}\right) \tag{5.9.1}$$

is significantly larger than the time scale of the atomic processes, see equation
(1.4.7),

$$\tau_a = \frac{1}{n_e\langle\sigma v\rangle} \tag{1.4.7}$$

(σ represents the cross section of the most frequent ionic processes in the plasma),
that is when,

$$t_{plasma} \gg \tau_a \tag{5.9.2}$$

If this inequality holds true, the charge and excitation state distributions have
enough time to adjust rapidly to the changing conditions of the plasma. In this
case one can use the local instantaneous temperature and ion density to compute
the charge and excitation state distributions by means of one of the above time-
independent models. In most laboratory plasmas and certainly in astrophysical
plasmas, this is the case.

   When condition (5.9.2) is not fulfilled, one has to use the full apparatus of the
rate equations to solve the local distributions, which now depend, in addition to
the local temperature and density, on the history of the plasma's evolution as well.
In this case one writes down a system of differential equations that describes the

rates of population and depopulation of all the charge and excited states. In the formalism of section 5.4 this is written as

$$\frac{dN_{\zeta,m}}{dt} = \underset{\substack{all \\ populating \\ processes}}{\sum} n_e^k N_{\zeta',m'} \mathcal{R}(\zeta',m' \to \zeta,m) - \underset{\substack{all \\ depopulating \\ processes}}{\sum} n_e^k N_{\zeta,m} \mathcal{R}(\zeta,m \to \zeta',m')$$

$$\zeta = 0,1,\dots,Z, \qquad \zeta' = \zeta, \zeta \pm 1, \qquad m = 0,1,\dots,M_\zeta,$$

$$m' = 0,1,\dots M_{\zeta'} \tag{5.9.3}$$

In a manner similar to that for the time-independent case, the complexity of the computations can be reduced by omitting from the equations the ionic charge and excited states that presumably have low abundance. Nevertheless, since in a time-dependent case every charge state is abundant at some stage of the plasma's evolution, many more states have to be considered than in the time-independent case.

A computation of the plasma evolution with time-dependent rate equations may exhaust the resources of even the most advanced present-day computers. To facilitate such computation, the atomic physics part is generally decoupled from the hydrodynamics part of the simulation, and is applied as a postprocessor on the results of the hydrodynamics code. This partially reduces the need for a huge memory and long computation times.

Assume that we know the local densities of all the ions at time t, $N_{\zeta,m}(\vec{r}, t)$, $\zeta = 0,1,\dots,Z$, $m = 0,1,\dots,M_\zeta$ (we recall that ζ is the charge and m is the state of excitation). The local density of these ions at time $t + \Delta t$ is computed by means of

$$N_{\zeta,m}(\vec{r}, t + \Delta t) = N_{\zeta,m}(\vec{r}, t) + \Delta N_{\zeta,m}(\vec{r}, t)$$

$$= N_{\zeta,m}(\vec{r}, t) + \Delta t \times \{\text{r.h.s. of } (5.9.3)\} \tag{5.9.4}$$

Two things should be pointed out with regard to this "simple" equation. First, the location $\vec{r}$ is understood to have the Lagrangian meaning, namely, $\vec{r}$ is attached to a constant group of particles, and moves together with them. Second, the temperatures and the ion densities in the rate coefficients that participate in the right hand side of equation (5.9.3) are time- and space-dependent,

$$T = T(\vec{r}, t), \qquad n_i = n_i(\vec{r}, t) \tag{5.9.5}$$

and their values are calculated by the hydrodynamics code.

A few computer programs are available for use in time-dependent problems. The most widely used are RATION and FLY, developed by R. W. Lee. The main approximation used in RATION and FLY is the use of only a limited number of excited states of lithiumlike, heliumlike, and hydrogenlike ions. The excited states of other charge states are neglected and only their ground states are accounted for.

6

The Emission Spectrum

The radiation emitted from hot plasmas is, perhaps, the most important diagnostics tool for these plasmas. Recently, more effort is being invested into the investigation of this radiation, mainly in the UV and x-ray regions, because some industrial applications have emerged that may use such high intensity, localized short wavelength radiation. For plasma diagnostics purposes, the important point is that the emitted spectrum carries information abut the local instantaneous plasma temperature and density conditions. For short life-time plasmas such as particle beam and nanosecond or femtosecond laser-produced plasmas, this is the only reliable diagnostic tool that is capable of giving information about the evolution of the plasma. In this chapter we shall consider the components of the emission spectrum. Its applications for purposes of plasma diagnostics will be discussed in chapter 8, while a brief summary of the industrial and other high-tech applications will be given in chapter 10.

The emission spectrum has two parts: the continuous spectrum and the line spectrum. Continuous spectra are important in the diagnosis of low-Z or fully ionized plasmas. The line spectra are more important in the analysis of intermediate and high-Z plasmas, particularly short lived ones.

The emitted radiation intensity and spectral distribution are determined by the plasma density and temperature. The intensities of both the continuous and the line spectra depend on the temperature and density conditions around the emitting ions, as well as the details of the emission mechanism. The spectral region into which the greater part of the emission takes place is determined mainly by the plasma temperature, and is centered approximately around the same spectral region where the emission from a black body radiator having the same temperature would have its maximum. Although the spectral structure of the emission from a real plasma is very much different from the Planckian shape, the maximum emission from such a plasma occurs nonetheless at energies around $\approx 2.8 T_e$, in a manner similar to black body radiation.

6.1 The Continuous Spectrum

Two kinds of processes contribute to the continuous spectrum. The first is the *bremsstrahlung* radiation. In plasma physics this is also called *free-free* radiation, to recall the initial and the final states of the radiating electron. The emission spectrum of this radiation is obtained by averaging the emission rate from a single electron over the Maxwellian distribution of velocities. A full derivation of the formula is described in Griem's (1964) book for a classical radiator as well as from quantum mechanical principles. We quote here only the results,

$$P_{ff}(\hbar\omega)\, d(\hbar\omega) = \frac{32}{3}\left(\frac{\pi}{3}\right)^{1/2} r_0^2 c \left(\frac{E_H}{T_e}\right)^{1/2} \bar{Z}^2 n_i n_e \, \exp\left(-\frac{\hbar\omega}{T_e}\right) d(\hbar\omega) \qquad (6.1.1)$$

where $r_0 = e^2/mc^2 = \alpha^2 a_0 = 2.818 \times 10^{-13}$ cm, is the electron electromagnetic radius, c is the speed of light, $E_H = 13.605$ eV is the binding energy of a ground state hydrogen, and n_i and n_e are the ion and electron densities, respectively. $P_{ff}(\hbar\omega)$ is the energy (in eV) emitted into a 1 eV spectral region around $\hbar\omega$, per unit time per unit volume. Its units are eV/(eV s cm^3). Equation (6.1.1) for the bremsstrahlung was derived from the radiation laws of classical physics. To account for more accurate quantum mechanical effects, a multiplicate *Gaunt factor*, G_{ff}, should be incorporated into equation (6.1.1). The Gaunt factor for the free-free radiation was calculated by W. J. Karzas and R. Latter (1961). They showed that the Gaunt factor is a slowly varying function that depends on the ionic species present in the plasma and the energy of the emitted photon, $\hbar\omega$. For practically interesting cases, the Gaunt factor is quite close to unity. The generally used assumption is that $G_{ff} = 1$. Equation (6.1.1) shows that the emission spectrum has a very simple spectral structure, an exponentially decreasing function of the photon energy.

The total bremsstrahlung intensity is obtained by integrating equation (6.1.1) over the whole spectrum,

$$P_{ff,tot} = \int_0^\infty P_{ff}(\hbar\omega)\, d(\hbar\omega) = \frac{16}{3}\left(\frac{\pi}{3}\right)^{1/2} r_0^2 c E_H \left(\frac{T_e}{E_H}\right)^{1/2} \bar{Z}^2 n_i n_e \qquad (6.1.2)$$

where $\bar{Z}$ is the average charge state. $P_{ff,tot}$ grows approximately in proportion to the square of the plasma density and the square root of the temperature. As $\bar{Z}(n_i, T)$ grows with the temperature and decreases with the plasma density, a power law dependence of $P_{ff,tot}$ on T and n_i is not rigorous.

In a low-Z plasma, such as a hydrogen or helium plasma, the free-free radiation is the main radiation mechanism. This is also true for higher Z plasmas at very high temperatures when the ions are fully stripped. In high-Z plasmas at lower temperatures, when the ions are partially "dressed" with several bound electrons, the main emission mechanism is the line emission, relative to which the bremsstrahlung provides a low continuous background radiation.

The second emission process of the continuous spectrum is the *recombination radiation*, also called the *free-bound* radiation. It is emitted during the recombination of a free electron with an ion. This radiation process occurs only above sharp so-called *edges*, which correspond to the binding energies of the bound states of the ions,

$$
P_{fb,\varsigma,m} = \begin{cases} \dfrac{64}{3}\left(\dfrac{\pi}{3}\right)^{1/2} cr_0^2 \varsigma N_\varsigma n_e [(E_{\varsigma-1,m} - \Delta\chi_{\varsigma-1})/T_e]^{3/2} \\ \qquad \times (1/n_p^3)\exp(-\hbar\omega/T_e)[1 - P_{\varsigma,m}] & \hbar\omega \ge E_{\varsigma-1,m} - \Delta\chi_{\varsigma-1} \\ \\ 0 & \text{otherwise} \end{cases}
$$

(6.1.3)

where m denotes the quantum numbers of the state, $m = (n_p, l)$, n_p and l are the principal and orbital quantum numbers of the bound state, $E_{\varsigma-1,m}$ is the binding energy of the ion after the recombination, $\Delta\chi_{\varsigma-1}$ is the continuum lowering, and $P_{\varsigma,m}$ is the population probability of the final bound state. The units of P_{fb}, too, are eV/(eV s cm^3). A multiplicative free-bound Gaunt factor, G_{fb}, can be incorporated into equation (6.1.3), but generally is assumed to be equal to 1 (Karzas and Latter, 1961). The last factor of equation (6.1.3), $(1 - P_{\varsigma,m})$, is included to account for the number of unoccupied states in the recombining ion.

The free-bound spectrum has edge structure, as indicated by equation (6.1.3). The various edges correspond to recombination into the atomic states of the abundant charge states. The recombination into the ground state has the highest probability, due to the $1/n_p^3$ dependence of the spectrum on the principal quantum number. Edges of excited states are only seldom observed in the spectrum. In high density plasmas the edges are broadened due to Stark broadening, as well as due to the fluctuations of the continuum lowering, and are, therefore, less prominent in high density plasmas than in lower density ones.

The total free-bound radiation is important at intermediate temperatures, particularly when the plasma consists mostly of hydrogenlike ions. At other temperatures its relative influence is small.

6.2 The Line Spectrum—Isolated Lines

The spectral line emission is the most important radiation process in hot plasmas, particularly in intermediate and high-Z plasmas in which the ions are not fully ionized. There are several very important effects related to line emission, which can be sorted, in general, according to the atomic number Z and the number of the electrons bound to the emitting ion. In this section we treat the simplest case of isolated spectral lines, which is characteristic of low-Z highly ionized species. The spectra of high-Z plasmas can be very complex, and the methods used to treat such spectra will be discussed in later sections of this chapter.

The emission rate of a spectral line is given by the simple formula

$$
P_{\varsigma,m'\to\varsigma,m} = N_{\varsigma,m'}\hbar\omega_{\varsigma,m'\to\varsigma,m}A(\varsigma,m'\to\varsigma,m)
$$

(6.2.1)

where ς is the charge of the emitting ion, m, m' are the ordinal numbers of the lower and upper ionic states in a list in which the ionic states are ordered according to their ascending energy ($m = 0$ corresponds to the ground state). In equation

(6.2.1), $N_{\zeta,m'}$ is the density of ions in the upper initial ionic state and $\hbar\omega_{m'\to m} = E_{\zeta,m'} - E_{\zeta,m}$ is the transition energy, which is also the energy of the emitted photon. Since the ionic transition occurs in the same ion, we will omit, in the following, the ionic charge, ζ, from the notation whenever this cannot lead to ambiguity. The units of $P_{m'\to m}$ are eV/(cm³ s). In accordance with the jargon used for the continuous spectrum, the line emission is termed also the *bound-bound* emission.

Equation (6.2.1) gives the total emission rate in a given spectral line. To get the spectral distribution of the emission within this line, equation (6.2.1) has to be multiplied by a *line profile* factor, $\mathcal{L}(\hbar\omega)$, which has the units of eV^{-1} and accounts for the finite width of the emission line. This factor has a unit area, $\int \mathcal{L}(\hbar\omega) \, d\hbar\omega = 1$. More about line profiles and line widths will be discussed in chapter 7. Here it will suffice to mention that in most cases this profile has either a Doppler or a Lorentzian shape, or a combination of the two. More complicated line shapes appear in complex situations of strong local electric microfields or photon reabsorption.

In a low density plasma in corona equilibrium, equation (6.2.1) can be further reduced. Using the definition of the population probability in terms of the total density of the charge state, (equation 5.2.9), then using equations (5.5.4–8) and (5.5.10) for the population probability of an excited state in a low density plasma, one finds

$$N_{\zeta,m'} = N_\zeta P_{\zeta,m'} = N_\zeta \frac{n_e}{\nu_{\zeta,m'}}$$

$$= N_\zeta n_e \frac{\mathcal{E}(0 \to m')(1 + d)}{\sum_{m'' < m'} A(m' \to m'')} \tag{6.2.2}$$

where N_ζ is the density of the charge state, $P_{\zeta,m'}$ is the population probability of state m', $\mathcal{E}(\ldots)$ is the electron impact excitation rate coefficient from the ground state to state m', and $\sum_{m'' < m'} A(\ldots)$ is the sum of all the Einstein coefficients for decay from state m' to all lower lying states. Finally, d is the ratio between the rate coefficients of the ionization and recombination processes that contribute to the population of state m' and the rate coefficient of the electron excitation, see equation (5.5.8). Substituting this last expression for the population probability into equation (6.2.1), one gets

$$P_{m'\to m} = N_\zeta n_e \mathcal{E}(0 \to m')(1 + d) \frac{A(m' \to m)}{\sum_{m'' < m'} A(m' \to m'')} \hbar\omega_{m'\to m} \tag{6.2.3}$$

This equation has two factors: The first (before the fraction) is the rate of excitations per unit volume from the ground state into the excited one (with corrections due to ionization and recombination processes). This exactly equals the rate of populations per unit volume of the excited state m'. The second is the relative number of transitions from state m' to state m, which equals the fraction of the probability in which this upper state decays to state m, divided by the probability of decays to all possible channels. The factor $\hbar\omega_{m'\to m}$ converts the *transition rate* into *emission intensity*. This formula is frequently used in the computation of the intensity of spectral lines in low density plasmas.

Hydrogenlike Ions

The spectra of hydrogenlike ions has a special status among the line spectra of low-Z and intermediate-Z spectra. This is so for several reasons: First, hydrogenlike ions are an abundant species in hot plasmas, particularly those found in magnetic confinement and astrophysical environments. Second, the spectral lines of hydrogenlike ions are strong and prominent features in the spectrum of hot plasmas, and are easily measurable. Finally, the spectra of H-like ions are known and well understood. In fact, the spectral lines have energies given by the famous Bohr formula,

$$\hbar\omega_{u,\ell} = \frac{Z^2 e^3}{2a_0} \left(\frac{1}{n_\ell^2} - \frac{1}{n_u^2} \right) \tag{6.2.4}$$

where n_ℓ and n_u are the principal quantum numbers of the lower and the upper states of the transition, respectively. The states are degenerate with respect to the orbital quantum number l, that is, all the levels within the same shell have the same energy.

The line spectra of ions, like those of neutral atoms, are organized into *series of lines*, which include all the emission lines from transitions that end at the same ionic state. The most important and famous is the *Lyman series* in hydrogenlike ions whose initial state is an $np\,^1P$ state and whose final state is the $1s\,^1S$ ground state,

$$\hbar\omega(\text{Lyman}) = \frac{Z^2 e^2}{2a_0} \left(1 - \frac{1}{n_u^2} \right) \tag{6.2.5}$$

The various lines in the Lyman series are denoted by the initial state of the transition, so that Ly_α, Ly_β, Ly_γ, etc., correspond to $n_u = 2, 3, 4, \ldots$, respectively. The energy of the Lyman series, as well as that of all other series of H-like ions, increases as the square of the atomic number Z. Thus the energy of the Lyman-α of hydrogen is $e^2/(2a_0) \times (3/4) = 10.20\,\text{eV}$, which is in the ultraviolet ($\lambda = 1215.8\,\text{Å}$), while the same line in hydrogenlike neon, $Z = 10$, is exactly 100 times more energetic, $1020\,\text{eV}$, and appears in the x-ray region ($\lambda = 1.2158\,\text{Å}$).

In the *Balmer series* the final lower state is the $2s$ state, $np\,^2P \to 2s\,^2S$, and the emitted photon energy is

$$\hbar\omega(\text{Balmer}) = \frac{Z^2 e^2}{2a_0} \left(\frac{1}{4} - \frac{1}{n_u^2} \right) \tag{6.2.6}$$

It is interesting to note that the Ly_α of an ion of charge Z exactly coincides with the Balmer-β ($n_u = 4$) line of ions of charge $2Z$.

The spectrum corresponding to a given series is located slightly below the recombination edge corresponding to the lower state. The higher members of the series are closer to the edge, and are more crowded together. As the

Einstein coefficients decrease for the higher initial states, the higher members of the series have a lower intensity than the lower ones.

The Inglis–Teller Limit

On the high limit side, the members of the series are found to be closer and closer together. When the finite width of the lines is also taken into account, then from some highest member and above the spacing between two neighboring members can be smaller than the lines' widths. Beyond this limit the lines cannot be resolved as separate lines any longer, but rather generate a smooth continuum. For the hydrogenlike Lyman series, the highest line that is still resolvable was calculated by Inglis and Teller assuming a Stark broadening. This is the so called *Inglis–Teller limit* (Inglis and Teller, 1939)

$$n_{p,max}^{15/2} \leq \frac{Z^{9/2}}{8 n_e a_0^3 \bar{Z}^{1/2}}$$

or, in its better known numerical form,

$$\log_{10} n_e \leq 23.93 + 4.5 \log_{10} Z - 0.5 \log_{10} \bar{Z} - 7.5 \log_{10} n_{p,max} \qquad (6.2.7)$$

In equation (6.2.7), $n_{p,max}$ is the principal quantum number of the uppermost state from which the transition is still resolvable as a separate line and n_e, as usual, is the electron density. In analyzing the spectrum from some high density astrophysical hydrogen plasma, the Inglis–Teller limit can be used to obtain an estimate of the electron density in the emitting volume.

Heliumlike Ions

The spectra of helium and heliumlike ions was investigated at the very beginning of the era of quantum mechanics (Heisenberg, 1926; Bethe and Salpeter, 1957). Unlike the hydrogen energy levels, the interaction between the two electrons splits levels with different orbital quantum numbers l. The energy level diagrams of helium and singly ionized lithium are shown in figure 6.1, and compared with the energy levels of hydrogen. The ground state is a $1s^2\,{}^1S$ state. An excitation of one of the electrons into the $n = 2$ state splits the resulting state into three LS states: $1s2s\,{}^1S$, $1s2p\,{}^1P$, and $1s2p\,{}^3P$. The large energy gap between the ground state and the first excited states is characteristic of all the inert gases. In the $1s2s$ excited state the inner $1s$ electron feels only a weak screening by the outer excited electron. However, the inner electron generates a strong screening of the nuclear potential at the location of the outer one, which therefore feels a potential that is very close to a hydrogenlike potential. The energies of the excited states of helium are therefore not very much different from those of hydrogen, see figure 6.1.

The radiative transition from the $1s2s$ to the ground state is, of course, strictly forbidden, while transitions from the two other excited $1s2p\,{}^1P$ and $1s2p\,{}^3P$ levels give rise to the two highest intensity lines in the spectrum, namely,

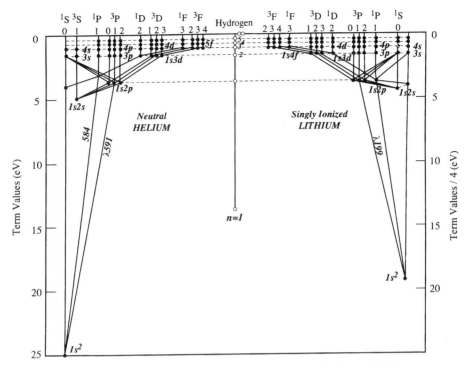

Figure 6.1 Energy level diagrams of helium and singly ionized lithium.

the $1s2p\,^1P \to 1s^2$ the so called *resonance line*, and the $1s2p\,^3P \to 1s^2$, the *intercombination line*. These two lines are frequently used for the purposes of diagnosing hot plasmas. Spectral lines from transitions among highly excited states are also used sometimes, but never as frequently as the two strongest lines.

The Structure of the Spectrum in the Isolated Line Case

To summarize this section, in figure 6.2 we show an emission spectrum of beryllium, as computed by Koniges et al. (1995) for purposes of wall shielding of a Tokamak machine. The figure shows "synthetic" beryllium spectra for an ion density of $10^{18}\,cm^{-3}$ and three temperatures, 1, 3, and 5 eV. In such density, a beryllium plasma is in LTE conditions and the charge state distributions can be calculated by the Saha equation (equation 5.2.7). The energies required to ionize the four ionization states of beryllium are 9.3, 18.2, 153.9, and 217.7 eV, respectively. At the given ion density, the average charges of the ions are $Z = 3.3, 2.2$, and 2.1 for $T_e = 1, 3$, and 5 eV; in other words, for the lowest of these temperatures the lithiumlike ions are the most abundant charge state, whereas for the other two heliumlike ions dominate.

For 1 eV there is no line emission present, because such a temperature is not sufficient to excite electrons to produce appreciable line emission in Li-like or

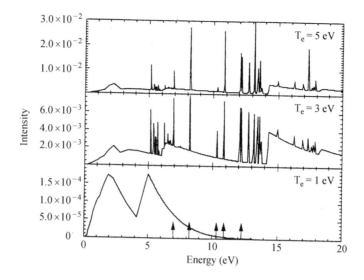

Figure 6.2 Computed beryllium spectrum for an ion density of 10^{18} cm^{-3}, and temperatures of 1, 3, and 5 eV. (Reprinted from Koniges, A. E., Eder, D. C., Wan, A. S., Scott, H. A., Dalhed, H. E., Mayle, R. W., and Post, D. E., 1995, *J. Nucl. Mat.*, **220**, 1116, with permission from Elsevier Science-NL.)

neutral beryllium. For 3 and 5 eV there is significant emission from Li-like species, although the majority of the radiation is emitted by He-like species. Figure 6.2 clearly shows the recombination edges into the ground and excited states; it shows the crowding of the higher members of the spectral line series and the reduction of their intensities near the recombination edges. These features are, in general, common to all the emission spectra of low-Z plasmas. In higher-Z hot plasmas, the density of spectral lines increases rapidly, and the simple isolated line structure, as shown in figure 6.2, is no longer valid.

6.3 Satellites

Satellites are low intensity features near strong spectral lines, in particular those of the hydrogenlike and heliumlike series. Their origin is dielectronic recombination accompanied by radiative stabilization (see section 4.9), in which the radiative transition between two ionic states occurs in the presence of a spectator electron. The extra screening generated by this spectator electron, which generally but not always occupies a higher quantum state, slightly modifies the ionic potential and thereby also the energy difference between the states of the transition. This modification, however, is not very large, so that the emitted photon energy is not too far from the energy of the original *parent transition*. The general effect of this screening is a reduction of the potential and of the emitted photon energy, which is shifted to the lower energy, longer wavelength side of the parent transition. The satellites that have been most widely used for the purposes of plasma diagnostics are the $1snl–2pnl$ satellites near the H-like Ly$_\alpha$ line, and the $1s^2nl–1s2pnl$ satellites near the He-like resonance and intercombination lines. With a spectator electron in the $n = 2$ state, the satellite line is well separated from the parent line, the satellites with the spectator electron at the $n = 3$ and $n = 4$ shells are much closer to the parent line, while satellites corresponding to spectator electrons in higher n states cannot, in general, be resolved from the main parent line.

The role of satellites in the spectra of hot plasmas has been fairly widely discussed in the literature. A partial list of papers includes those of A. H. Gabriel (Gabriel 1972; Gabriel and Jordan, 1972; Gabriel and Page, 1972), which give the basic formulas for the intensity ratios between the satellites and the parent line as function of the plasma temperature. Gabriel and Jordan (1972) assigned LS coupled terms to the satellite lines near the He-like resonance line by comparing observed wavelengths with energies computed using Hartree–Fock techniques. This classification is reproduced in table 6.1. F. Bely-Dubau and S.

Table 6.1 Annotation of the Individual Lines of the Satellite Spectrum of Helium-like Ion Resonance Lines

Array	Multiplet	Line	Key letter[a]
$1s^22p–1s2p^2$	$^2P^0–^2P$	3/2–3/2	a
		1/2–3/2	b
		3/2–1/2	c
		1/2–1/2	d
	$^2P^0–^4P$	3/2–5/2	e
		3/2–3/2	f
		1/2–3/2	g
		3/2–1/2	h
		1/2–1/2	i
	$^2P^0–^2D$	3/2–5/2	j
		1/2–3/2	k
		3/2–3/2	l
	$^2P^0–^0S$	3/2–1/2	m
		1/2–1/2	n
$1s^22p–1s2s^2$	$^2P^0–^2S$	3/2–1/2	o
		1/2–1/2	p
$1s^22s–1s2s2p$	$^2S–(^1P)^2P^0$	1/2–3/2	q
		1/2–1/2	r
	$^2S–(^3P)^2P^0$	1/2–3/2	s
		1/2–1/2	t
	$^2S–^4P^0$	1/2–3/2	u
		1/2–1/2	v
$1s^2–1s2p$	$^1S–^1P^0$	0–1	w
	$^1S–^3P^0$	0–2	x
		0–1	y
$1s^2–1s2s$	$^1S–^3S$	0–1	z

[a] See text.

Volonté (1980) have published a review paper on this subject, with emphasis on the analysis and interpretation of satellite lines in terms of the plasma parameters; see also (Bely-Dubau et al., 1979). In their paper one can find a full derivation of the satellite intensity theory and the related atomic parameters, such as emission wavelengths and transition probabilities.

Satellite to Parent Line Ratio

Assume a low density plasma that is in corona equilibrium (CE), that is, the ionization is in a steady state and most of the ions are in their ground state. We shall focus on the satellite lines near the resonance line in heliumlike ions. The emission of photons into these satellites occurs in a lithiumlike ion, whose electronic configuration is that of an excited heliumlike ion with an additional spectator electron. The combined energy level diagram of heliumlike and lithium-like ions is shown in figure 6.3. This figure illustrates that the doubly excited states of lithiumlike ions are above the ionization limit of the heliumlike ions, and can, therefore, autoionize spontaneously. The LS coupling approximation is very successful scheme for these ions, at least as far along the sequence as oxygenlike ions. In this scheme the two most intense satellites are

$$1s2p(^1P)2s(^2P^0) \rightarrow 1s^22s(^2S), \qquad 1s2p^2(^2D) \rightarrow 1s^22p(^2P^0)$$

which have approximately equal intensities.

In the following, we will illustrate our arguments by the first of these transitions and calculate the ratio of the intensities of the $1s2p\,(^1P)2s\,(^2P^0) \rightarrow 1s^22s(^2S)$ satellite of the lithium-like ions (lines q and r in table 6.1), relative to the parent $1s2p\,(^1P) \rightarrow 1s^2\,(^1S)$ line in the heliumlike ion (line w in table 6.1). As explained above, this satellite arises when a dielectronic recombination occurs between a free electron and a ground state heliumlike ion, $1s^2\,(^1S)$ (which, for simplicity of

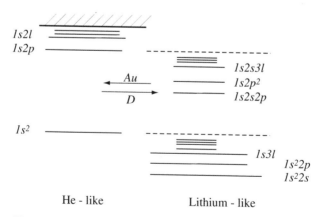

Figure 6.3 Schematic energy level diagram for the explanation of the origin of the satellites of the He-like lines.

notation, we shall call state α), resulting in a doubly excited state in the lithiumlike ion, $1s2p\,(^1P)2s\,(^2P^0)$ (state β), in which the excited $2p$ electron decays spontaneously to the ground state of the lithiumlike ion, $1s^2 2s\,(^1S)$ (state γ). We shall compare the intensity of this transition to the parent transition in the helium-like species from an excited state, $1s2p\,(^1P)$ (state π) to the ground state, $1s^2$ (state α). The intensity of the satellite line, that is, the number of photons emitted per unit volume per unit time, equals the rate of the dielectronic recombination, whose rate was found in section 4.9, see equation (4.9.6),

$$P_s = N_\alpha n_e \mathcal{R}^{(d)}(\alpha \to \beta \to \gamma; T_e)$$

$$= N_\alpha n_e \frac{1}{2}\frac{g_\beta}{g_\alpha}\left(\frac{2\pi(\hbar c)^2}{mc^2 T_e}\right)^{3/2}\frac{A_{\pi\to\alpha}Au(\beta\to\alpha)}{Au(\beta\to\alpha)+A_{\pi\to\alpha}}\exp\left(-\frac{E_\beta}{T}\right) \qquad (6.3.1)$$

Here N_α is the density of ground state heliumlike ions, g_β, g_α are the statistical weights of the corresponding states, and Au and A are the rate coefficient for autoionization and the Einstein coefficient of the transition, respectively. The emission rate of the parent transition in heliumlike ions in low density plasmas was calculated in equation (6.2.3),

$$P_{\pi\to\alpha} = N_\alpha n_e[\mathcal{E}(\alpha\to\pi; T_e)+\mathcal{R}^{(2)}(H-like\ g.s. \to\pi; T_e)]$$

$$= N_\alpha n_e\mathcal{E}(\alpha\to\pi; T_e)(1+d) \qquad (6.3.2)$$

where d is the ratio between the rate of the recombination processes into the H-like ions that populate the He-like excited state π and the rate coefficients of electron impact excitation. The ratio of the satellite to the parent lines is, therefore, independent of the density, and depends solely on the electron temperature. Substituting van Regemorter's formula for the electron impact excitation, one obtains

$$\frac{P_s(\text{satellite})}{P(\text{line }w)} = 3.125\times10^6\sqrt{2}\pi^{3/2}\frac{g_\beta}{g_\alpha}\left(\frac{(\hbar c)^2}{mc^2 E_H}\right)^{3/2}\frac{E_\pi}{T}$$

$$\times e^{(E_\pi-E_\beta)/T}\frac{A_{\pi\to\alpha}Au(\beta\to\alpha)}{Au(\beta\to\alpha)+A_{\pi\to\alpha}}\times\frac{1}{f_{\pi\to\alpha}G(E_\pi/T)(1+d)}$$

$$= 4.48\times10^{-10}\left(\frac{1}{eV^2}\right)\frac{g_\beta}{g_\alpha}e^{(E_\pi-E_\beta)/T}\frac{E_\pi^3}{T}\frac{Au(\beta\to\alpha)}{Au(\beta\to\alpha)+A_{\pi\to\alpha}}$$

$$\times\frac{1}{G(E_\pi/T)(1+d)} \qquad (6.3.3)$$

where E_π is the energy of state π, E_β is the energy of the doubly excited state with respect to the ground state of the He-like ion, $f_{\pi\to\alpha}$ is the oscillator strength of the parent transition, and $G(y)$ is defined in equation (4.6.7). In the second of the above equations, we have used equation (4.5.4) to express the ratio of the Einstein A-coefficient to the oscillator strength in terms of E_π and atomic constants. The intensity ratio in equation (6.3.3) is now independent of the density.

The temperature and Z dependences of the satellite/parent intensity ratio can be deduced from the following considerations: Usually $E_\pi - E_\beta$ is less than T, so that the exponential in equation (6.3.3) does not depend too strongly on T. The main temperature variations come, therefore, from the $1/T$ term. Regarding the dependence on the atomic number, Z, we recall that in corona equilibrium the ratio between the ionization energy of the most abundant charge state and the temperature is more or less constant (see section 5.3). If we are considering plasmas in which heliumlike ions are the most abundant species, regardless of their atomic number, the term E_π/T is a constant with respect to Z. It turns out that only the Einstein coefficient, $A_{\pi \to \alpha}$, determines the Z dependence of the satellite-to-parent line ratio (see the first equation in (6.3.3)), which therefore scales approximately as Z^4.

Examples of satellite spectra are shown in figures 6.4 and 6.5. In figure 6.4 a typical experimental spectrum of the satellites near the Lyman-α of hydrogenlike aluminum is shown (Audebert et al., 1984). This spectrum was obtained from a laser-irradiated plasma at an electron density of about 1×10^{22} cm^{-3} and a temperature around 600 eV. In figure 6.5 are shown high resolution spectra of heliumlike iron from the coronal region of the sun (Feldman et al., 1988). These spectra were recorded by a Soviet spacecraft. The notation of the satellites is the same as in table 6.1. The temperature derived from this spectrum for the sun is 25×10^6 K ≈ 2000 eV.

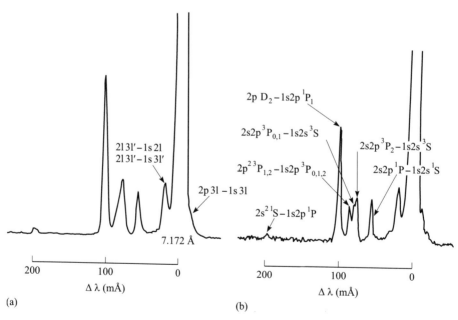

Figure 6.4 Comparison of two experimental spectra of H-like aluminium, obtained under two different irradiation conditions: (a) 100 J, 1.06 μm; (b) 55 J, 1.06 μm, spot 100 μm. (From Audebert, 1984.)

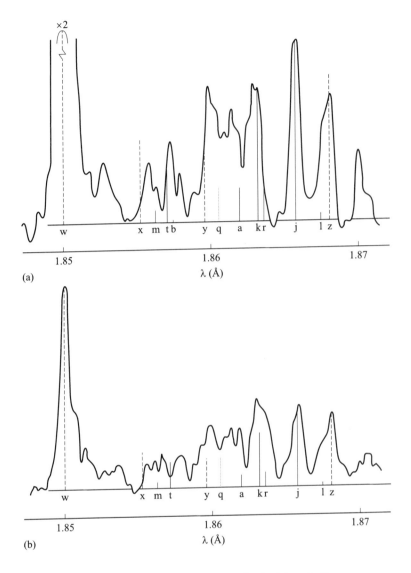

Figure 6.5 High resolution spectra of highly ionized iron in the sun, recorded by a Soviet spacecraft. (a) $T = 24 \times 10^6$ K. (b) $T = 25 \times 10^6$ K. (From Feldman et al., 1988.)

6.4 Unresolved Transition Arrays (UTAs)

The simple line structure of the bound-bound transitions prevails only for rather low-Z plasmas consisting of ions with a small number of bound electrons. As the number of the bound electrons increases, the bound-bound spectrum becomes very complex and includes more and more closely spaced lines. Eventually, when the number of the bound electrons becomes large, the density of lines in some parts of the spectrum becomes so high that their spacing is smaller than their width. The lines then become unresolvably close to each other and the original line spectrum gives place to, what is called, an *unresolved transition array* (*UTA*).

An example illustrates the complexity of the problem. An electronic configuration of w electrons in a configuration nl (generally denoted as nl^w), has $2(2l+1)$ different quantum states. The w electrons of the configuration can be put into these states in $\binom{2(2l+1)}{w}$ ways. For example, a $3d^4$ configuration can have $(2 \cdot 5)!/[(10-4)!4!] = 210$ different states. Each of these states has a slightly different energy, but altogether these states are quite close to each other. Similarly, a configuration of $3d^3 4f^1$ has

$$\binom{2(2 \cdot 2 + 1)}{3}\binom{2(2 \cdot 3 + 1)}{1} = 120 \cdot 14 = 1680$$

states. The maximum number of transitions between the configurations $3d^3 4f^1 \rightarrow 3d^4$ is $210 \times 1680 \approx 3.5 \times 10^5$, but the actual number of transitions, taking into account all the selection rules, is "only" 5523, all of which are closely spaced in the same spectral region.

To make this situation even more complicated, one also has to take into account these same transitions in the neighbouring ion, that is, the satellites of each transition, with one or more spectator bound electrons is outer orbitals, originating from dielectronic recombination into these ions. These transitions, too, span about the same spectral region as the parent transitions. For many-electron ions the computations of such UTAs becomes an unmanageable task.

An interesting attempt to make a full computation of the UTAs for a bromine plasma was carried out by A. Goldberg, B. Rozsnyai and P. Thompson (1986). These authors used a *collective vector method*, developed by A. Goldberg and S. Bloom, (Bloom and Goldberg, 1986), by which they could compute the shape of UTAs. Their results are reproduced in figure 6.6. Such detailed computations are possible, however, only for a limited number of cases. These figures show the complexity of the problem; at the same time they also give indications in which direction one should look if wishing to treat such spectra.

The first serious approach to reducing the problem of the computation of UTAs was carried out by C. Bauche-Arnoult, J. Bauche and M. Klapisch (1979). They described the closely spaced energy levels of many-electron ions and the transitions among their levels by means of band structures, that is, as statistical entities that are characterized by their first two moments, namely, their mean energy and width.

The average energy and the standard deviation of the energy distribution of a given electronic configuration are the first and second moments of the Hamiltonian,

$$E_{av} = [\langle \phi_i | \mathcal{H} | \phi_i \rangle]_{av} \tag{6.4.1}$$

and

$$\sigma^2 = [\langle \phi_i | \mathcal{H} | \phi_i \rangle^2]_{av} - [\langle \phi_i | \mathcal{H} | \phi_i \rangle]_{av}^2 \tag{6.4.2}$$

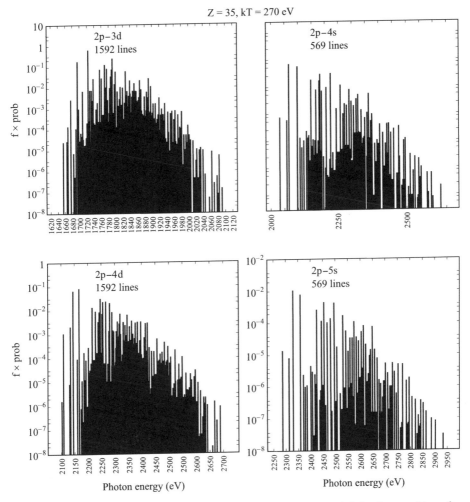

Figure 6.6 Computed emission spectrum of the $2p$–$3d$, $2p$–$4s$, $2p$–$4d$, $2p$–$5s$ transitions for a bromine plasma at $T = 270\,\text{eV}$ and $n_i = 2 \times 10^{20}\,\text{cm}^{-3}$. (From Goldberg et al., 1986.)

The meaning of this last equation is that the standard deviation, σ^2, is the average value of the squares of the energies of the *eigenstates* ϕ_i minus the square of the average value of their energies. The interactions included in the nonrelativistic Hamiltonian, $\mathcal{H}$, are the electrostatic and the spin–orbit operators,

$$\mathcal{H} = \sum_{i>j=1}^{N} \frac{e^2}{r_{ij}} + \sum_{i=1}^{N} \zeta(r_i)\vec{s}_i \cdot \vec{l}_i + \alpha L(L+1) \tag{6.4.3}$$

where the last term is a second order energy effect, first proposed by Trees (1951a,b), which accounts for correlation effects between the bound electrons (Cowan, 1981). The computations of the average quantities in equations (6.4.1–

2) are cumbersome and lengthy, and the interested reader is encouraged to read the original paper (Bauche-Arnoult et al., 1979).

Using a j–j coupling scheme, and the Hermitian properties of the Hamiltonian, C. Bauche-Arnoult, J. Bauche, and M. Klapisch expressed the width of the energy distribution in a configuration nl^N in the form

$$\sigma^2 = D_1 + D_2 + D_3 + D_4 \qquad (6.4.4)$$

The values of the Ds are shown in table 6.2, reproduced from their paper. In that paper there are also similar formulas for configurations of the sort of $nl^N n'l'^{N'}$. In table 6.2 the $F^k(ll')$ and the $G^k(ll')$ are the *Slater direct* and *exchange integrals* (Cowan, 1981; Condon and Shortley, 1987). The other quantities in table 6.2 incorporate the $3j$ and the $6j$ symbols. The summations are limited by the selection rules of these symbols.

The shape of the emission spectrum receives a similar statistical treatment. One considers transitions from a group of closely spaced levels, A, to another group of levels B, and denotes the individual levels by $a(\in A)$ and $b(\in B)$. The standard deviation of the weighted line wavenumber distribution is the square root of the variance,

$$\sigma^2 = \mu_2 - \mu_1^2 \qquad (6.4.5)$$

where now

$$\mu_n = \frac{1}{W} \sum_{a,b} w_{ab} [\langle a|\mathcal{H}|a\rangle - \langle b|\mathcal{H}|b\rangle]^n \qquad (6.4.6)$$

In (6.4.6) w_{ab} is the strength of an individual transition from $a \in A$ to $b \in B$, and $W = \sum_{ab} w_{ab}$ is the sum of the weights. The weight of the transition is proportional to the matrix element of the dipole operator,

$$w_{ab} = |\langle a|Z|b\rangle|^2 \qquad (6.4.7)$$

where $Z = \sum_i z_i$ is, within a constant factor, the total dipole operator, and the z_i are the z components of the position vectors of the bound electrons. The weighting of the matrix elements of the Hamiltonian by the appropriate transition probabilities gives the correct width of the UTA.

These authors' formula for the width of UTAs of transition of the kind $nl^N n'l' \to nl^{N+1}$ is expressed in the form

$$\sigma^2 = \sum_{i=1}^{7} E_i \qquad (6.4.8)$$

where the quantities E_i are shown in table 6.3. The E_i are given in terms of the $3j$, $6j$, and $9j$ coefficients, as well as the two-particle Slater direct and exchange radial integrals, $F^k(ll')$ and $G^k(ll')$. The fact that they could reduce these very complicated expressions to computable forms is the central importance of their work. In their paper there are also similar formulas for transitions of the sort of $nl^N n'l'$–$nl^N n''l''$.

Table 6.2 Different parts of the formula giving the variance σ^2 of the distribution of the level energies belonging to a given configuration. $a = N(N-1)(4l-N+1)(4l-N+2)$; $b = N(4l-N+2)N'(4l'-N'+2)$.

(a) Configuration nl^N: $\sigma^2 = \sum_{i=1}^{4} D_i$, where

$$D_1 = \sum_{k\neq0}\sum_{k'\neq0}\left(\frac{2\delta(k,k')}{(2k+1)} - \frac{1}{(2l+1)(4l+1)} - (-1)^{k+k'}\begin{Bmatrix} l & l & k \\ l & l & k' \end{Bmatrix}\right)$$

$$\times \frac{(2l+1)^3}{(4l-1)8l(4l+1)}\begin{pmatrix} l & k & l \\ 0 & 0 & 0 \end{pmatrix}^2\begin{pmatrix} l & k' & l \\ 0 & 0 & 0 \end{pmatrix}^2 aF^k(ll)F^{k'}(ll)$$

$$D_2 = \sum_{k\neq0}\left(-\frac{1}{(2l+1)(4l+1)} + (-1)^k\begin{Bmatrix} l & l & k \\ l & l & 1 \end{Bmatrix}\right)\frac{(l+1)(2l+1)^2}{2(4l-1)(4l+1)}\begin{pmatrix} l & k & l \\ 0 & 0 & 0 \end{pmatrix}^2 a\alpha F^k(ll)$$

$$D_3 = \left(-\frac{1}{(2l+1)(4l+1)} + \frac{2}{3} - \begin{Bmatrix} l & l & 1 \\ l & l & 1 \end{Bmatrix}\right)\frac{l(l+1)^2(2l+1)}{2(4l-1)(4l+1)} a\alpha^2$$

$$D_4 = \frac{l(l+1)}{4(4l+1)}N(4l-N+2)\zeta_{nl}^2$$

(b) Configuration $nl^N n'l'^{N'}$: $\sigma^2 = \sum_{i=1}^{7} D_i + \sum_{i=1}^{4} D_i'$ (see text)

$$D_5 = \sum_{k\neq0}\sum_{k'\neq0}\frac{\delta(k,k')(2l+1)(2l'+1)}{(2k+1)(4l+1)(4l'+1)}\begin{pmatrix} l & k & l \\ 0 & 0 & 0 \end{pmatrix}^2\begin{pmatrix} l' & k & l' \\ 0 & 0 & 0 \end{pmatrix}^2 bF^k(ll')F^{k'}(ll')$$

$$D_6 = \sum_{k}\sum_{k'}\left(\frac{\delta(k,k')}{(2k+1)} - \frac{1}{4(2l+1)(2l'+1)}\right)\frac{(2l+1)(2l'+1)}{(4l+1)(4l'+1)}$$

$$\times \begin{pmatrix} l & k & l' \\ 0 & 0 & 0 \end{pmatrix}\begin{pmatrix} l & k' & l' \\ 0 & 0 & 0 \end{pmatrix} \times bG^k(ll')G^{k'}(ll')$$

$$D_7 = \sum_{k\neq0}\sum_{k'}(-1)^{k+k'+1}\begin{Bmatrix} l' & l' & k' \\ l & l & k \end{Bmatrix}\frac{(2l+1)(2l'+1)}{(4l+1)(4l'+1)}\begin{pmatrix} l & k & l \\ 0 & 0 & 0 \end{pmatrix}\begin{pmatrix} l' & k & l' \\ 0 & 0 & 0 \end{pmatrix}$$

$$\times \begin{pmatrix} l & k' & l' \\ 0 & 0 & 0 \end{pmatrix}^2 bF^k(ll')G^{k'}(ll')$$

(From Bauche-Arnoult et al., 1979.)

Some care should be taken when computing the average wavenumber of a UTA, because the strong individual emission lines of the transition are not necessarily located at the center of the distribution, and an off-centered strong line may distort the shape of the UTA, so that its center is slightly shifted from its average position. For transitions of the kind of $l^N l''-l^N l'$ there is no shift from the expected location,

$$\hbar\omega_{av}(l^N l'' - l^N l') = E_{av}(l^N l'') - E_{av}(l^N l') \tag{6.4.9}$$

Table 6.3 Different parts of the formula giving the variance σ^2 of the energy distribution of the transitions between $nl^N n'l'$ and $nl^N nl''$ configurations. $\Delta F^k = F^k(ll)(\text{in } l^N l') - F^k(ll)(\text{in } l^N l'')$, $\Delta\zeta_{nl} = \zeta_{nl}(\text{in } l^N l') - \zeta_{nl}(\text{in } l^N l'')$; $y = N(N-1)$ $\times(4l - N + 1)(4l - N + 2)$, $t = N(4l - N + 2)$

$$E_1 = \sum_{k \neq 0}\sum_{k' \neq 0}\left(\frac{2\delta(k,k')}{(2k+1)} - \frac{1}{(2l+1)(4l+1)} - (-1)^{k+k'}\left\{\begin{matrix} l & l & k \\ l & l & k' \end{matrix}\right\}\right)$$

$$\times \frac{(2l+1)^3}{(4l-1)8l(4l+1)}\begin{pmatrix} l & k & l \\ 0 & 0 & 0 \end{pmatrix}^2\begin{pmatrix} l & k' & l \\ 0 & 0 & 0 \end{pmatrix}^2 y\,\Delta F^k\,\Delta F^{k'}$$

$$E_2 = \sum_{k \neq 0}\sum_{k' \neq 0}\frac{\delta(k,k')}{(2k+1)}\frac{(2l+1)(2l'+1)}{(4l+1)}\begin{pmatrix} l & k & l \\ 0 & 0 & 0 \end{pmatrix}^2\begin{pmatrix} l' & k & l' \\ 0 & 0 & 0 \end{pmatrix}^2 tF^k(ll')F^{k'}(ll')$$

E_2': same as E_2 with l'' replacing l'

$$E_3 = \sum_{k}\sum_{k'}\left(\frac{\delta(k,k')}{(2k+1)} - \frac{1}{4(2l+1)(2l'+1)}\right)\frac{(2l+1)(2l'+1)}{(4l+1)}\begin{pmatrix} l & k & l' \\ 0 & 0 & 0 \end{pmatrix}^2$$

$$\times\begin{pmatrix} l & k' & l' \\ 0 & 0 & 0 \end{pmatrix}^2\times tG^k(ll')G^{k'}(ll')$$

E_3': same as E_3 with l'' replacing l'

$$E_4 = \sum_{k \neq 0}\sum_{k'}(-1)^{k+k'+1}\left\{\begin{matrix} l' & l' & k \\ l & l & k' \end{matrix}\right\}\frac{(2l+1)(2l'+1)}{(4l+1)}\begin{pmatrix} l & k & l \\ 0 & 0 & 0 \end{pmatrix}\begin{pmatrix} l' & k & l' \\ 0 & 0 & 0 \end{pmatrix}$$

$$\times\begin{pmatrix} l & k' & l' \\ 0 & 0 & 0 \end{pmatrix}^2 tF^k(ll')G^{k'}(ll')$$

E_4': same as E_4 with l'' replacing l'

$$E_5 = \sum_{k \neq 0}\sum_{k' \neq 0}\frac{2(-1)^k\delta(k,k')}{(2k+1)}\left\{\begin{matrix} l'' & l'' & k \\ l' & l' & 1 \end{matrix}\right\}\frac{(2l+1)(2l'+1)(2l''+1)}{(4l+1)}$$

$$\times\begin{pmatrix} l & k & l \\ 0 & 0 & 0 \end{pmatrix}^2\begin{pmatrix} l' & k & l' \\ 0 & 0 & 0 \end{pmatrix}\begin{pmatrix} l'' & k & l'' \\ 0 & 0 & 0 \end{pmatrix}tF^k(ll')F^{k'}(ll'')$$

$$E_6 = \sum_{k}\sum_{k'}\left(-2\left\{\begin{matrix} k & k' & 1 \\ l'' & l' & l \end{matrix}\right\}^2 + \frac{1}{2(2l+1)(2l'+1)(2l''+1)}\right)\frac{(2l+1)(2l'+1)(2l''+1)}{(4l+1)}$$

$$\times\begin{pmatrix} l & k & l' \\ 0 & 0 & 0 \end{pmatrix}^2\begin{pmatrix} l & k' & l'' \\ 0 & 0 & 0 \end{pmatrix}^2 tG^k(ll')G^{k'}(ll'')$$

$$E_7 = \sum_{k \neq 0}\sum_{k'}(-1)^{k'+1}\left\{\begin{matrix} l'' & l'' & k \\ l' & l' & 1 \end{matrix}\right\}\left\{\begin{matrix} l'' & l'' & k \\ l & l & k' \end{matrix}\right\}\frac{(2l+1)(2l'+1)(2l''+1)}{(4l+1)}$$

$$\times\begin{pmatrix} l & k & l \\ 0 & 0 & 0 \end{pmatrix}\begin{pmatrix} l' & k & l' \\ 0 & 0 & 0 \end{pmatrix}\begin{pmatrix} l & k' & l'' \\ 0 & 0 & 0 \end{pmatrix}^2 tF^k(ll')G^{k'}(ll'')$$

E_7': same as E_7 with l' and l'' interchanged

For the spin-orbit contribution:

$$E_8 = \frac{l(l+1)}{4(4l+1)}t(\Delta\zeta_{nl})^2 + \frac{l'(l'+1)}{4}\zeta_{n'l'}^2 + \frac{l''(l''+1)}{4}\zeta_{n''l''}^2 - \frac{l'(l'+1) + l''(l''+1) - 2}{4}\zeta_{n'l'}\zeta_{n''l''}$$

(From Bauche-Arnoult, 1979.)

Here $E_{av}(l^N l')$ is the *configuration averaged energy*; see for example Cowan (1981, chapter 6-2). In transitions of the sort $l^{N+1}-l^N l'$ there is a shift in the center of the energy distribution, whose value is

$$\delta\hbar\omega(l^{N+1} - l^N l') = N\,\frac{(2l+1)(2l'+1)}{4l+1}\left(\sum_{k\neq 0} f_k F^k(ll') + \sum_{k\geq 0} g_k G^k(ll')\right) \quad (6.4.10)$$

where

$$f_k = \begin{pmatrix} l & k & l \\ 0 & 0 & 0 \end{pmatrix}\begin{pmatrix} l' & k & l' \\ 0 & 0 & 0 \end{pmatrix}\begin{Bmatrix} l & k & l \\ l' & 1 & l' \end{Bmatrix}$$

$$g_k = \begin{pmatrix} l & k & l' \\ 0 & 0 & 0 \end{pmatrix}^2 \left(\frac{2}{3}\delta_{k,1} - \frac{1}{2(2l+1)(2l'+1)}\right) \quad (6.4.11)$$

so that for transitions of this kind the center of the UTA is

$$\hbar\omega_{av}(l^{N+1} - l^N l') = E_{av}(l^N l') - E_{av}(l^{N+1}) + \delta\hbar\omega(l^{N+1} - l^N l') \quad (6.4.12)$$

When the average energy or wavenumber of the distribution and its width are known, one may use a Lorentzian distribution function to obtain an approximate shape of the spectral band. Repeating this procedure for all the UTAs expected in a given spectral region yields the emission spectrum and related quantities of complex spectra of many electron multiply ionized ions.

The capability of the theory of the UTAs to simulate complex experimental spectra that had been intractable by other means is quite impressive. In figure 6.7 is shown a comparison of an experimental spectrum of molybdenum between 20 and 55 Å, and theoretical computations, using the theory of UTAs. The experimental spectrum was obtained by means of a high power vacuum spark. The ion density is about $10^{18}\,\mathrm{cm}^{-3}$, and $T = 200\,\mathrm{eV}$. In this density and temperature domain the main ionization states are those between MoXV and MoXXIII (Ni to Ca-like). The relative intensities of the different transition arrays were fitted to the experiment. The comparison shows that the location and the width of the computational transition arrays reproduce the experimental spectrum to very good accuracy.

Recent developments of the UTAs and the statistical treatments of atomic spectra can be found in Bauche and Bauche-Arnoult (1990).

6.5 Super Transition Arrays (STAs)

The UTA model is a powerful method for characterizing the emission spectrum of hot plasmas. It is an efficient solution to the problem of unresolved spectral structures as long as the number of the populated atomic configurations is relatively small. In high density plasmas, however, the number of populated configurations increases rapidly due to collisional excitations among the various states, and the number of UTAs grows to an intractable quantity. For example, for an

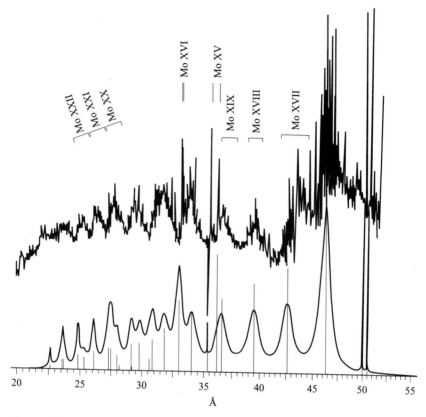

Figure 6.7 Comparison of experimental and theoretical spectrum of MoXV–MoXXII. The transition arrays are $3d^{N+1}$–$3d^N 4p$, $-3d^N 4f$, $-3d^N 5p$, $-3d^N 5f$. (From Bauche-Arnoult et al., 1979.)

ion in LTE that has $n_{max} = 10$ bound ionic shells, the number of individual transitions is about 10^{15} and the number of UTAs may go up to 10^9.

An interesting attempt to reduce the number of the UTAs for these complicated cases was made by A. Bar-Shalom, J. Oreg and their collaborators (Bar-Shalom et al., 1989), by introducing the model of *super transition arrays* (STAs). The central point in their model is the definition of *superconfigurations*, which are closely spaced electronic configurations grouped together. Such grouping reduces significantly the number of objects that one has to consider, with only a minor sacrifice in accuracy. The method of grouping the electronic configurations into superconfigurations is, roughly speaking, a matter of personal experience, fortified by an iterative computational procedure. The population of the various superconfigurations is assumed to have a thermal Boltzmann-type distribution.

The *super transition arrays* are transitions between the superconfigurations. As the number of configurations decreases, the number of the transitions is reduced by orders of magnitudes relative to the UTA method. In an example in their paper, the authors reproduced the results for the total transition array $3d_{5/2}-2p_{3/2}$ in iron in a plasma of $T = 200\,\mathrm{eV}$ and $n_i = 8.5 \times 10^{22}\,\mathrm{cm}^{-3}$. An explicit UTA calculation for this special case would require the computation of 40 000 configurations. With only eight super transition arrays Bar-Shalom et al. could reproduce the UTA results with remarkable accuracy.

Since its first publication, the STA model has been applied successfully to compute opacities of several materials of astrophysical interest. The method is being developed to account for non-LTE conditions as well. The original formulas and recipes of the STA model would be too lengthy to repeat here. The interested reader is encouraged to go back to the original paper (Bar-Shalom et al., 1989).

Line Broadening

7.1 Introduction

The width of spectral lines emitted from astrophysical and laboratory plasmas is an extremely valuable diagnostic tool for the study of the plasma conditions surrounding the emitting ion. The line width is a parameter that is very sensitive to the local microfield around the emitting ions, and can be used, therefore, to provide information about the local temperature and density conditions that cannot be obtained by other diagnostic means.

A spectral line is broadened by several mechanisms:

- *Natural line broadening*, which is the width due to the finite lifetime of ionic excited states.
- *Doppler broadening*, whose origin is the thermal velocity distribution of the emitting ions.
- *Electron impact broadening*, caused, as its name indicates, by electron collisions.
- *Quasi-static Stark broadening*, whose origin is the relatively slowly changing local electrostatic fields, generated by nearby ions, which split and shift the energy levels of the radiating ion.

These broadening effects add up, and the total line width is a combination of all of them. However, in general, only one of these effects dominates, depending on the local plasma temperature and density, while the others are regarded as smaller corrections.

The shape of a spectral line is conveniently described by means of the *line profile function*, $\mathcal{L}(\omega)$, which is defined such that

$$I(\omega)\,d\omega = I_0 \mathcal{L}(\omega)\,d\omega \qquad (7.1.1)$$

is the number of photons emitted in the range of frequencies $[\omega, \omega + d\omega]$, and I_0 is the total number of photons emitted in the line. The units of $\mathcal{L}(\omega)$ are *seconds* and it is normalized to unity,

$$\int \mathcal{L}(\omega)\, d\omega = 1 \qquad (7.1.2)$$

In this chapter, unlike the previous ones, we characterize the spectrum by the photon's frequency, ω, rather than its energy $\hbar\omega$.

The shapes of $\mathcal{L}(\omega)$ are, in most cases, either as Lorentzian distribution,

$$\mathcal{L}(\omega) = \frac{\Gamma}{2\pi} \frac{1}{(\omega - \omega_0)^2 + (\Gamma/2)^2} \qquad (7.1.3)$$

or a Gaussian distribution,

$$\mathcal{L}(\omega) = \frac{1}{\sqrt{2\pi}\sigma} \exp\left[-\frac{1}{2}\left(\frac{\omega - \omega_0}{\sigma}\right)^2\right] \qquad (7.1.4)$$

See figure 7.1 for a comparison between these two distributions. Both distributions attain their maximum at $\omega = \omega_0$. The Lorentzian line shape attains a maximum of $\mathcal{L}(0) = 2/\pi\Gamma$, while for a Gaussian distribution $\mathcal{L}(0) = 1/\sqrt{2\pi}\sigma$. Γ and σ determine the width of the distributions. $\Gamma/2$ is the half-width of the spectral line at half its peak value, $\mathcal{L}(\omega) = \frac{1}{2}\mathcal{L}(0)$ for $\omega - \omega_0 = \pm\Gamma/2$. In a similar manner, in a Gaussian distribution $\mathcal{L}(\omega)$ decreases to $0.606\ldots$ of its maximum value when $\omega - \omega_0 = \pm\sigma$, and the intensity of the line reduces to one-half of its value at the line center when $\omega - \omega_0 = \pm\sigma\sqrt{2\log 2} = \pm 1.177\sigma$. Generally, both $\Gamma \ll \omega_0$ and $\sigma \ll \omega_0$, so that $\mathcal{L}(\omega)$ is a function which takes nonzero values only in a very narrow part of the spectrum. As can be seen from figure 7.1, the Gaussian is concentrated more around the line center, whereas the Lorentzian has larger *line*

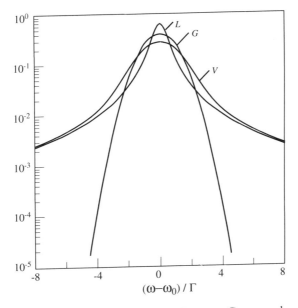

Figure 7.1 Comparison between Lorentz, Gauss, and Voigt line profiles.

wings; that is, its decrease for frequencies far from the line center is much slower than the Gaussian's.

7.2 What Is Line Broadening?

Assume an atomic oscillator emitting a continuous radiation of amplitude $A(t)$ such that

$$A(t) = A_0 \exp(i\omega_0 t) \qquad (7.2.1)$$

The Fourier transform of this amplitude is given by the simple formula

$$\tilde{A}(\omega) = \frac{1}{\sqrt{2\pi}} \int_{-\infty}^{\infty} A(t) e^{-i\omega t} \, dt = A_0 \delta(\omega - \omega_0) \qquad (7.2.2)$$

In this simple case the oscillation, and the emitted photons, have only one Fourier component, that is, one sharply defined frequency, ω_0. The emission rate of the radiative energy is proportional to the square of the amplitude,

$$I(\omega) \propto |\tilde{A}(\omega)|^2 \qquad (7.2.3)$$

which, too, has a sharp line at the same frequency, and the line profile is actually a Dirac function,

$$L(\omega) = \delta(\omega - \omega_0) \qquad (7.2.4)$$

Assume now that due to interaction with a second particle, which will be called the *perturber*, the *radiator* emits the radiation only for a finite time, say between $-t_0$ and $+t_0$. The Fourier transform of the amplitude now is

$$\begin{aligned}
\tilde{A}(\omega) &= \frac{1}{\sqrt{2\pi}} \int_{-t_0}^{t_0} A(t) \exp(-i\omega t) \, dt \\
&= A_0 \frac{1}{\sqrt{2\pi}} \int_{-t_0}^{t_0} \exp[i(\omega_0 - \omega)t] \, dt \\
&= A_0 \sqrt{\frac{2}{\pi}} \frac{\sin[(\omega_0 - \omega)t_0]}{\omega_0 - \omega}
\end{aligned} \qquad (7.2.5)$$

The profile function now acquires the form

$$L(\omega) = \frac{I(\omega)}{I_0} = \frac{2}{\pi} \frac{\sin^2[(\omega_0 - \omega)t_0]}{(\omega_0 - \omega)^2} \qquad (7.2.6)$$

Now more frequencies take part in the emitted spectrum. Although the maximum emission still occurs at the original frequency ω_0, other frequencies on both sides of the emission peaks also show up. Figure 7.2 shows a schematic line shape of equation (7.2.6).

The more frequent the perturbations felt by the radiating ion, the shorter is the emission period t_0, and the profile of the line becomes broader and broader. In contrast, in low density plasmas, where the perturbations are rare, the lifetime t_0

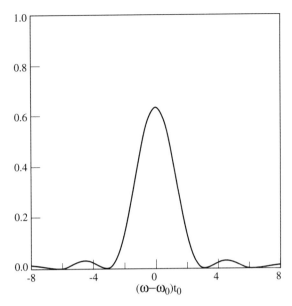

Figure 7.2 The shape of the fictitious line profile defined by equation (7.2.6).

becomes longer and the line becomes narrower. In principle, for $t_0 \to \infty$ the line profile acquires a sharp δ-function type profile of the kind given by equation (7.2.4).

In a real plasma the origin of the perturbations of the line emissions, and the corresponding line broadenings, is the interactions between the radiator and its neighboring particles. Interpretation of the emission line widths can, in principle, provide valuable information about the local microfields around the radiator and the interaction of the radiating ion with its plasma environment.

7.3 Natural Line Broadening

The simplest broadening mechanism of spectral lines is due to the finite lifetime of the excited states, see for instance Heitler (1954). An atomic excited state decays spontaneously to one of the lower states within a time period that is characterized by the Einstein coefficient $A_{u \to \ell}$, where u symbolizes the quantum numbers of the *u*pper state, and ℓ those of the *l*ower state. The probability to find the ion at the instant t in the upper state is, in fact, an exponentially decreasing function,

$$P_u(t) = P_u(0)e^{-\Gamma t} = \psi_u(\vec{r})\psi_u^*(\vec{r})e^{-\Gamma t} \tag{7.3.1}$$

where $\psi_u(\vec{r})$ is the eigenfunction of the upper state, and $\Gamma = \sum_{\ell < u} A_{u \to \ell}$ is the sum of the Einstein coefficients for transitions into all possible lower states. The amplitude of the photon emission during the transition from the upper to the lower state has, therefore, a damped oscillatory time dependence,

$$A(t) \propto e^{i\omega_0 t} e^{-\Gamma t/2} \tag{7.3.2}$$

where $\omega_0 = (E_u - E_\ell)/\hbar$ is the transition energy. The Fourier transform of this amplitude is proportional to

$$\tilde{A}(\omega) \propto \frac{1}{\sqrt{2\pi}} \int_{-\infty}^{\infty} dt \, \exp\left[-\left(\frac{\Gamma}{2} + i(\omega - \omega_0)\right)t\right] = -\frac{1}{\sqrt{2\pi}} \frac{1}{\Gamma/2 + i(\omega - \omega_0)} \tag{7.3.3}$$

The natural line profile has, therefore, a Lorentzian shape,

$$\mathcal{L}(\omega) = \frac{\Gamma}{2\pi} \frac{1}{(\omega - \omega_0)^2 + (\Gamma/2)^2} \tag{7.3.4}$$

where we have added a normalization factor. If the lower state is also an excited state, then the line width Γ has to account also for the broadening of this state, and Γ assumes the form (Griem, 1964)

$$\Gamma = \sum_{m<u} A(u \rightarrow m) + \sum_{m'<\ell} A(\ell \rightarrow m') \tag{7.3.5}$$

The natural broadening is the minimum width possible for a spectral line. In laboratory plasmas this natural line width is much smaller than the broadenings caused by other mechanisms, and is only seldom measurable.

7.4 Doppler Broadening

A different type of broadening mechanism is Doppler broadening. Owing to the motion of the emitting ions, the frequency of the detected photon in the laboratory frame, ω, differs from the frequency emitted in the frame of the moving ion, ω_0, by an amount equal to

$$\Delta\omega = \omega - \omega_0 = \omega_0 \frac{v_x}{c} \tag{7.4.1}$$

where v_x is the component of the velocity of the emitter along the line of sight to the detector. This velocity has a Maxwell–Boltzmann distribution, which in our case is the one-dimensional version of equation (1.3.12),

$$f_{v_x}(v_x) \, dv_x = \left(\frac{M}{2\pi T_e}\right)^{1/2} \exp\left(-\frac{Mv_x^2}{2T_e}\right) dv_x \tag{7.4.2}$$

Here M is the mass of the radiating ion. Combining equation (7.4.2) with (7.4.1), one finds the probability that the emission line will be detected at frequency ω, shifted by an amount $\Delta\omega$ from the emitted frequency ω_0,

$$Prob\{\omega - \omega_0 = \Delta\omega\} \, d(\Delta\omega) = Prob\left(v_x = c\frac{\Delta\omega}{\omega_0}\right) d(\Delta\omega) = f_{v_x}\left(c\frac{\Delta\omega}{\omega_0}\right) dv_x$$

$$= \left(\frac{M}{2\pi T_e}\right)^{1/2} \exp\left[-\frac{Mc^2}{2T_e}\left(\frac{\Delta\omega}{\omega_0}\right)^2\right] \frac{c}{\omega_0} \, d(\Delta\omega) \tag{7.4.3}$$

Denote

$$\sigma^2 = \frac{T\omega_0^2}{Mc^2} \qquad (7.4.4)$$

Then the resulting line profile is

$$\mathcal{L}(\Delta\omega)\,d(\Delta\omega) = \frac{1}{\sqrt{2\pi}\sigma}\exp\left[-\frac{1}{2}\left(\frac{\Delta\omega}{\sigma}\right)^2\right]d(\Delta\omega) \qquad (7.4.5)$$

which means that the effect of the ions' motion in the plasma is the broadening of the emission line into a Gaussian shape whose half-width at half-maximum (*HWHM*) is

$$HWHM = \sigma\sqrt{2\log 2} = \omega_0\sqrt{\frac{T\,2\log 2}{Mc^2}} \qquad (7.4.6)$$

The width of the Doppler broadening depends on the plasma temperature only; the higher the temperature the larger is the line width. The relative Doppler line width, however, is very small. In fact, for hydrogen plasmas the ratio of the *HWHM* to the frequency of the line center is of the order of

$$\frac{HWHM}{\omega_0} = \sqrt{\frac{T\,2\log 2}{Mc^2}} \approx 3\times 10^{-5}\text{--}3\times 10^{-3} \qquad (7.4.7)$$

for temperatures between 1 and 10^4 eV, and is even smaller for plasmas of higher Z.

When the line is subject to two kinds of broadenings, of which one generates a Lorentzian and other a Gaussian line shape, the combined line profile is a combination of both distribution functions. Such a line profile is called a *Voigt distribution*,

$$\mathcal{L}_V(\omega;\Gamma,\sigma)\,d\omega = d\omega\int_{-\infty}^{\infty}d\omega'\,\mathcal{L}_L(\omega';\Gamma)\mathcal{L}_G(\omega-\omega';\sigma) \qquad (7.4.8)$$

Here, $\mathcal{L}_V(\omega;\Gamma,\sigma)$, $\mathcal{L}_L(\omega;\Gamma)$, and $\mathcal{L}_G(\omega;\sigma)$ are the Voigt, the Lorentzian, and the Gaussian profile functions, respectively. Using equations (7.1.3) and (7.1.4), after some manipulations, equation (7.4.8) is reduced to the form

$$\mathcal{L}_V(\omega;\Gamma,\sigma)\,d\omega = \frac{\Gamma}{2\pi^{3/2}}\int_{-\infty}^{\infty}dy\,\frac{e^{-y^2}}{(\omega-\omega_0+\sqrt{2}\sigma y)^2 + (\Gamma/2)^2} \qquad (7.4.9)$$

The Voigt profile is shown in figure 7.1 for the same Γ and σ (both equal to 1) as the other profiles. It can be seen that the Voigt shape is approximately Gaussian near the line center, and has a $1/(\omega-\omega_0)^2$ decrease in the line wings, far from the line center, like the Lorentzian profile. This property of the Voigt line shape is very important to the understanding of the background from emission/absorption lines in the spectral regions between the lines.

7.5 Electron Impact Broadening

The broadenings caused by the interactions of a radiator with its plasma environment are closely related to the theory of atomic collisions. These collisions may be very rapid, with a duration much shorter than the natural lifetime of the decaying level, or alternatively, may be slow relative to the level's lifetime. The rapid scattering processes are mainly due to the rapidly moving free electrons in the plasma. In this case we speak of *electron impact broadening*. At the other extreme are the perturbations of the radiating ion by its slow-moving neighboring ions. In this case we speak of *quasi-static broadening*. In this section we concentrate on the electron impact broadening.

Assume a radiator approached by a perturber, which can be either an electron or an ion. The general interaction pattern between the two particles can be very complicated. We shall make a few simplifying approximations that will help in the understanding of the underlying physical picture without significantly sacrificing the accuracy of the results.

1. It is assumed that the motion of the perturbing particles can be described by classical trajectories, and the wave nature of the perturber is neglected.
2. For simplicity, it is also assumed that the trajectory of the perturbing particle is rectilinear. A full calculation that takes full account of the hyperbolic motion of the perturber improves the accuracy of the results only slightly, but greatly obscures the physical picture. For a full treatment see, for instance, Sahal-Brechot (1969) or Griem (1974).
3. It is also assumed that the greatest influence on the radiator is due to its interaction with its nearest neighbor.
4. Finally, the perturbation is assumed to be elastic and not to induce transitions among the energy levels of the radiator.

Within the framework of these assumptions, the perturbing particle generates an external potential on the radiator of

$$V(R) = V\left(\sqrt{\rho^2 + [v(t - t_0)]^2} \right) \tag{7.5.1}$$

where R is the distance between the perturber and the radiator at time t, ρ is the impact parameter, and t_0 is the time of the nearest approach. As a result of the interaction, the atomic levels of the radiator are perturbed by an amount

$$\Delta E_n = \langle n| - eV[R(t)]|n\rangle \tag{7.5.2}$$

During the encounter with the perturber, there is a continuous modification of the energy levels, and thereby of the emission frequency, by an amount

$$\Delta \omega(t) = \frac{1}{\hbar}[\Delta E_n(t) - \Delta E_m(t)] = -\frac{e}{\hbar}[\langle n|V[R(t)]|n\rangle - \langle m|V[R(t)]|m\rangle] \tag{7.5.3}$$

This perturbation of the line energy, which has a statistical nature, is the origin of the corresponding line broadening.

In the presence of this external potential, the oscillations of the atomic oscillator are

$$A(t) = \exp\left(i\omega_0 t + i \int_{-\infty}^{t} \Delta\omega(t') \, dt' \right) = \exp(i\omega_0 t + i\eta(t)) \tag{7.5.4}$$

where we have denoted by

$$\eta(t) = \int_{-\infty}^{t} \Delta\omega(t') \, dt' \tag{7.5.5}$$

the total change of the phase of the oscillations caused by the interaction up to time t. The emitted line profile is

$$I(\omega) = \lim_{T \to \infty} \frac{1}{T} \left| \frac{1}{\sqrt{2\pi}} \int_{-T?2}^{+T/2} A(t) \exp(-i\omega t) \, dt \right|^2 \tag{7.5.6}$$

After an appropriate change of the variables in equation (7.5.6), this equation can be rewritten as

$$I(\omega) = \frac{1}{\pi} \mathcal{R}\left(\int_{0}^{\infty} \Phi(\tau) \exp[-i(\omega - \omega_0)\tau] \, d\tau \right) \tag{7.5.7}$$

where

$$\Phi(\tau) = \lim_{T \to \infty} \frac{1}{T} \int_{-T/2}^{T/2} A^*(t) A(t + \tau) \, dt \tag{7.5.8}$$

is the *autocorrelation function* of the oscillations. This function is the time average of the product of the amplitude of the oscillations at two instants separated by a time difference τ. If τ is smaller than the *correlation time* between the two collisions, this average is positive (as can be tested, for example, by taking $\tau = 0$). If, however, the time difference τ is so large that at time $t + \tau$ the radiator has enough time to "forget" the influence of the phase of the oscillations at time t, then the phases of the oscillations at these times are independent of each other, so that the time average, and the corresponding function, vanishes.

The nature of the interaction is imbedded in the function $\Delta\omega(t)$. As this factor is proportional to the potential between the perturber and the radiator, one can assume that schematically it has the form

$$\Delta\omega = \frac{C_m}{R^m} \tag{7.5.9}$$

with $m = 2$ for a dipole interaction, $m = 3$ for a quadrupole interaction, and $m = 6$ for van-der-Waals forces between the two particles. C_m is a constant, depending on the multipolarity of the interaction but independent of R. If both the radiator and the perturber are charged, then the order of magnitude of C_m is $Z_p Z_r e^2 a_0^{m-1}/\hbar$, where eZ_p and eZ_r are the charges of the perturber and the radiator. We shall not try to give a more elaborate form for C_m. It depends on the accurate shape of the interaction between the particles and has to be calculated for each type of interaction separately. Even without having the exact form of C_m we shall be able to derive important information about the line broadening.

Substituting equation (7.5.9) for $\Delta\omega$ into equation (7.5.5), one gets

$$\eta(\rho, v) = C_m \int_{-\infty}^{\infty} \frac{dt}{\{\rho^2 + [v(t - t_0)]^2\}^{m/2}} = \alpha_m \frac{C_m}{v\rho^{m-1}} \qquad (7.5.10)$$

where

$$\alpha_m = \sqrt{\pi} \frac{\Gamma[(m - 1)/2]}{\Gamma(m/2)} \qquad (7.5.11)$$

or, numerically, $\alpha_2 = \pi$, $\alpha_3 = 2$, $\alpha_4 = \pi/2$ $\alpha_5 = 4/3$, and $\alpha_6 = 3\pi/8$. Although the integral in equation (7.5.10) should be done over the time of interaction only, extending its upper limit to infinity results in a sufficiently good approximation.

The explicit appearance of the impact parameter, ρ, in the equation of η automatically splits the interactions into two categories: strong interactions between the radiator and the perturber, corresponding to a small impact parameter, and weak interactions when the impact parameter is large. The boundary between these two cases is defined by the impact parameter ρ_0 for which $\eta(\rho_0) = 1$,

$$\rho_0 = \left(\frac{\alpha_m C_m}{v}\right)^{1/(m-1)} \qquad (7.5.12)$$

This ρ_0 is called the *Weisskopf radius* and it depends on the plasma temperature (through the velocity v) and on the multipolarity of the interaction. The Weisskopf radius determines the limit between close and distant encounters inside a plasma.

Consider an ion in the plasma. The motion of the other plasma particles generates a randomly varying potential on this ion. This random pattern of the potential fluctuations consists of a rather slow component due to the ion–ion interactions and a high frequency component due to the electron impacts. If the plasma density is not too high, the duration of the electron collisions is much shorter than the average time between two consecutive electron–ion collisions. Under this condition, the sharp potential fluctuations due to the electron impacts can be very closely approximated by δ-function-like spikes as a function of time, and one can consider the collisions to be instantaneous. This is the condition for *electron impact broadening*, which will be the subject of the rest of this section.

Next we turn to the calculation of the autocorrelation function $\Phi(\tau)$ (equation 7.5.8). One of the useful assumptions in statistical systems is that time averaging can be replaced by an averaging over the statistical assembly. This is the so-called *ergodic hypothesis*. This assumption is equivalent to the fact that over a long time one ion undergoes the same statistical distribution of perturbations as the whole assembly of ions does at a given instant. Denoting by angle brackets the statistical distribution over the assembly, one gets (taking $t = 0$),

$$\Phi(\tau) = \langle A(0)A^*(\tau)\rangle$$

$$= \langle \exp[i\eta(0)] \exp[i\eta(\tau)]\rangle = \langle \exp[i\eta(\tau)]\rangle \qquad (7.5.13)$$

The change in the correlation function during time $\Delta\tau$ is given by

$$\Delta\Phi = \Phi(\tau + \Delta\tau) - \Phi(\tau) = \langle\exp[i\eta(\tau + \Delta\tau)]\rangle - \langle\exp[i\eta(\tau)]\rangle$$

$$= \langle\exp[i\eta(\tau)]\exp[i\Delta\eta]\rangle - \langle\exp[i\eta(\tau)]\rangle \qquad (7.5.14)$$

when $\Delta\eta$ is the phase shift produced by the collisions during the time interval $\Delta\tau$. If the collisions are instantaneous, occuring in practically zero time, then $\Delta\eta$ is not affected by the instantaneous value of $\eta(\tau)$, and these two quantities are statistically independent. Therefore

$$\Delta\Phi = \langle\exp[i\eta(\tau)]\rangle\{\langle\exp[i\Delta\eta]\rangle - 1\}$$

$$= -\Phi(\tau)\langle 1 - \exp[i\Delta\eta]\rangle \qquad (7.5.15)$$

This formula can further be simplified. If, as usual, n_e is the electron density, and $\bar{v}$ is the electron mean velocity, then the number of impacts with the radiator with impact parameter between ρ and $\rho + d\rho$ during the time period $\Delta\tau$ is $n_e\bar{v}\Delta\tau 2\pi\rho\, d\rho$. Since the phase shift $\Delta\eta$ during $\Delta\tau$ due to a collision is a function of the impact parameter, summing over all possible collisions during $\Delta\tau$ gives

$$\langle 1 - \exp[i\Delta\eta]\rangle = n_e\bar{v}\,\Delta\tau\int_0^\infty 2\pi\rho\, d\rho[1 - \exp(i\Delta\eta)]$$

$$= n_e\,\Delta\tau\langle v(\sigma' - i\sigma'')\rangle \qquad (7.5.16)$$

where,

$$\sigma' = 2\pi\int_0^\infty (1 - \cos\Delta\eta)\rho\, d\rho \qquad (7.5.17)$$

$$\sigma'' = 2\pi\int_0^\infty \sin\Delta\eta\,\rho\, d\rho$$

The quantities of σ' and σ'' are called the *width and shift effective cross sections*. Substituting equation (7.5.16) back into equation (7.5.15), one gets

$$\Delta\Phi = -n_e\Delta\tau\langle v(\sigma' - i\sigma'')\rangle\Phi(\tau) \qquad (7.5.18)$$

or

$$\Phi(\tau) = \exp[-\tau n_e\langle v(\sigma' - i\sigma'')\rangle] \qquad (7.5.19)$$

which is the autocorrelation function of the oscillations of the radiator. Finally, inserting equation (7.5.19) into the equation of the line intensity (equation 7.5.7) and normalizing the result to unit integral, one finds

$$\mathcal{L}(\omega) = \frac{w}{2\pi}\frac{1}{(\omega - \omega_0 - \Delta)^2 + (w/2)^2} \qquad (7.5.20)$$

which is a Lorentzian line shape. The width of this line shape, w, is related to the real part of the width-shift cross section,

$$w = 2n_e\langle v\sigma'\rangle \qquad (7.5.21)$$

The line is also shifted from its original position by an amount

$$\Delta = n_e \langle v\sigma'' \rangle \tag{7.5.22}$$

which depends on the imaginary part of the width-shift cross section.

By using equation (7.5.9) for Δw, the integrals in equation (7.5.10) can be carried out analytically for $m \geq 3$. We list the results:

$$m = 3 \qquad w = 2\pi^2 n_e C_3$$

$$m = 4 \qquad w = 11.4 n_e C_4^{2/3} \langle v^{1/3} \rangle \qquad \Delta = \frac{\sqrt{3}}{2} w \tag{7.5.23}$$

$$m = 6 \qquad w = 7.16 n_e C_6^{2/5} \langle v^{3/5} \rangle \qquad \Delta = 0.36 w$$

For the most important case, namely that of dipole interaction, $m = 2$, the phase $\eta(\rho) \propto 1/\rho$ and the integrals in equation (7.5.17) diverge logarithmically. To overcome this difficulty one has to insert a cutoff impact parameter, ρ_{max}, which accounts for the screening of the interaction by the plasma medium when the radiator–perturber distance is large. The result is

$$m = 2 \qquad w = \frac{n_e \pi^3 C_2^2}{\langle v \rangle} \left[0.923 - \ln\left(\frac{\pi C_2}{\rho_{max}\langle v \rangle}\right) + \ldots \right] \tag{7.5.24}$$

This result does not depend critically on ρ_{max}, and generally one takes the Debye radius D for the cutoff radius ρ_{max}.

As the average velocity of the electrons, $\langle v \rangle$, is proportional to the square root of the electron temperature, T_e, equation (7.5.24) shows a very important scaling property of the electron impact broadening, namely, that this width is approximately proportional to the electron density and inversely proportional to the square root of the electron temperature

$$w \propto \frac{n_e}{\sqrt{T_e}} \tag{7.5.25}$$

The simple form obtained for the electron impact broadening (equations 7.5.23–24), needs some important but rather complicated corrections to include several other effects:

- Screening corrections of the surrounding plasma particles.
- Higher multipole contributions.
- Inelastic collisions.

Griem proposed a series of formulas that already include corrections for these effects. For the electron impact broadening formula for neutral hydrogen and singly ionized helium ions, see equation (417) in Griem (1974). Other formulas suggested by Griem are for the Lyman series (see his equations (494) and (496)) and a general formula for transition from nP states (see his equation (526), reproduced in table 7.1. These formulas are not too simple to use, and are valid only for weakly coupled low density plasmas.

Table 7.1 Electron impact broadening of nP states

$$w \approx 3\pi n_e \left(\frac{2m}{\pi kT}\right)^{1/2} \left(\frac{\hbar}{mZ}\right)^2 n^2 \left[\frac{l+1}{2l+1}\right] [n^2 - (l+1)^2] \times \ln\left\{5 - \frac{4.5}{\sqrt{Z}} + \xi_{l,l+1}^{-1}\left[1 + \frac{kTn^2}{E_H(Z-1)Z}\right]^{-1}\right\}$$

$$+ \frac{l}{2l+1}(n^2 - l^2)\ln\left\{5 - \frac{4.5}{\sqrt{Z}} + \xi_{l,l-1}^{-1}\left[1 + \frac{kTn^2}{E_H(Z-1)Z}\right]^{-1}\right\} + \frac{n^2}{3} + \frac{n^2}{3}\frac{kT}{E_H}\left(1 + \frac{kT}{E_H} + \frac{Z^2}{n^4}\right)^{-1}$$

$$+ \frac{1}{9}(n^2 + 3l^2 + 3l + 11)\ln\left\{1.4 + \left(\frac{kT}{E_H}\right)^{3/2}\frac{n^3}{(Z-1)Z^2}\left[1 + \frac{kTn^2}{E_H(Z-1)Z}\right]^{-1}\right\}\right]$$

(From Griem, 1974.)

7.6 Quasi-Static Stark Broadening

The other extreme model for the line broadening is the quasi-static or Stark broadening, which is valid when the interaction time between a perturber and the radiator is longer than the time between two collisions. This is the case when the perturber is an ion whose velocity is much less than that of the electrons.

The central mechanism for the Stark broadening is the electrostatic field generated by the perturber on the radiator. This interaction between the perturber and the radiator adds to the Hamiltonian a term

$$\mathcal{H}_{elec} = -\vec{F} \cdot \vec{d} \qquad (7.6.1)$$

where $\vec{F}$ is the electrostatic field (we keep the letter E for energies) and $\vec{d}$ is the electric dipole moment of the ion. Due to this interaction, the ionic levels are split according to the absolute value of the *magnetic quantum number, m,* which means that the magnetic sublevels with $+m$ and $-m$ remain degenerate.

The Stark splitting of the spectral lines has a complex nature. For weak electrostatic fields, it is proportional to $|\vec{F}|^2$; this is the region of the so-called *quadratic Stark effect*. This quadratic behavior for weak fields is explained by the polarization of the ion by the electric field, so that the dipole moment, too, increases linearly with the local electric field. At higher fields, the dipole moment attains its limiting constant value, and the Stark splitting and shift increase only linearly with the increasing electric field. This is the region of the *linear Stark effect*. At very high electric fields, the effect behaves in a complex nonlinear manner. For hydrogenlike ions the quadratic effect at low fields is very small, and one can fairly assume that these ions display the linear effect only (Condon and Shortley, 1987).

Assume an ion is located at $r = 0$. The motions of nearby ions generate a fluctuating electrostatic field on this ion. These fluctuations split and shift the levels according to the local instantaneous electrostatic field. The distribution of the Stark broadening is, therefore, closely related to the distribution of the electrostatic field on this ion.

We assume a linear Stark effect, namely, that the frequency shift is proportional to the local electrostatic field,

$$\Delta\omega = \omega - \omega_0 \propto F \propto \frac{1}{R^2} \tag{7.6.2}$$

The line profile is therefore proportional to the distribution of the electric field strength in the plasma,

$$\mathcal{L}(\Delta\omega)\,d(\Delta\omega) = P_F(F)\,dF \tag{7.6.3}$$

Here $P_F(F)\,dF$ is the probability that the strength of the local instantaneous electrostatic field on the ion is between $F = |\vec{F}|$ and $F + dF$. The calculation of the line profiles in the case of quadratic and higher multipole Stark effects will not be treated here. These can be found, for instance, in the book by Mihalas (1970).

The Nearest Neighbor Approximation

First, we shall find the electrostatic field distribution, $P_F(F)$, in the *nearest neighbor approximation*, namely, assuming that the electric field strength on the radiator is produced solely by the nearest neighbor.

In terms of the distance to the nearest neighbor, R, the electric field strength is written as

$$F = \frac{|Z_p e|}{R^2} = F_N \left(\frac{R_i}{R}\right)^2 \tag{7.6.4}$$

where we have denoted $F_N = |Z_p e|/R_i^2$, and $R_i = (3/4\pi n_i)^{1/3}$ is the ion sphere radius, equation (1.2.7). Using the definition of R_i, F_N takes the form

$$F_N = (4\pi/3)^{2/3}|Z_p e|n_i^{2/3} = 2.599|Z_p e|n_i^{2/3} \tag{7.6.5}$$

Define the *reduced electric field*, β_N, by

$$\beta_N = \frac{F}{F_N} = \left(\frac{R_i}{R}\right)^2 \tag{7.6.6}$$

or

$$R = R_i\beta_N^{-1/2} \tag{7.6.7}$$

The probability that the nearest neighbor will be at a distance R was calculated in a weakly correlated plasma in section 2.7 (equation (2.7.1). Substituting equation (7.6.7) into (2.7.1) for R, one finds for the electric field strength distribution

$$P_F(\beta_N)\,d\beta_N = P_{NN}(R_i\beta_N^{-1/2})\,\frac{dR}{d\beta_N}\,d\beta_N$$

$$= \frac{3}{2}\frac{1}{\beta_N^{5/2}}\exp(-\beta_N^{-3/2})\,d\beta_N \tag{7.6.8}$$

Denote by ω^* the proportionality factor between $\Delta\omega$ and β_N (see equation (7.6.2)), then the line profile, equation (7.6.3), is calculated as

$$\mathcal{L}(\Delta\omega)\,d(\Delta\omega) = P_F\left(\frac{\Delta\omega}{\omega^*}\right)\frac{d\beta_N}{d(\Delta\omega)}\,d(\Delta\omega) = \frac{3}{2}\frac{\omega^{*3/2}d(\Delta\omega)}{(\Delta\omega)^{5/2}}\exp\left[-\left(\frac{\omega^*}{\Delta\omega}\right)^{3/2}\right]$$

$$(7.6.9)$$

The distribution of the electrostatic field, $P_F(\beta_N)$, and the line profile, $\mathcal{L}(\Delta\omega)$, both vanish for $\beta_N \to 0$ and $\Delta\omega \to 0$, as well as for $\beta_N \to \infty$ and $\Delta\omega \to \pm\infty$. The line profile attains its maximum at $\Delta\omega = \omega^*(3/5)^{2/3} = 0.711\omega^*$. The asymptotic expansion of the line profile at the line wings is

$$\mathcal{L}(\Delta\omega) \approx \frac{3}{2}\frac{d(\Delta\omega)}{\omega^*}\left(\frac{\omega^*}{\Delta\omega}\right)^{5/2}\left[1 - \left(\frac{\omega^*}{\Delta\omega}\right)^{3/2} + \frac{1}{2}\left(\frac{\omega^*}{\Delta\omega}\right)^3 + \cdots\right], \qquad \frac{\Delta\omega}{\omega^*} \gg 1$$

$$(7.6.10)$$

which decreases as $(\Delta\omega)^{-5/2}$ in the line wings, a slightly stronger decrease than in a Lorentzian profile.

The Holtsmark Distribution

The calculation of the electric field strength distribution in the linear Stark case was studied by Holtsmark as early as 1919. In his derivation, which is reproduced below, he took into account the total electrostatic field generated by all the plasma particles, not only the nearest neighbor. The calculation has several assumptions. First, it corresponds to the electrostatic field only, $F \propto 1/R^2$, and neglects higher multipole terms of the electrostatic interactions. Second, the particles are assumed to be uncorrelated, thereby constraining the validity of his treatment to weakly correlated plasmas. Finally, the distribution of the electrostatic field inside the plasma is assumed to be isotropic, with no preferential direction, which limits the validity to plasmas without external magnetic or electric fields.

Again, assume that the radiator is located at $r = 0$. Denote by M the number of all the other ions in the plasma, which are located at points $\vec{r}_1, \vec{r}_2, \ldots, \vec{r}_j, \ldots, \vec{r}_M$. Each of these ions generates an electric field of

$$\vec{F}_j = \frac{Z_p e}{r_j^2}\,\vec{\epsilon}_j \tag{7.6.11}$$

where $\vec{\epsilon}_j = \vec{r}_j/r_j$ is the unit vector in the direction of the jth ion and $Z_p e$ is the charge of the perturber. The total electric field generated by all these ions is

$$\vec{F} = \sum_{j=1}^{M}\vec{F}_j \tag{7.6.12}$$

Denote by $P(\vec{r}_1, \vec{r}_2, \ldots, \vec{r}_M)\,d^3r_1\,d^3r_2\cdots d^3r_M$ the probability to find the first of these ions at an infinitely small volume d^3r_1 around $\vec{r}_1$, the second in d^3r_2 around $\vec{r}_2$, and so on. If the particles are uncorrelated, the probability to find a particle in a small volume element d^3r is simply d^3r/V, where V is the plasma volume. For the total probability one obtains, therefore,

$$P(\vec{r}_1, \vec{r}_2, \ldots, \vec{r}_M) \, d^3r_1 \, d^3r_2 \cdots d^3r_M = \frac{1}{V^M} d^3r_1 \, d^3r_2 \cdots d^3r_M \qquad (7.6.13)$$

The probability that the total electric field on the test ion at $r = 0$ equals $\vec{F}$ is

$$P_F(\vec{F}) \, d^3F = \int \cdots \int \delta \left(\vec{F} - \sum_{j=1}^M \vec{F}_j \right) P(\vec{r}_1, \vec{r}_2, \ldots, \vec{r}_M) \, d^3r_1 \, d^3r_2 \cdots d^3r_M$$

$$= \frac{1}{V^M} \int \cdots \int \delta \left(\vec{F} - \sum_{j=1}^M \vec{F}_j \right) d^3r_1 \, d^3r_2 \cdots d^3r_M \qquad (7.6.14)$$

To continue the calculations, we first look for the Fourier transform of this expression

$$A(\vec{\xi}) = \int \exp(i\vec{\xi} \cdot \vec{F}) P_F(\vec{F}) \, d^3F$$

$$= \frac{1}{V^M} \int \cdots \int d^3r_1 \, d^3r_2 \cdots d^3r_M \left[\int d^3F \, \exp(i\vec{\xi} \cdot \vec{F}) \delta \left(\vec{F} - \sum_{j=1}^M \vec{F}_j \right) \right]$$

$$= \frac{1}{V^M} \int \cdots \int d^3r_1 \, d^3r_2 \cdots d^3r_M \, \exp \left(i\vec{\xi} \cdot \sum_{j=1}^M \vec{F}_j \right)$$

$$= \left[\frac{1}{V} \int d^3r_j \, \exp(i\vec{\xi} \cdot \vec{F}_j) \right]^M \qquad (7.6.15)$$

Here ξ is the parameter of the Fourier transform. The units of ξ are reciprocal to those of the electric field, that is, they are $\mathrm{cm}^2/\mathrm{esu}$. In equation (7.6.15), the second line is obtained by inserting equation (7.6.14) for P_F into the first line and rearranging the order of the integrals. The third line is derived by carrying out the integration over the components of the total electric field, $\vec{F}$, exploiting the special properties of the δ-function. The last line is obtained by separation of the combined integral into a product of M equal factors.

The last integral in equation (7.6.15) can be further reduced. Denoting by θ the angle between $\vec{\xi}$ and $\vec{F}_j$, and accounting for the isotropy of the plasma medium, one gets

$$\frac{1}{V} \int d^3r_j \, \exp(i\vec{\xi} \cdot \vec{F}_j) = \frac{2\pi}{V} \int r^2 \, dr \, \exp(i\xi F_j \cos \theta) \, d(\cos \theta)$$

$$= \frac{4\pi}{V} \int r^2 \, dr \, \frac{\sin[\xi F_j(r)]}{\xi F_j(r)}$$

$$= \frac{4\pi}{V} \left[\int r^2 \, dr - \int r^2 \, dr \left(1 - \frac{\sin[\xi F_j(r)]}{\xi F_j(r)} \right) \right] \qquad (7.6.16)$$

The last line in equation (7.6.16) is obtained by adding and subtracting $\int r^2 \, dr$ from the integral in the third line. The first integral in the last line of equation (7.6.16) is just $V/4\pi$.

The second integral is solved by using a substitution,

$$Y = \xi F_j = \xi \frac{eZ_p}{r^2}, \qquad r^2 = \xi \frac{eZ_p}{Y}$$

$$dr = -(\xi eZ_p)^{1/2} \frac{dY}{2Y^{3/2}}$$

(7.6.17)

by which the second integral becomes

$$\int_0^\infty r^2 \, dr \left(1 - \frac{\sin[\xi F_j(r)]}{\xi F_j(r)}\right) = \frac{1}{2}(\xi eZ_p)^{3/2} \int_0^\infty \frac{dY}{Y^{5/2}} \left(1 - \frac{\sin Y}{Y}\right)$$

$$= \frac{1}{2}(\xi eZ_p)^{3/2} \times \frac{4(2\pi)^{1/2}}{15}$$

(7.6.18)

The result of the last integral is obtained by several consecutive integrations by parts. Substituting equations (7.6.16) and (7.6.18) into (7.6.15) one gets

$$A(\xi) = \left(1 - \frac{4\pi n_i}{M} \frac{2(2\pi)^{1/2}}{15}(\xi eZ_p)^{3/2}\right)^M$$

(7.6.19)

where we have used the relationship $n_i = M/V$ to replace the volume V by the ion density. Finally, if M is very large, as is usually the case, the right hand side of equation (7.6.19) tends to the exponential function

$$A(\xi) = \exp[-(\xi F_0)^{3/2}]$$

(7.6.20)

where F_0 is the *Holtsmark field strength*,

$$F_0 = 2\pi \left(\frac{4}{15}\right)^{2/3} |eZ_p| n_i^{2/3} \approx 2.603 |eZ_p| n_i^{2/3}$$

(7.6.21)

This field strength deviates only very slightly from the result of the nearest neighbor approximation, equation (7.6.5).

We define the reduced field strength as

$$\beta = \frac{F}{F_0}$$

(7.6.22)

The *Holtsmark distribution function*, $H(\beta)$ for the reduced electric field strength is the inverse Fourier transform of equation (7.6.20),

$$H(\beta) \, d\beta = P_F(F) \frac{dF}{d\beta} \, d\beta$$

$$= \frac{1}{2\pi} F_0 \int_0^\infty d\xi \, \exp(-i\xi F) A(\xi)$$

$$= \frac{2\beta}{\pi} \, d\beta \int_0^\infty dx \, \sin(\beta x) \exp(-x^{3/2})$$

(7.6.23)

There is no analytical solution for the integral in terms of known elementary or special functions; therefore, this integral can be computed only numerically. Its asymptotic expansion for large β (which means large F and large $\Delta\omega$, i.e., line wings) is (Griem, 1974)

$$H(\beta) \sim \frac{2}{\pi} \sum_{n=1}^{\infty} \frac{(-1)^{n+1}}{n!} \Gamma\left(\frac{3n+4}{2}\right) \sin\left(\frac{3n\pi}{4}\right) \beta^{(3n+2)/2}$$

$$= \frac{1.496}{\beta^{5/2}} \left(1 + \frac{5.107}{\beta^{3/2}} + \frac{14.43}{\beta^3} + \cdots\right) \qquad (7.6.24)$$

The leading term is very close to the nearest neighbor approximation, equation (7.6.10), which indicates that the nearest neighbor approximation gives a reasonably correct approximation for the calculation of the Stark line profile in weakly coupled plasmas, particularly at the line wings.

The Corrections of Hooper

The first attempt to include ion–ion correlations in the total electric field distribution was made in a series of papers by C. F. Hooper (1966, 1968) who carried out such calculations to second order. He assumes that the ions are pointlike positive charges that are moving in a uniform neutralizing negatively charged background. His starting point is a Boltzmann distribution for the positive charge distribution function,

$$P(\vec{r}_1, \vec{r}_2, \ldots, \vec{r}_M)\, d^3r_1\, d^3r_2 \cdots d^3r_M = \frac{\exp[(-eV(\vec{r}_1, \vec{r}_2, \ldots, \vec{r}_M)/T)]}{Z(T)}$$

$$\times \frac{1}{V^M}\, d^3r_1\, d^3r_2 \cdots d^3r_M \qquad (7.6.25)$$

where $Z(T)$ is the partition function, equation (1.3.3). Substituting this into equation (7.6.14), and following the same reasoning as above, he found corrections to second order to the Holtsmark distribution. His results are expressed in the form

$$A(\xi) = \exp[-\gamma L^2(\xi) + I_1(\xi) + I_2(\xi)] \qquad (7.6.26)$$

where $L = \xi e/R_i^2$ and R_i is the ion sphere radius. $I_1(\xi)$ and $I_2(\xi)$ are the first and second order corrections, respectively. These functions, as well as the form of γ, are too lengthy to be repeated here, and the interested reader is encouraged to go back to the original papers.

Results from this theory are in remarkably good agreement with Monte-Carlo type calculations, see figure 7.3. It can be shown that Hooper's results approach the Holtsmark distribution for very high temperature and low density plasmas, for which the assumption of uncorrelated plasma is a good approximation. At higher densities or lower temperatures, the particle correlations play an important role and the Hooper functions are the more correct ones.

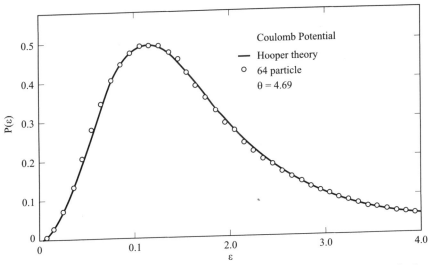

Figure 7.3 Comparison of the electric microfield distribution by Hooper's method (solid line) with the results of a Monte Carlo computation (points). (From Hooper, 1966.)

7.7 Line Broadening: Lyman Series

We end this chapter by applying the above formulas to the very important case of the line widths of the Lyman lines, $n_p p \to 1s$, of hydrogenlike ions (n_p is the principal quantum number). We assume that the plasma consists of one material only, that is, it is not a mixture. The following formulas are taken from the book of Griem; more specifically, they are displayed in equations (490)–(512) in Griem (1974). These formulas also incorporate corrections for effects that were not discussed in our previous sections.

The Stark Line Profile of the Lyman Series

First we give the quasi-static Stark width for these lines. Let

$$\Delta\omega_i = (10\pi\sqrt{3})^{2/3} \frac{\hbar}{Zm} \left(\sum_{\varsigma=0}^{Z} \varsigma^{3/2} N_\varsigma \right)^{2/3} = (10\pi\sqrt{3})^{2/3} \frac{\hbar}{Zm} n_i \overline{(Z^{3/2})}^{2/3} \quad (7.7.1)$$

then the shape of the *Lyman-α* line is proportional to

$$\text{Ly}_\alpha: \quad \mathcal{L}_s(\omega) \propto \begin{cases} 1/(10\Delta\omega_i) = \text{const.} & \text{for } |\Delta\omega| \le \Delta\omega_i \\ (\pi\sqrt{3}/|\Delta\omega|^{5/2})(\hbar/Zm)^{3/2}\overline{Z^{3/2}}n_i & \text{for } |\Delta\omega| \ge \Delta\omega_i \end{cases} \quad (7.7.2)$$

where $\Delta\omega = \omega - \omega_0$ is the frequency difference from the line center. In equations (7.7.1–2), as usual, N_ς is the density of ions of charge state ς, and $n_i \overline{Z^{3/2}} = \sum_{\varsigma=0}^{Z} \varsigma^{3/2} N_\varsigma$. The other symbols have their usual meaning. By the way, $\mathcal{L}_s(\omega)$ in equation (7.7.2) has still to be multiplied by a constant factor to be normalized to unit integral.

For the higher Lyman terms we use a slightly different definition for $\Delta\omega_i$,

$$\Delta\omega_i = 4(n_p^2 - 1)\frac{\hbar}{Zm} n_i (\overline{Z^{3/2}})^{2/3} \tag{7.7.3}$$

where we recall that n_p is the principal quantum number of the upper state. Using this definition, one obtains for $n_p \geq 3$,

$$\text{Ly}_{\beta,\gamma,...}: \quad \mathcal{L}_s(\omega) \propto \begin{cases} 3/(10\Delta\omega_i) = \text{const.} & \text{for } |\Delta\omega| \leq \Delta\omega_i \\ 3(\Delta\omega_i)^{3/2}/10|\Delta\omega|^{5/2} & \text{for } |\Delta\omega| \geq \Delta\omega_i \end{cases} \tag{7.7.4}$$

To account for the radiator–perturber correlations, multiply the N_ζ in the summations by

$$\exp\left[-2\left(\frac{\zeta Z\hbar\,\Delta\omega}{6E_H}\right)^{1/2}\frac{(Z-1)E_H}{T}\right] \tag{7.7.5}$$

before carrying out the averaging over $\zeta^{3/2}$. The other parts of the formula remain unchanged. Equation (7.7.5) corresponds to the Lyman-α only. To adjust it for the Lyman-β and Lyman-γ, replace the "6" in equation (7.7.5) by 12 and 24, respectively. To account for the dynamic effects, multiply the formulas for the Lyman-α by

$$\left(1 - 3 \times 10^{-5}\frac{ZT}{\hbar|\Delta\omega|}\right) \tag{7.7.6}$$

For Lyman-β and Lyman-γ divide the correction term by 2 and 5 respectively.

The Electron Impact Width of the Lyman Series

The electron impact line profile of the Lyman lines has a Lorentzian form,

$$\mathcal{L}_e(\omega) = \frac{w}{\pi}\frac{1}{(\Delta\omega)^2 + w^2} \tag{7.7.7}$$

The line width in this case has a rather complicated shape,

$$w \approx \pi n_e \left(\frac{2m}{\pi T}\right)^{1/2}\left(\frac{\hbar}{Zm}\right)^2 (A + B + C) \tag{7.7.8}$$

In this equation, n_e is the electron density, A, B, and C are given below, and the other symbols have their usual meaning.

In equation (7.7.8), A is a term that corresponds to the line broadening by the weak collisions of the electrons with large impact parameter,

$$A = (3n_p^4 - 9n_p^2)\log\left[\frac{T^{3/2}}{(Z-1)E_H^{1/2}\hbar\omega_c}\left(1 + \frac{Tn_p^2}{Z(Z-1)E_H}\right)^{-1}\right] \tag{7.7.9}$$

In this formula, ω_c is the Debye or Lewis cutoff frequency,

$$\omega_c = \max(|\Delta\omega|, \omega_p, \omega_F, \Delta\omega_i) \tag{7.7.10}$$

where $\Delta\omega_i$ is given in equations (7.7.1) and (7.7.3), $\omega_p = (4\pi n_e e^2/m)^{1/2}$ is the plasma frequency, and $\omega_F = 2Z^2\alpha^2\omega_0/3n_p$ (here α is the fine structure constant).
   The B term in equation (7.7.8) stems for the strong collisions with electrons,

$$B = n_p^4\left[1 + \frac{2T}{E_H}\left(1 + \frac{2T}{E_H} + \frac{Z^2}{n_p^4}\right)^{-1}\right] \tag{7.7.11}$$

Finally, the C term is due to inelastic scatterings,

$$C = \left(\frac{n_p^4}{3} + \frac{17n_p^2}{3}\right)\log\left[1.4 + \left(\frac{T}{E_H}\right)^{3/2}\frac{n_p^3}{Z^2(Z-1)}\left(1 + \frac{Tn_p^2}{Z(Z-1)E_H}\right)^{-1}\right] \tag{7.7.12}$$

The total line shape is obtained by the combination of the electron impact line profile, equation (7.7.7), with the Stark profile, equation (7.7.2) or (7.7.4). In most cases, however, only one of these two broadening mechanisms determines the line profile.

8

Experimental Considerations: Plasma Diagnostics

The most important diagnostic tool of hot plasmas is perhaps the emission spectrum, which carries information about the local instantaneous plasma temperature and density conditions. For short lifetime plasmas such as plasmas generated by nanosecond or femtosecond lasers or by particle beams, this is the only reliable means of studying the plasma evolution. This chapter is devoted to the description of some of the methods that use the emission spectrum for purposes of plasma diagnostics. We will not give a full account of the instrumental part, but instead emphasize the *principles* underlying these methods. From this point of view, this chapter would perhaps best be called the *theory of experimental plasma spectroscopy.*

8.1 Measurements of the Continuous Spectrum

Spectroscopy of the Continuum Radiation

Several experimental techniques take advantage of the fact that the continuous spectrum has an exponentially decreasing structure

$$P_{f-f}, P_{f-b} \propto \exp\left(-\frac{\hbar\omega}{T_e}\right) \tag{8.1.1}$$

whose slope depends on the local electron temperature, see equations (6.1.1) and (6.1.3). From a semilog plot of the continuous spectrum one can, therefore, immediately infer the electron temperature, see below. This diagnostic is very convenient in fully ionized plasmas, and particularly in high temperature hydrogen plasmas, which have no line spectra. These methods are less convenient in plasmas consisting of partially ionized species, in which the emission lines distort the simple structure of the continuous spectrum.

An experimental spectrum from a polyethylene surface is shown in figure 8.1. Above the carbon Lyman-α line the spectrum is exponentially decreasing with a slope corresponding to $T_e = 170 \pm 20$ eV. Similar behavior is routinely seen above the spectral region of the lines.

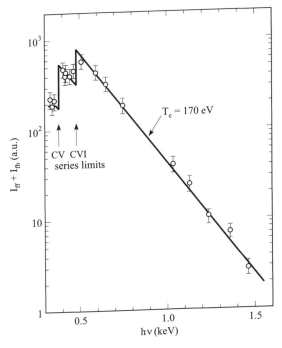

Figure 8.1 The continuum emission spectrum from a polythylene surface. (From Galanti and Peacock, 1975.)

The Absorption Foil Technique

A widely used experimental method is the *absorption foil technique*. This is a low accuracy method that gives some space- and time-integrated estimate of the electron temperature in a plasma. The great advantage of this technique is the relative ease of implementing it in an experiment: being based on electronic detectors, it gives prompt results. The experimental setup consists of several x-ray detectors (generally four or more PIN diodes), behind beryllium foils of increasing thickness (see figure 8.2). Aluminum or higher Z material foils (gold or silver) can be used as well, but are not recommended for reasons that will be made clear below.

This technique exploits the fact that the absorption properties of the foils depend on their material and thickness. The mean free path (m.f.p.) of a photon of energy $\hbar\omega$ in the foil is $\lambda_{mfp}(\hbar\omega) = 1/n\sigma(\hbar\omega)$, where n is the density of the atoms in the foil, and $\sigma(\hbar\omega)$ is the absorption cross section. Let D be the foil thickness (generally a few μm), then all the photons whose energy, $\hbar\omega$, is such that

$$\frac{D}{\lambda_{mfp}(\hbar\omega)} = n\sigma(\hbar\omega)D > 1 \qquad (8.1.2)$$

will, with high probability, be absorbed in the foil and not reach the detector.

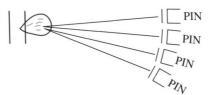

Figure 8.2 Schematic description of the absorption foil technique.

The absorption cross section and m.f.p. of beryllium are shown in figure 8.3. It has an edge structure at the energies corresponding to the ionization energies of the various atomic shells, with the *K-edge* at $\hbar\omega = 118.36\,\text{eV}$ being the highest. Beyond the edges, the absorption coefficient is a monotonically and rather steeply decreasing function of the incident photon energy,

$$\sigma(\hbar\omega) = \sigma_0 \left(\frac{\hbar\omega_0}{\hbar\omega}\right)^{\alpha} \tag{8.1.3}$$

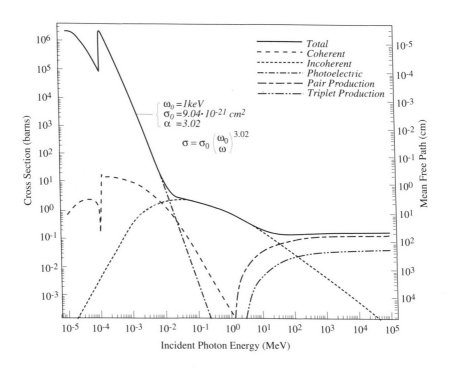

Figure 8.3 Photoabsorption coefficient and mean free path of beryllium. (From Cullen et al., 1989.) (Courtesy of the University of California, Lawrence Livermore National Laboratory and the Department of Energy.)

whose values for beryllium are $\alpha = 3.02$, and $\sigma_0 = 9.04 \times 10^{-21}$ cm^2 if $\hbar\omega$ is measured in keV ($\hbar\omega_0 = 1$ keV). The absorption coefficients of other materials are similar in structure, but with different locations of the edges and a different value of σ_0. It should be noted, however, that the value of the slope, α, is close to 3 for all the materials and all absorption edges.

Inserting equations (8.1.3) into (8.1.2), one finds that low energy photons, up to some *critical energy*, $\hbar\omega_c$, will be absorbed by the foil with high probability, and only higher energy photons will cross the foil and arrive at the detector. This limit between the transmitted and absorbed photons is quite sharp. The critical energy is defined by the condition $n\sigma(\hbar\omega_c)D = 1$, and depends on the foil thickness,

$$\hbar\omega_c(D) = \hbar\omega_0(n\sigma_0 D)^{1/\alpha} \tag{8.1.4}$$

The above discussion suggests that the absorption factor of the photons in the foil can be written to a very good approximation as

$$\exp[-n\sigma(\hbar\omega)D] \approx \begin{cases} 1 & \hbar\omega \geq \hbar\omega_c \\ 0 & \hbar\omega < \hbar\omega_c \end{cases} \tag{8.1.5}$$

The radiation intensity impinging on the detectors is

$$\begin{aligned} I(D; T_e) &= \int_0^\infty P_c(\hbar\omega)\, \exp[-n\sigma(\hbar\omega)D]\, d(\hbar\omega) \\ &= P_0 \int_{\hbar\omega_c(D)}^\infty \exp\left(-\frac{\hbar\omega}{T_e}\right) d(\hbar\omega) \\ &= P_0 \exp\left(-\frac{\hbar\omega_c(D)}{T_e}\right) \end{aligned} \tag{8.1.6}$$

In the first line of equation (8.1.6), $P_c(\hbar\omega)$ is the total, free-free + free-bound, continuous spectrum, expressed in the form of equation (8.1.1). In the second line, P_0 is the coefficient of the exponential factor of the continuous spectrum, and we have used the approximation in equation (8.1.5) to convert the integration over the complicated exponential factor into a lower limit of the integration. Equation (8.1.6) suggests that if one plots the pulse height, $I(D; T_e)$, in the detectors as a function of the critical energy, $\hbar\omega_c(D)$, on a semilog scale, the slope will give the reciprocal of the electron temperature.

The absorption foil technique is widely used to find the temperatures of low-Z and fully ionized plasmas. For high-Z plasmas the emission spectrum has not only the continuous but also the line component, and the interpretation of the results may become more complicated as the penetration of the line spectrum through the foils must also be accounted for. To avoid this difficulty, the thicknesses of the foils are adjusted so that the critical energies are all above the spectral line energies.

A different type of difficulty of this technique is the penetration of radiation through the transparency "windows" below the K-edge. This may be a problem when the absorbing foil is aluminum or other high-Z material, whose K-edge is in the keV region. In this case a full calculation of the first integral in equation (8.1.6) must be carried out. Such calculations can provide very good agreement with the experimental results (see figure 8.4).

8.2 Measurements of the Line Spectrum

The line spectrum provides much more flexibility for plasma diagnostics than the continuous spectrum. The wealth of the emission lines in the spectrum enables the application of more sophisticated and more accurate methods for the determination of the plasma temperature and density. Only the standard and most frequently used methods for plasma diagnostics will be reviewed below. The more tricky and intriguing methods, which have been used in some of the more interesting experiments, must, unfortunately, be left out.

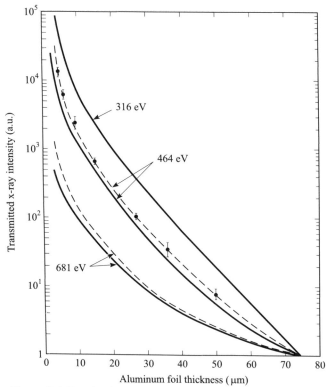

Figure 8.4 Results from an absorption foil experiment. Solid circles (●) represent experimental data. The dashed and solid lines are computational results for $n_i = 10^{20}$ and $n_i = 10^{21}$ cm^{-3}, respectively. (Reprinted with permission from Gupta, P. D., Popil, R., Fedoseyevs, R., Offenberger, A. A., Salzmann, D., and Capjack, C. E., 1986, *Appl. Phys. Lett.*, **48**, 103. Copyright 1986 American Institute of Physics.)

X-Ray Spectrometers

The standard instrument for measuring the x-ray spectrum is the *spectrometer* or spectrograph. The simplest of these is the *crystal spectrometer* (figure 8.5). It is based on Bragg's law, which states that if a beam of x-rays impinges on a crystal at an angle θ, only components whose wavelength satisfy the Bragg condition,

$$n_B\lambda = 2d \sin \theta \qquad (8.2.1)$$

will undergo constructive interference and have an efficient specular reflection from the crystal. In equation (8.2.1), n_B is the order of the dispersion, and d is the lattice spacing of the crystal. From equation (8.2.1), it follows that the largest wavelength that is measurable by means of a given crystal is $2d$. For a parallel beam of x-rays, the crystal selects only one of the wavelengths, namely the one that complies with equation (8.2.1). From a point source, however, the rays arrive at the crystal at various angles, and the crystal disperses the wavelengths according to the angles of arrival. One therefore gets the whole spectrum, see figure 8.5a.

Crystal spectrometers were initially used in the hard x-ray region. Later, using crystals of large lattice spacing, this technique became applicable, with reasonably good resolution, for soft x-rays of wavelengths as long as 25 Å. Using soap-film crystals and other more sophisticated microstructures, the range of applicability of this technique was extended to 100 Å.

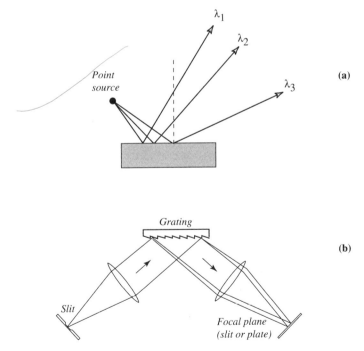

Figure 8.5 The principle of (a) flat crystal and (b) grating spectrometers.

Photons with the same wavelength from the various parts of the plasma impinge on the crystal at various transverse angles, and are dispersed to different transverse locations on the photographic film. Thus, the dispersion in the transverse direction—perpendicular to the spectral dispersion direction—is proportional to the intensity of the x-rays emitted from the various portions of the plasma. The crystal spectrometer therefore also gives some spatial resolution of the diagnosed plasma.

As its name indicates, the *grating spectrometer* uses a grating as the dispersive element (figure 8.5b). The condition for constructive interference is

$$n_B \lambda = d(\sin \alpha + \sin \beta) \tag{8.2.2}$$

where d is the separation between the grooves, and α and β are the angles of incidence and reflection, respectively. These angles satisfy the relationship $\alpha - \phi = \phi - \beta$, where ϕ is the blaze angle of the grating. The grating spectrometer is efficient for long wavelength radiation, and can be applied in the vacuum ultraviolet region (VUV), above 100 Å, where crystal spectrometers are no longer available. The resolving power of the grating spectrometer is $\mathcal{R} = \lambda / \Delta \lambda = M n_B$, where $\Delta \lambda$ is the wavelength difference between two lines which are on the limit of resolvability, and M is the number of illuminated grooves.

The Simplest Way to Estimate the Plasma Temperature

Perhaps the simplest way to get some information about the plasma temperature is by inspection of the charge states whose emission lines take part in the spectrum. As each charge state has its maximum abundance in a rather limited temperature region, such an initial observation of the line spectrum immediately gives an estimate, albeit of very low accuracy, of the plasma temperature. For instance, if one generates an aluminum plasma and finds spectral lines of heliumlike ions but no lines from hydrogenlike species, one can safely assume that the temperature of the plasma is between 100 and 400 eV. This is because the heliumlike ions have nonzero abundance above approximately 100 eV, while hydrogenlike ions start to show up only above 400 eV (see figures 5.1 and 5.3). Such simple considerations are sometimes useful for getting an initial low accuracy estimate of the plasma temperature.

Plasma Temperature Measurements from Line Intensity Ratios

The standard spectroscopic method consists of measuring the ratio of intensities of two lines within the same series of a given charge state, ζ, such as the ratio of the Lyman-α and β lines, or the He-α and He-β lines. From equation (6.2.1), one finds that this ratio is given by

$$\frac{P(m' \to m)}{P(m'' \to m)} = \frac{N_{\zeta,m'}}{N_{\zeta,m''}} \frac{\hbar \omega_{m' \to m}}{\hbar \omega_{m'' \to m}} \frac{A(m' \to m)}{A(m'' \to m)} \tag{8.2.3}$$

where we recall that $P(m' \to m)$ and $P(m'' \to m)$ are the experimental intensities of the two lines, whose initial excited states are m' and m'' (we omit the notation of the charge state, ζ, since it is the same for all three states m, m', and m''), $N_{\zeta,m'}$ and $N_{\zeta,m''}$ are the densities of the initial excited states, $\hbar\omega_{m' \to m}$ and $\hbar\omega_{m'' \to m}$ are the transitions energies, and $A(m' \to m)$ and $A(m'' \to m)$ are the corresponding Einstein coefficients. In most cases, the lower state is the ground state, $m = 0$, but sometimes transitions to an excited state, such as the Balmer series (lower state $= 2s$), are used as well. The ratio of the intensities on the left hand side of equation (8.2.3) is the quantity measured in the experiment; the values of the Einstein coefficients and the transition energies are found in the literature. By inserting these into equation (8.2.3), one finds the ratio of the partial densities of the initial excited states, $N_{\zeta,m'}/N_{\zeta,m''}$, which depends on the plasma temperature and density, as explained in chapter 5.

If the plasma is in the LTE condition, the ratio of the densities of the two initial states is determined by the Boltzmann distribution,

$$\frac{N_{\zeta,m'}}{N_{\zeta,m''}} = \frac{g_{\zeta,m'}}{g_{\zeta,m''}} \exp\left(-\frac{E_{\zeta,m'} - E_{\zeta,m''}}{T_e}\right) \tag{8.2.4}$$

Inserting this into equation (8.2.3) provides an equation from which T_e can be solved for directly. In low density plasmas, where corona equilibrium prevails, this density ratio is

$$\frac{N_{\zeta,m'}}{N_{\zeta,m''}} = \frac{\nu_{\zeta,m''}(T_e)}{\nu_{\zeta,m'}(T_e)} \tag{8.2.5}$$

where the ν are defined in equations (5.5.5–8) in terms of the rate coefficients of the ionic processes in the plasma. Inserting equation (8.2.5) into equation (8.2.3) results in an equation, whose left hand side is measured in the experiment, and whose right hand side depends on the electron temperature.

If the plasma is in an intermediate condition, in which neither LTE nor CE are valid, one has to apply the whole apparatus of the collisional–radiative model to find the temperature and density conditions that reproduce the experimentally measured line intensity ratio. It may happen that a whole contour of temperatures and densities may be in agreement with the experimental values, in which case a comparison with a third line or even more lines is necessary to resolve the ambiguity.

In principle, one can also measure the intensity ratios of lines emitted from two different charge states. For instance, one may measure the ratio of the intensities of the Lyman-α and the He-α lines, which are, in general, the strongest and most easily measurable lines. Such a comparison, however, can be valid only for long lifetime plasmas that are under constant temperature and density conditions for periods that are much longer than the times at which these parameters are changing. For plasmas fulfilling such conditions, one can safely assume that the bulk of the spectral intensities was emitted while the plasma was in constant temperature and density conditions. Such an assumption is not valid for short lifetime plasmas, such as laser generated ones, in which at any given instant the heliumlike and hydrogenlike lines are emitted from different spatial portions of the plasma. The lines from the hydrogenlike species are emitted from regions that are at a higher temperature than the lines from heliumlike ions. In addition, at every given point in the plasma, the spectral lines from hydrogenlike species are emitted at different times than those from heliumlike ions, according to the variation in time of the local temperature. In plasmas with steep temperature gradients and/or rapid time variations, the comparison of the intensities of lines from two different charge states may lead to incorrect results.

We conclude this section by showing in figures 8.6, 8.7, and 8.8 several realistic x-ray spectra that have been used for the diagnostics of plasmas. Figure 8.6 shows the densitometer trace of the solar flare spectrum, recorded by the NRL slitless spectrograph on Skylab in the 200–250 Å region. Analysis of the spectrum enabled the identification of the presence of a large number of ions in the solar atmosphere and their relative abundances (Feldman, 1992).

Figure 8.7 shows spectra of highly ionized gold and tantalum plasmas in the 4–6 Å spectral region. The raw spectra exhibit the complex nature of the emission spectrum of these high-Z materials, which under the experimental conditions includes mainly lines of Ni-, Cu-, and Zn-like ions. The analysis of such complex spectra was carried out by comparing the experimental spectra to synthetic simulational spectra. Such a comparison for gold showed the best fit when the electron temperature in the simulation was 240 eV.

Figure 8.8 shows argon spectra measured from an indirectly driven laser-irradiated capsule, after background subtraction. The time- and space-integrated plasma temperature was inferred from the line intensities, whereas the line shapes were used to calculate the plasma density, see section 8.5 below. In fact, from the ratio of the intensities of the $Ly_\beta/He_\beta = 0.4$ an electron temperature $T_e = 1000 \pm 50$ eV was found, and from the width of the He_β line an ion density of $n_i = (1.2 \pm 0.3) \times 10^{24}$ cm^{-3} was inferred.

Satellite-to-Parent Line Intensities

In a large number of experiments the intensity ratio of the satellite to the parent line was used to measure the plasma temperature. In section 6.3 we discussed in detail the origin of the satellites and their intensities. Equation (6.3.3), reproduced here for the reader's convenience, is an explicit formula for the ratio of the satellite/parent line intensities,

$$\frac{P_s(\text{satellite})}{P(\text{line } w)} = 4.48 \times 10^{-10} \left(\frac{1}{\text{eV}^2}\right) \frac{g_\beta}{g_\alpha} e^{(E_\pi - E_\beta)/T} \frac{E_\pi^3}{T}$$

$$\times \frac{Au(\beta \to \alpha)}{Au(\beta \to \alpha) + A_{\pi \to \alpha}} \frac{1}{G(E_\pi/T)(1d)} \qquad (6.3.3)$$

The notation is explained in section 6.3. The left hand side of this equation incorporates the satellite/parent intensity ratio, which is measured from the emission spectrum. On the right hand side of the equation, only the electron temperature, T_e, is unknown, whereas all the other terms are constant parameters,

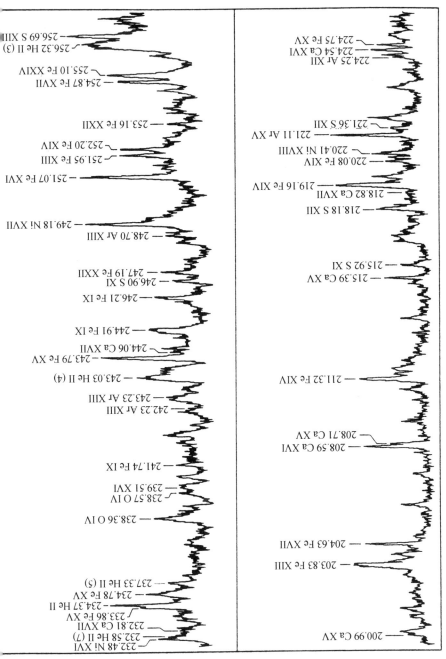

Figure 8.6 Densitometer trace of the solar flare spectrum recorded by the NRL slitless spectrograph on Skylab. (From Feldman et al., 1992.)

Wavelength (Å)

Photographic density

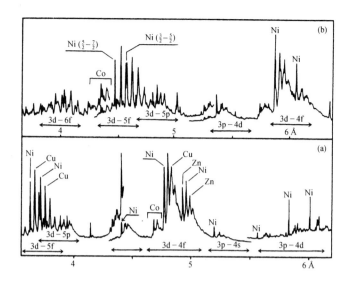

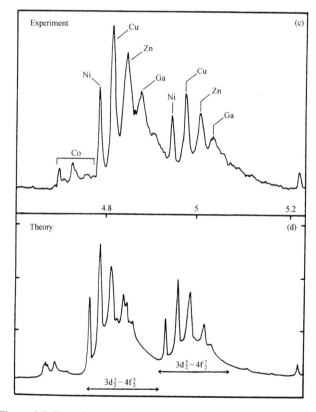

Figure 8.7 Experimental gold (a), and tantalum (b) spectra from laser-produced plasmas. Comparison of the 4.7–5.2 Å region of the spectrum from the gold plasma (c) with computational results (d) yields the best fit for a 240 eV ionization temperature. (From Gauthier et al., 1986.)

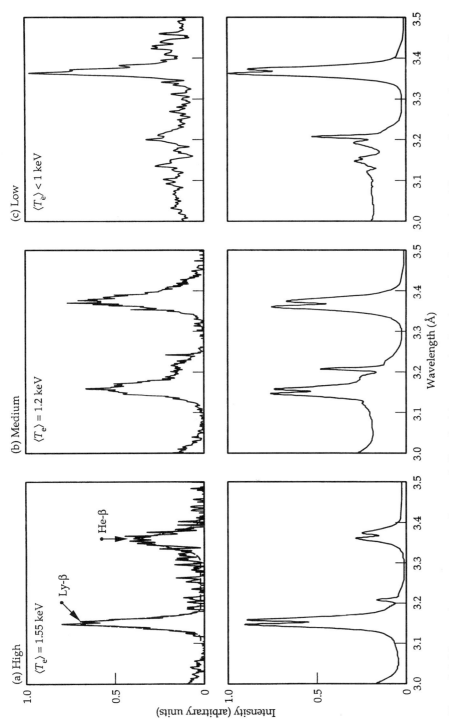

Figure 8.8 Measured argon spectrum from an indirectly driven capsule. Continuum is subtracted and lines are fitted to determine line widths and intensity ratios. (From Hammel et al., 1993.) (Courtesy of the University of California, Lawrence Livermore National Laboratory and the Department of Energy.)

independent of the plasma conditions, which can be found in the literature or computed separately. Equation (6.3.3) can be regarded, therefore, as an equation from which the electron temperature can be inferred. In fact, the solar temperature in figure 6.5 was deduced from the spectrum by means of such a method. In a similar manner, the experimental plasma temperature was inferred from figure 6.4 using equation (6.3.3).

8.3 Space-Resolved Plasma Diagnostics

All the techniques described above measure a space- and time-averaged temperature or density. For some purposes, however, one is interested in obtaining information about the spatial distribution or the temporal evolution of the plasma temperature or density. In this section we discuss several of the most frequently used methods to obtain space-resolved information by using the plasma radiation.

The Pinhole Camera

One of the techniques for obtaining a spatially resolved x-ray picture of the plasma is the *pinhole camera* (figure 8.9). The principle of the operation of such a camera is quite simple: The plasma radiation goes through a pinhole whose diameter is much smaller than the plasma dimensions but larger than the thickness of the surrounding foil. A $25\,\mu$m diameter hole in a $10\,\mu$m thick aluminum foil are the approximate dimensions used in the standard cameras. A detector, which is either a photographic film or (more frequently) a charge-coupled device (CCD) or a microchannel plate detector (MCP), is placed behind the foil and records the radiation penetrating through the pinhole. To prevent visible and other long wavelength radiation from reaching the detector, the pinhole is covered by a thin foil of $\sim 10\,\mu$m beryllium or $\sim 1\,\mu$m aluminum.

Each point on the detector is exposed to the radiation only from a small portion of the plasma. The spatial resolution of the detector is determined by the diameter of the pinhole: The smaller is the hole, the better is the resolution. On the other hand, a pinhole of too small a diameter reduces the intensity of the radiation impinging on the detector.

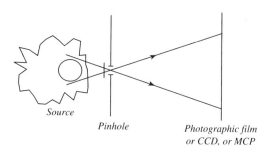

Source

Pinhole

Photographic film
or CCD, or MCP

Figure 8.9 The principle of the pinhole camera.

The magnification, $\mathcal{M}$, of the camera is determined by the ratio

$$\mathcal{M} = \frac{\text{pinhole} - \text{detector distance}}{\text{pinhole} - \text{plasma distance}} \tag{8.3.1}$$

In the simplest form of the procedure, one plots the contours of equal intensity arriving at the detector. These contours provide information about the (time-integrated) spatial distribution of the plasma temperature and density. Figure 8.10 is an example showing the raw images of a laser-compressed deuterium pellet at five different phases of the compression and the contours of equal intensity (Hammel et al., 1992).

By using two or more pinholes, with foils of different thicknesses, one obtains a *space-resolved absorption foil system*. Using equation (8.1.6) for the data processing, one can obtain a space-resolved plasma temperature distribution (time integrated) inside the plasma. The apparatus by which figure 8.10 was obtained also has some time resolution capability. The five pictures in the sequence are, in fact, taken at different times, about 60 ps apart. The figures show the compression of a pellet under the irradiation of a high intensity laser (the first four pictures), and its disassembling in the final image.

A different method of using x-ray pinhole cameras, with a CCD recording device, is shown in figures 8.11 and 8.12. Figure 8.11 is a schematic description of the experimental setup used in the Institute of Laser Engineering (ILE) GEKKO XII system in a laser-induced compression equipment. Three pinhole cameras at angles of 120° from each other viewed the compressed pellet at the

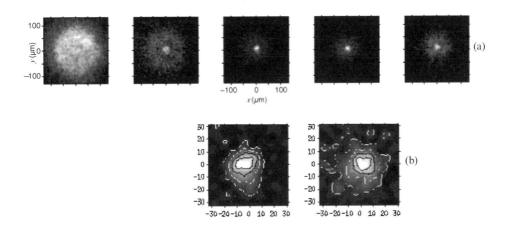

Figure 8.10 (a) Sequence of five pinhole images of a constricting deuterium plasma. The temporal interframe spacing is ≈ 60 ps. (b) The equal-emission contours at peak compression time on an enlarged scale. (From Hammel et al., 1992.) (Courtesy of the University of California, Lawrence Livermore National Laboratory and the Department of Energy.)

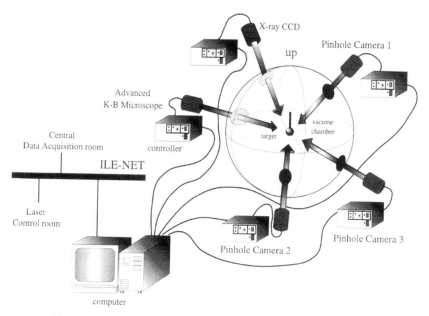

Figure 8.11 Experimental setup for a three pinhole camera imaging system. (From Matsushita et al., 1995.)

moment of peak compression. By using processing methods borrowed from tomography, a three-dimensional picture of the x-ray emitting regions has been reconstructed (figure 8.12). The constriction of the pellet and the radiation from the outward-propagating halo can be seen clearly. These figures indicate how powerful the pinhole camera techniques can be for plasma diagnostics purposes.

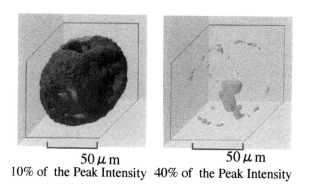

50 μm   50 μm
10% of the Peak Intensity 40% of the Peak Intensity

Figure 8.12 Reconstructed 3D image. (From Matsushita et al., 1995.)

Space-Resolved Spectroscopy

Space-resolved spectroscopy is based on the fact that spectrometers disperse the incident spectrum in one direction and the spatial image of the source in the perpendicular direction. A representative example of a space-resolved spectrum from a laser-produced aluminum plasma is shown in figure 8.13. The bending of the spectral lines is a geometrical effect of the spectrometer. Such spectra display the regions in the plasma wherefrom the various lines are emitted. From figure 8.13 it can be seen that some lines are emitted only from the dense parts of the plasma whereas others are emitted also from the low density plume. From the line intensity ratios as function of the distance, one can infer a space-resolved mapping of the temperature in the plasma.

A more sophisticated method is based on the fact that curved crystals can be used as selective imaging optics for the diagnostics of laser-produced plasmas. A full explanation of the principle of operation would take us too far away from the mainstream of this book; a full description can be found in Förster (1991). In a very interesting experiment carried out by a German–Japanese collaboration (Vollbrecht et al., 1997), a five-channel x-ray monochromatic camera was applied to observe the compression of a fusion pellet. The pellet consisted of spherical "pusher" enriched with chlorine and filled with deuterium mixed with a small amount (0.3%) of argon. The results are shown in figure 8.14. Figure 8.14a and b shows the images of two chlorine lines (Ly_β and 3.631 Å and He_β at 3.800 Å) and two argon lines (Ly_β at 3.275 Å and He_β at 3.366 Å), as well as the image of the continuum around 3.294 Å. The images in figure 8.14a show a symmetric compression. The second set, figure 8.14b, was obtained in a different experiment, and exhibits strong inhomogeneities. In the last set of images, figure 8.14c, the spatial distribution of the temperature was inferred from the ratio of the intensities of the argon Ly_β and He_β lines. The map of the intensity ratios was converted into a map of temperatures using the CRSS model.

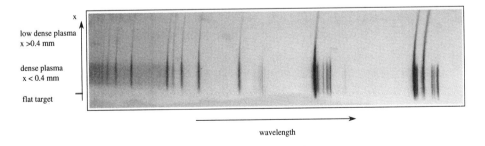

Figure 8.13 Space-resolved spectrum of a laser-produced aluminum plasma. (Reprinted from Förster, E., Fill, E. E., Gäbel, K., He, H., Missalla, T., Renner, O., Uschmann, I., and Wark, J., 1994, *J. Quant. Spectrosc. Rad. Trans.*, **51**, 101, with permission from Elsevier Science Ltd.)

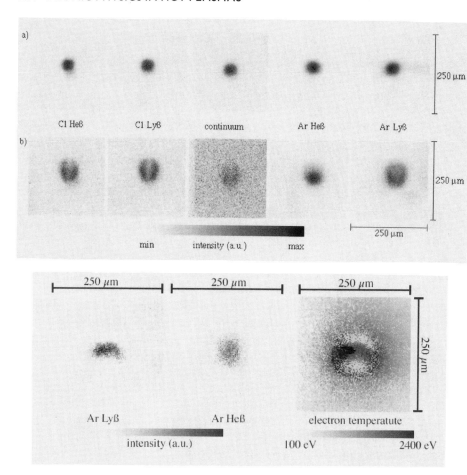

Figure 8.14 Space-resolved temperature measurement. (Reprinted from Vollbrecht, M., Uschmann, I., Förster, E., Fujita, K., Ochi, Y., Nishimura, H., and Mima, K., 1988, *J. Quant. Spectrosc. Rad. Trans.*, in press, with permission from Elsevier Science Ltd.)

8.4 Time-Resolved Spectra

The techniques for the measurement of time-resolved spectra are divided into two groups according to the plasma lifetime. The emission spectra of long lived plasmas, with lifetimes longer than 10^{-8} s are measured, in general, by commercially available photosensitive devices such as photodiodes or photomultipliers. As the speed of ordinary electronic circuits is of the order of 10^{-9} s, such devices are fast enough to provide a sufficiently accurate time-resolved scan of the evolution of long lived plasmas. However, other methods have to be used for plasmas whose lifetime is of the order of a few nanoseconds or less. The main devices used for such plasmas are the *streak camera* and the *framing camera*, which can provide time resolutions down to several picoseconds (10^{-12} s).

Photodiodes and Photomultipliers—Long Lived Plasmas

Photodiodes are used to record the radiation emerging from long lifetime plasmas for all spectral regions: visible, near UV, extreme UV, and x-rays, depending on the nature of the analysed plasma. The recording device is, in most cases, an oscillograph or magnetic memory. The pulse shape gives information about the evolution of the plasma temperature, the period of persistence of a constant temperature and density inside the plasma, and the plasma lifetime.

The most popular technique measures the time evolution of several spectral lines in the plasma. In this technique the photodiodes are placed on the focal plane of a spectrometer, at locations into which some spectrally interesting lines are emitted. From the ratio of the temporal shape of the pulse heights in the diodes, one can then unfold the evolution in time of the plasma temperature by using the methods of section 8.2.

An example of such an experiment is shown in figure 8.15. In this experiment a gaseous plasma source was investigated for purposes of plasma opening switches. CH_4 was used as the plasma source, seeded with a small amount of lithium for diagnostics purposes. The emerging radiation was observed by means of a spectrograph in the visible region. A fiber bundle array was used to transfer the collected light from the spectrograph into photomultiplier tubes. The temporal behavior of the emission lines from the neutral lithium LiI $2p$–$2s$

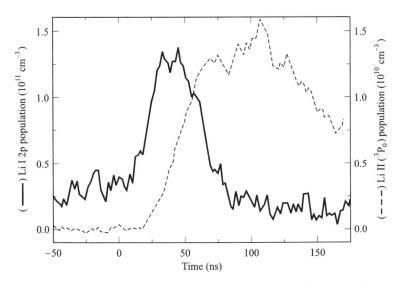

Figure 8.15 Time-resolved spectrum for the observation of the evolution of the charge state distribution. (Reprinted with permission from Sarfaty, M., Maron, Y., Krasik, Ya. E., Weingarten, A., Arad, R., Shpitalnik, R., Fruchtman, A., and Alexiou, S., 1995, *Phys. Plasmas*, **2**, 2122. Copyright 1995 American Institute of Physics.)

at 6708 Å, and LiII $2p(^3P^0)$–$2s(^3S)$ at 5485 Å was recorded. From the absolute intensity of these lines, the population densities of neutral and singly ionized lithium was inferred.

The Principle of Operation of the Streak Camera

Photodiodes cannot give time resolutions better than 10^{-9} s or so. To resolve the time evolution of plasmas of nanosecond or shorter lifetimes, one needs a streak camera.

The principle of operation of the streak camera is similar to that of an oscillograph. We shall first describe the camera for visible light, and then explain the differences found in the x-ray streak camera. The light from the plasma source falls through a narrow slit onto a photocathode (figure 8.16). The electrons emitted from the photocathode are accelerated through a very high potential difference, about 20–40 kV, to the anode. Due to the high voltage, the electron beam, whose initial shape is rectangular, keeps it original shape when hitting the anode with very little blurring effect.

On their way toward the anode, the electrons pass between two deflecting electrodes. A linear voltage ramp, which grows at a rate of about 20 kV in a 1 ns time period, is applied to these electrodes. Electrons arriving in the region between the electrodes undergo different deflections, depending on the instantaneous value of the deflecting voltage. Electrons at early times go through with small deflection, whereas electrons at later times are more and more deflected as the applied voltage increases. The variation in time of the intensity of the incident radiation is, therefore, imaged slice by slice in the vertical direction. The intensity in the horizontal direction is proportional to the spatial distribution of the emitted radiation, whereas the intensity in the vertical direction is proportional to the

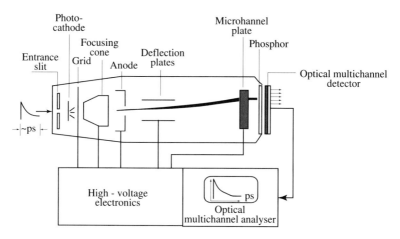

Figure 8.16 The principle of operation of the streak camera.

development in time. A microchannel plate is used as an image intensifier to increase the intensity on the anode.

The amplified radiation impinges on a phosphorescent material that emits radiation in the visible in proportion to the number of electrons hitting the anode. A detector, which is a photographic plate or a CCD device or an optical multichannel analyser (OMA), is used to record the radiation from the phosphor.

The x-ray streak camera works on the same principles as that in the visible. The difference stems from the different mean free paths of x-ray and visible photons in the cathode. To get an efficient photons-to-electrons conversion, the photons in an x-ray streak camera have to hit the photocathode sideways. Other parts of the camera are the same as for visible photons.

Streak cameras in the visible, particularly suitable for measuring the evolution of low temperature hydrogen plasmas, are commercially available with a 2 ps time resolution. Time resolutions of x-ray streak cameras are not quite as good, about 10 ps, but 0.5 ps cameras are being designed.

For *time-resolved spectroscopy*, an x-ray streak camera is located at the focal plane of an x-ray spectrometer. The line intensities are recorded as a function of time. One can follow the time evolution of the ratio of intensities of two lines, and therefrom calculate the evolution of the temperature and density in the plasma. The interpretation of a streak camera picture is never simple. An example of a time-resolved spectrum is shown in figure 8.17. Such pictures have, in general, low resolution in both the temporal and the spectral directions. Nonetheless, they provide an insight into the time evolution of a plasma that cannot be obtained by any other means.

Another example of time-resolved diagnostics is displayed in figures 8.18 and 8.19. Figure 8.18 describes the experimental setup, which incorporates a crystal spectrometer in conjunction with an x-ray streak camera (Ochi et al., 1995). The

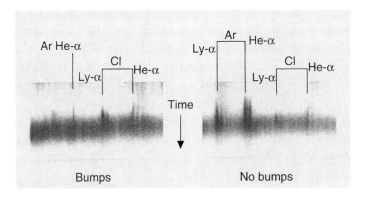

Figure 8.17 Steak record of chlorine-doped laser-heated capsule. (Reprinted from Hammel, B. A., Keane, C. J., Dittrich, T. R., Kania, D. R., Kilkenny, J. D., Lee, R. W., and Levedahl, W. K., 1994, *J. Quant. Spectrosc. Rad. Trans.*, **51**, 113, with permission from Elsevier Science Ltd.)

time evolution of the Lyman-β and He-β, γ lines from a laser-produced argon plasma are shown in figure 8.19. The periods of emission of the various excited species in the plasma are seen from the figure. From the ratio of the intensities at a given moment (which is the ratio along a horizontal line in the figure) one can derive the instantaneous temperature in the plasma.

The development of time- and/or space-resolved plasma diagnostics has undergone great advances in recent years, and more is expected in the coming ones. It must be pointed out, however, that the more advanced space- or time-resolved techniques are all based on the detection of the radiation emitted from the plasma, since this radiation is the only agent arriving at the outside detectors that carries information about the instantaneous plasma conditions in real time.

8.5 The Line Width

All the methods reviewed so far, which measure the slope of the continuous spectrum or the intensities of the spectral lines, characterize mainly the plasma temperature, and to a lesser extent the plasma density. For the measurement of the plasma density, one is better off using a different parameter. The property that is most affected by the plasma density is the width of the spectral lines, particularly in high density plasmas. In such plasmas the lines' widths are determined by the electron impact or the quasi-static broadenings caused by the local electron and ion microfields. The relevant formulas have been developed in chapter 7. These formulas can be used to find the plasma density from the measured line width.

We will not repeat here the formulas for the computation of the line width. Instead, we describe a few experiments in which this technique has been used to estimate the plasma density. An example has already been shown in figure 8.8 in which the line intensity ratios were used to find the local temperature and the line widths were used to estimate the local density in the plasma. Two more experiments are shown in figures 8.20 and 8.21. In figure 8.20 the $n = 4 \rightarrow n = 5$ emission line in a triply ionized (Li-like) carbon plasma is analyzed. The plasma was generated by a gas-liner pinch discharge. The driver gas was hydrogen with small amounts of methane, nitrogen and carbon dioxide as test gases. The line widths were measured at two times: at the moment of maximum pinch, and

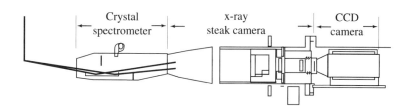

Figure 8.18 Time-resolved spectrograph. (From Ochi et al., 1995.)

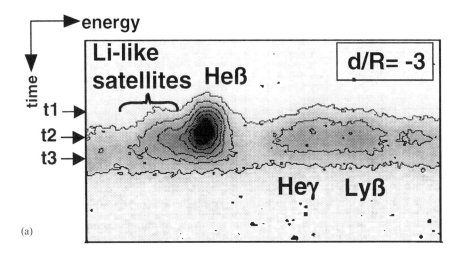

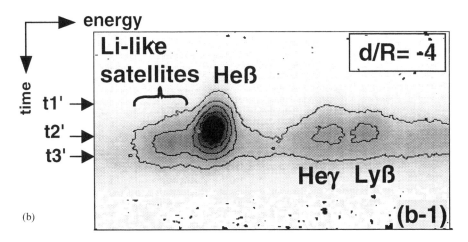

Figure 8.19 Time evolution of the Ly_β, He_β, and He_γ lines in a laser-produced argon plasma. (From Ochi et al., 1995.)

150 ns later. The temperatures were obtained by measuring two ratios of line intensities: $(2p_{1/2}-2p_{3/2})(/(2s_{1/2}-2p_{1/2})$ and $(3s_{1/2}-3p_{3/2})/(3s_{1/2}-3p_{1/2})$. The theoretical line profiles were simulated for several electron densities and the best fit with the experimental line profile was obtained for $n_e = 1.81 \times 10^{18}\,\mathrm{cm}^{-3}$ at the moment of the maximum pinch and slightly less, $n_e = 1.05 \times 10^{18}\,\mathrm{cm}^{-3}$, 150 ns later.

A similar experiment is shown in figure 8.21 carried out by the same group. Here the neutral helium 2^3P-3^3D transition line was measured. From the

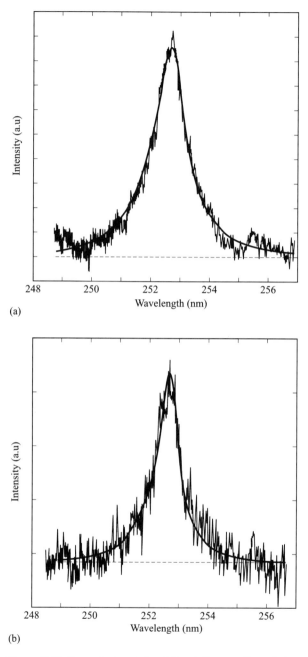

Figure 8.20 Recorded and fitted Stark line profiles of the $n = 5 \rightarrow n = 4$ transition in lithiumlike carbon (in (a) $n_e = 1.81 \times 10^{18}\,\mathrm{cm}^{-3}$, $kT_e = 8.6\,\mathrm{eV}$; in (b) $n_e = 1.05 \times 10^{18}\,\mathrm{cm}^{-3}$, $kT_e = 9.0\,\mathrm{eV}$). The calculated profile incorporates the Doppler and the apparatus profiles. (From Glenzer et al., 1994.)

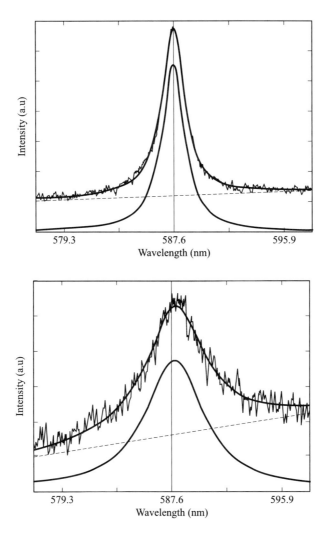

Figure 8.21 Recorded and fitted Stark line profiles of the $2^3P–3^3D$ transition in neutral helium (in (a) $n_e = 0.49 \times 10^{18}$ cm^{-3}, $kT_e = 3.84$ eV; in (b) $n_e = 2.49 \times 10^{18}$ cm^{-3}, $kT = 5.47$ eV). (Reprinted from Büscher, S., Glenzer, S., Wrubel, Th., and Kunze, H.-J., 1995, *J. Quant. Spectrosc. Rad. Trans.*, **54**, 73, with permission from Elsevier Science Ltd.)

comparison of the theoretical and experimental line profiles, electron densities of $n_e = 0.49 \times 10^{18}$ cm^{-3} at electron temperature $T = 3.84$ eV, and $n_e = 2.49 \times 10^{18}$ cm^{-3} at electron temperature $T = 5.47$ eV were deduced.

The Absorption Spectrum and Radiation Transport

9.1 Basic Definitions of the Radiation Field

The radiation field in an infinitesimal volume in space, at any given moment, is characterized by three parameters: (a) the direction of the propagation of the radiation, (b) the energy spectrum of the photons (equivalently, one can characterize the spectrum by means of the frequency, or the wavelength, or the wavenumber), (c) the polarization of the radiation. In the case of unpolarized radiation, which we shall assume in the following, only the first two of these parameters are meaningful. If the radiation is characterized in terms of the photons in the radiation field, one can obtain a full description of this field by means of the *photon distribution function*, which is defined such that

$$f_{ph}(\vec{r}, t; \vec{e}, \hbar\omega) \, d(\hbar\omega) \, d^3r \, d\Omega \qquad (9.1.1)$$

is the number of photons (at a given moment t) in a volume element d^3r around the point $\vec{r}$, whose energies are between $\hbar\omega$ and $\hbar\omega + d(\hbar\omega)$ and which are propagating into the infinitesimal solid angle $d\Omega$ around direction $\vec{e}$ ($|\vec{e}| = 1$). The units of $f_{ph}(\vec{r}, t; \vec{e}, \hbar\omega)$ are photons/(eV sr cm^3), (sr $\equiv$ steradian).

The *spectral radiation intensity*, or *specific intensity*, $I(\vec{r}, t; \vec{e}, \hbar\omega)$, characterizes the *energy* intensity of the radiation field, namely,

$$I(\vec{r}, t; \vec{e}, \hbar\omega) \, d(\hbar\omega) \, dS \, d\Omega \qquad (9.1.2)$$

is the radiative energy around the point $\vec{r}$ whose spectral distribution is within an infinitesimal spectral energy range $[\hbar\omega, \hbar\omega + d(\hbar\omega)]$ and which propagates into a solid angle $d\Omega$ around a direction defined by the unit vector $\vec{e}$ through an area dS perpendicular to the direction of propagation. The units of the radiation intensity are ergs/(cm^2 s sr eV). We have made a distinction between the total energy in the beam, which is measured in ergs (in cgs units), and the spectral region of the photons, which is measured in eV, but of course, these two quantities can be expressed in the same units, in which case the units of $I(\vec{r}, t; \vec{e}, \hbar\omega)$ become simply (cm^2 s sr)$^{-1}$. Other definitions of the specific intensity characterize the radiation intensity in a frequency interval $[\nu, \nu + d\nu]$, or a wavelength interval $[\lambda, \lambda + d\lambda]$, or a wavenumber interval $[k, k + dk]$, with the appropriate changes of the units.

Every photon in the given energy interval carries energy $\hbar\omega$ and moves with the speed of light c. The relation between the spectral intensity and the photon distribution function is, therefore, given by the simple relation

$$I(\vec{r}, t; \vec{e}, \hbar\omega) = \hbar\omega c f_{ph}(\vec{r}, t; \vec{e}, \hbar\omega) \tag{9.1.3}$$

The *spectral radiant energy density*, $U(\vec{r}, t; \hbar\omega)$ is the amount of radiant energy in the above energy range and spatial volume, regardless of the direction of the photon propagation. This energy density is obtained by integrating $f_{ph}(\vec{r}, t; \vec{e}, \hbar\omega)$ or $I(\vec{r}, t; \vec{e}, \hbar\omega)$ over all possible directions of the photon paths,

$$U(\vec{r}, t; \hbar\omega) = \hbar\omega \int_{4\pi} f_{ph}(\vec{r}, t; \vec{e}, \hbar\omega)\, d\Omega = \frac{1}{c} \int_{4\pi} I(\vec{r}, t; \vec{e}, \hbar\omega)\, d\Omega \tag{9.1.4}$$

$U(\vec{r}, t; \hbar\omega)$ is measured in units of erg/(cm^3 eV). Finally, imagine a unit area with normal unit vector $\vec{n}$ at point $\vec{r}$. Photons are traversing this area in both directions, from left to right and from right to left. The radiant energy crossing this area from left to right within an infinitesimal cone of solid angle $d\Omega$ around the direction vector $\vec{e}$, per unit time in the given energy range, is given by $\hbar\omega c f_{ph}(\vec{r}, t; \vec{e}, \hbar\omega) \cos\theta\, d\Omega$ (see figure 9.1). The total amount of energy crossing this area from left to right is given by

$$\hbar\omega c \int_{2\pi} f_{ph}(\vec{r}, t; \vec{e}, \hbar\omega) \cos\theta\, d\Omega = \int_{2\pi} I(\vec{r}, t; e, \hbar\omega) \cos\theta\, d\Omega \tag{9.1.5}$$

where θ is the angle between the direction of the photons $\vec{e}$ and the normal $\vec{n}$, $\cos\theta = \vec{e} \cdot \vec{n}$. This quantity is called the *one-sided radiant energy flux*. The two one-sided radiant fluxes through the plane, from left to right and from right to left, have opposite directions. This difference in their sign is reflected by the

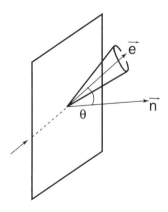

Figure 9.1 Schematic description of the quantities in equation (9.1.4).

change in sign of the cosine function of the integral. The total *spectral radiant energy flux* through the area, with a normal unit vector $\vec{n}$, is

$$F(\vec{r}, t; \hbar\omega, \vec{n}) = \hbar\omega c \int_{4\pi} f_{ph}(\vec{r}, t; \vec{e}, \hbar\omega) \cos(\vec{e} \cdot \vec{n}) \, d\Omega$$

$$= \int_{4\pi} I(\vec{r}, t; \vec{e}, \hbar\omega) \cos(\vec{e} \cdot \vec{n}) \, d\Omega \qquad (9.1.6)$$

This is the component of a vector whose direction is $\vec{n}$. The *total radiant energy flux* is defined as

$$\vec{F}(\vec{r}, t; \hbar\omega, \vec{n}) = \int_{4\pi} I(\vec{r}, t; \vec{e}, \hbar\omega) \vec{e} \, d\Omega \qquad (9.1.7)$$

The units of $\vec{F}$ are erg/(cm^2 s eV). In addition to the radiation field, $\vec{F}$ depends also on the direction of the area, $\vec{n}$.

The four quantities f_{ph}, I, U, and $\vec{F}$, are the most useful in the characterization of the radiation field.

If the radiation has an isotropic distribution, that is, it has the same intensity in all directions, then f_{ph} and I are independent of the propagation direction $\vec{e}$, and the angular integrations in equations (9.1.4–7) can be carried out trivially. In this case,

$$U(\vec{r}, t; \hbar\omega) = 4\pi\hbar\omega f_{ph}(\vec{r}, t; \hbar\omega) = \frac{4\pi}{c} I(\vec{r}, t; \hbar\omega) \qquad (9.1.8)$$

and

$$\vec{F} = 0 \qquad (9.1.9)$$

because the two one-sided fluxes exactly cancel each other. The one-sided flux does not vanish, and its value is

$$F(\text{one-sided}; \hbar\omega) = \int_{2\pi} I(\vec{r}, t; \hbar\omega) \cos\theta \, d\Omega = \pi I(\vec{r}, t; \hbar\omega) \qquad (9.1.10)$$

The *integrated energy density, intensity,* and *flux* of the radiation are obtained from their spectral counterparts by integrating over the whole energy spectrum:

$$U(\vec{r}, t) = \int_0^\infty U(\vec{r}, t; \hbar\omega) \, d(\hbar\omega)$$

$$I(\vec{r}, t; \vec{e}) = \int_0^\infty I(\vec{r}, t; \vec{e}, \hbar\omega) \, d(\hbar\omega) \qquad (9.1.11)$$

$$\vec{F}(\vec{r}, t) = \int_0^\infty \vec{F}(\vec{r}, t; \hbar\omega) \, d(\hbar\omega)$$

It must be emphasized again that the definitions in equations (9.1.1–11) refer to a spectral region that is defined in terms of the photon energy, $\hbar\omega$. Other useful definitions of the above four quantities, f_{ph}, I, U, and $\vec{F}$, refer to spectral regions that are expressed in terms of the photon frequency $[\nu, \nu + d\nu]$, or wavelength $[\lambda, \lambda + d\lambda]$, or wavenumber $[k, k + dk]$ $(k = 1/\lambda)$.

9.2 The Radiation Field in Thermodynamic Equilibrium:
The Black Body Radiation

Assume a medium in which both the material particles and the radiation are in thermodynamic equilibrium. Such systems have a well-defined temperature, T, which is the same for the electrons, the ions, as well as the radiation field. By Kirchhoff's law, under these considerations, the number of photons absorbed by the medium from the group of photons whose energy is in the spectral range between $\hbar\omega$ and $\hbar\omega + d(\hbar\omega)$, and which propagates into a given solid angle $d\Omega$ per unit volume per unit time, exactly equals the number of photons emitted by the medium into the same photon group. This equality between the absorption and the emission of radiation is the condition for *black body radiation*. The black body radiation is isotropic, because in thermodynamic equilibrium there is no preferred direction in space.

The spectral energy density function in a system in thermodynamic equilibrium is the famous *Planck distribution function*, which was the first step in the development of quantum theory,

$$U_{TE}(\hbar\omega; T) = \frac{8\pi h\nu^3}{c^3} \frac{1}{\exp(\hbar\omega/T) - 1}$$

$$= \frac{2\hbar\omega^3}{\pi c^3} \frac{1}{\exp(\hbar\omega/T) - 1} \tag{9.2.1}$$

where the subscript TE denotes thermodynamic equilibrium. We recall that the temperature is in energy units, that is, we work in units in which the Boltzmann constant $k_B = 1$.

The limiting cases of $U_{TE}(\hbar\omega; T)$ for low and high photon energies are given by

$$U_{TE}(\hbar\omega; T) = \begin{cases} 8\pi T\nu^2/c^3 & \text{Rayleigh-Jeans radiation law,} \\ & \hbar\omega \ll T \\ \\ 8\pi h\nu^3 \exp(-\hbar\omega/T)/c^3 & \text{Wien radiation law, } \hbar\omega \gg T \end{cases} \tag{9.2.2}$$

The Rayleigh–Jeans law is the result of the classical electromagnetic theory. It leads to the paradox of the ultraviolet catastrophe, when applied to a too high energy spectral region. It was the conjecture of Planck that solved this paradox, thereby establishing the beginning of the modern quantum theory. As the black body radiation is isotropic, the spectral intensity, by equation (9.1.8), becomes

$$I_{TE}(\hbar\omega; T) = \frac{c}{4\pi} U_{TE} = \frac{\hbar\omega^3}{2\pi^2 c^2} \frac{1}{\exp(\hbar\omega/T) - 1} \tag{9.2.3}$$

The *integrated equilibrium radiant energy density* is obtained by integrating equation (9.2.3) over all possible energies. The result is

$$U_{TE}(T) = \int_0^\infty U_{TE}(\hbar\omega; T)\, d(\hbar\omega) = \frac{4\sigma_{SB} T^4}{c} \tag{9.2.4}$$

where

$$\sigma_{SB} = \frac{2\pi^5}{15h^3c^2} = 1.0283 \times 10^{12}\,\mathrm{erg}/(\mathrm{cm}^2\,\mathrm{s}\,\mathrm{eV}^4) = 5.6705 \times 10^{-5}\,\mathrm{erg}/(\mathrm{cm}^2\,\mathrm{s}\,\mathrm{deg}^4)$$

(9.2.5)

is the *Stefan–Boltzmann constant*. In a similar manner, the equilibrium value of the specific intensity leads to the form

$$I_{TE}(T) = \int_0^\infty I_{TE}(\hbar\omega; T)\,d(\hbar\omega) = \frac{1}{\pi}\sigma_{SB}T^4$$

(9.2.6)

In thermodynamic equilibrium, the total radiant energy flux is of course zero, because the radiation is isotropic. The one-sided flux, however, does not vanish, and can be obtained by integrating the flux over one hemisphere,

$$F_{TE}(T) = \int_{2\pi} I_{TE}\cos\theta\,d\Omega = \pi I_{TE} = \frac{c}{4}U_{TE}(T) = \sigma_{SB}T^4$$

(9.2.7)

This is the radiation flux emanating from a semi-infinite medium in thermodynamic equilibrium and flowing into an empty half-space.

9.3 Absorption of Photons by a Material Medium

Absorption of photons emitted from a plasma may take place either within the same plasma, in which case we are speaking about *reabsorption*, or in some other material medium between the emitting plasma and a detector. The problems of photon absorption has two independent parts: The first is the *basic absorption mechanisms* through which a photon is absorbed by the individual plasma particles, namely, the electrons and the ions. The second part is related to the *radiative transport* of a photon beam inside a plasma, that is, to the interplay of successive emissions and reabsorptions of photons by the plasma material in any given volume element. This is a *collective effect* of the plasma constituents on the intensity of a photon beam propagating in the plasma. First we shall discuss the basic processes by which electrons and ions absorb photons from the radiation field, and shall consider the radiation transport in later sections.

We omit from our discussion processes of *photon scattering*, in which a photon having energy $\hbar\omega$ and propagating in some direction $\vec{e}$ is transformed by some process to a photon of energy $\hbar\omega^*$ propagating in direction $\vec{e}^*$. Most of these processes can be separated into two consecutive parts: The first is the absorption of the original photon and the second is the reemission of the outgoing one. Such separation can be justified if the absorbing particle had enough time to interact with the local plasma microfields between the times of absorption and reemission and thereby to 'forget" its initial state when emitting the photon. This is the case, for instance, in resonant photoabsorption followed by spontaneous decay, in which the emitted photon has the same energy as the absorbed one but moves in a different direction. The lifetimes of the excited states of ions are, in general, much longer than the time between two successive collisions undergone by the ion, except, perhaps, in very low density plasmas. The direction of the emitted photon is, therefore, uncorrelated with the direction of the initial one. On the other hand, *Rayleigh scattering* and *Compton scattering* are pure scattering processes, in which there are strong correlations between the energy and direction of propagation of the incident and outgoing photons. These scattering processes however, play an important role only in a very few experimentally or astrophysically interesting plasmas. The problem of photon scattering in stellar atmospheres is explained in detail in, for example Mihalas (1970).

When a beam of photons traverses a material medium whose thickness is D, some of the photons are absorbed and some are transmitted (see figure 9.2). The number of photons absorbed in an infinitesimally thin slice of material of thickness dx is proportional to the number of photons in the beam and to the slice thickness,

$$df_{ph}(x, t; \vec{e}, \hbar\omega) = -\kappa(x; \hbar\omega)f_{ph}(x, t; \vec{e}, \hbar\omega)\,dx$$
$$dI(x, t; \vec{e}, \hbar\omega) = -\kappa(x; \hbar\omega)I(x, t; \vec{e}, \hbar\omega)\,dx$$

(9.3.1)

Here $\kappa(x; \hbar\omega)$ is the *absorption coefficient* of the material at coordinate x (units cm^{-1}). In a homogeneous medium, where $\kappa(x; \hbar\omega)$ is independent of x, the solution of equation (9.3.1) yields an exponentially decreasing intensity,

$$f_{ph}(x = D, t; \vec{e}, \hbar\omega) = f_{ph,0}(x = 0, t; \vec{e}, \hbar\omega)\,\exp[-\kappa(\hbar\omega)D]$$
$$I(x = D, t; \vec{e}, \hbar\omega) = I_0(x = 0, t; \vec{e}, \hbar\omega)\,\exp[-\kappa(\hbar\omega)D]$$

(9.3.2)

where $f_{ph,0}(x = 0, t; \vec{e}, \hbar\omega)$ and $I_0(x = 0, t; \vec{e}, \hbar\omega)$ are the photon distribution and specific intensity in front of the material slab; $f_{ph}(x = D, t; \vec{e}, \hbar\omega)$ and $I(x = D, t; \vec{e}, \hbar\omega)$ are the corresponding quantities behind the slab; and D is the slab thickness (see figure 9.2).

Equation (9.3.2) indicates that, regarding their absorption properties, plasmas can be divided into two limiting cases:

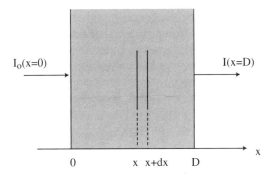

Figure 9.2 Schematic description of the quantities in equation (9.3.1–2).

- *Optically thin plasmas*, $\kappa(\hbar\omega)D \ll 1$, which absorb only a small part of the photons in the radiation field.
- *Optically thick plasmas*, $\kappa(\hbar\omega)D \geq 1$, which absorb a significant portion of the propagating photons.

The absorption coefficient, $\kappa(\vec{r}; \hbar\omega)$, determines the distance at which the intensity reduces to e^{-1} of its initial value and is thus the reciprocal of the mean free path $(\lambda_{mfp}(\vec{r}; \hbar\omega)$; see section 4.2)

$$\kappa(\vec{r}; \hbar\omega) = \frac{1}{\lambda_{mfp}(\vec{r}; \hbar\omega)} \tag{9.3.3}$$

By virtue of equation (4.2.2), $\kappa(\vec{r}; \hbar\omega)$ is proportional to the local density of the absorbing particles, $n(\vec{r})$, and the *total absorption cross section*, $\sigma(\vec{r}; \hbar\omega)$ of the medium,

$$\kappa(\vec{r}; \hbar\omega) = n(\vec{r})\sigma(\hbar\omega) \tag{9.3.4}$$

where $\sigma(\hbar\omega)$ is the sum of the cross sections of *all* the absorbing processes by the plasma ions and electrons. The term *opacity* is generally used for $\kappa(\vec{r}; \hbar\omega)/\rho(\vec{r})$, where $\rho(\vec{r})$ is the local *mass density* of the plasma in g/cm^3. The units of the opacity are cm^2/g.

In a manner similar to the discussion in chapter 6 about the photon emission processes, the various absorption mechanisms of photons in a plasma can also be classified according to whether the initial and/or final state of the absorbing electron is bound or free. In a different method of classification, we distinguish between *continuous* and *line absorption* mechanisms. Using the terminology of the emission spectrum, in the continuous absorption we speak about

- *free-free absorption* which is *inverse bremsstrahlung* (see section 4.1);
- *bound-free absorption*, which is the *photoionization* process (see section 4.8).

The only line absorption mechanism is the

- *bound-bound absorption*, or *resonant photoabsorption* process.

In the following we first introduce the characteristics of the individual processes and in later sections discuss the radiation transport effects.

9.4 The Continuous Photoabsorption Cross Section

Photoionization

Photoionization was discussed extensively in section 4.8. Its cross section is characterized by a series of absorption edges, corresponding to absorption by bound electrons in the various ionic levels. The absorption cross section by the electrons in each shell behaves approximately according to a power law above the absorption threshold energy, called also the *absorption edge*,

$$\sigma_{ph-i}(\hbar\omega) = \begin{cases} \sigma_0(\hbar\omega_0/\hbar\omega)^\alpha & \hbar\omega \geq \hbar\omega_0 \\ 0 & \hbar\omega < \hbar\omega_0 \end{cases} \tag{9.4.1}$$

where $\hbar\omega_0$ is the edge energy, σ_0 is the cross section at the edge tip, and α is a constant, generally close to 3. The height of the edges scales approximately as $Z^{4.5}$, where Z is the atomic number of the ion, see equation (4.8.2), and diminishes approximately as $1/n_p^5$, where n_p is the principal quantum number (q.n.) of the initial state of the absorbing bound electron.

The total photoionization cross section is the weighted sum of the absorption coefficients by electrons in all possible charge and excitation states, with the partial densities, $N_{\zeta,m}$ (ζ is the ion charge and m is the excitation state) as the weighting factors,

$$\sigma_{phi,total}(\hbar\omega) = \frac{1}{n_i} \sum_{\zeta,m} N_{\zeta,m}\sigma_{phi,\zeta,m}(\hbar\omega) \tag{9.4.2}$$

A typical cross section is shown in figure 9.3 for an aluminium plasma at an ion density of $10^{21}\,\text{cm}^{-3}$ and electron temperatures of 32, 100, 320, and 1000 eV (Salzmann and Wendin, 1978). The edge structure of the absorption spectrum can be seen clearly. It is interesting to note that slightly below the absorption edges there is an energy "window" where the absorption coefficient is low. Photons with energies in this spectral region have much longer mean free paths than those above the edges. It must be emphasized, however, that the bound-

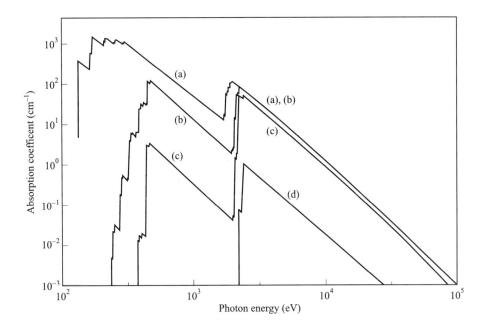

Figure 9.3 The photoionization absorption cross section of aluminium, at $n_i = 10^{21}\,\text{cm}^{-3}$ and electron temperatures of (a) 32, (b) 100, (c) 320, and (d) 1000 eV. (Reprinted with permission from Salzmann, D. and Krumbein, A., 1978, *J. Appl.. Phys.*, **49**, 3229. Copyright 1978 American Institute of Physics.)

bound resonant absorption lines are located exactly in this region and they provide the principal absorbing mechanisms in the window area.

In a real plasma, the sharp absorption edges are shifted due to continuum lowering, and are broadened both by the Stark broadening of the initial bound state and by the fluctuations of the continuum lowering around the absorbing ion.

Inverse Bremsstrahlung

The *inverse bremsstrahlung* process consists of the absorption of a photon by a free electron whose energy is increased accordingly. The momentum conservation law requires that this process can proceed only in the presence of an ion, which carries the extra momentum. This absorption process is important particularly at very low energies, below the lowest absorption edge, and is most important for photons having energies of a few eV and less (see figure 9.4). The absorption coefficient, κ_{ib} ($\kappa_{ib} = n_i \sigma_{ib}$, units cm^{-1}) decreases inversely as the third power of the photon energy, $\hbar\omega$, (Zeldovitch and Raizer, 1966).

$$\kappa_{ib} = \frac{32\pi^3}{3\sqrt{3}} \frac{e^6(\hbar c)^2}{mc^2} \frac{1}{(2\pi mc^2 T_e)^{1/2}} \bar{Z}^2 n_i n_e \frac{1}{(\hbar\omega)^3} g_{ib}(\hbar\omega) \qquad (9.4.3)$$

where $g_{ib}(\hbar\omega)$ is the Gaunt factor, which in most cases can be taken as unity. Since at such low energies the wavelength is rather long, collective absorption by the plasma electrons has to be taken into account. A correction factor which accounts for collective effects was calculated by T. Johnston and J. Dawson (1973). Their modified expression for the transition rate has the form

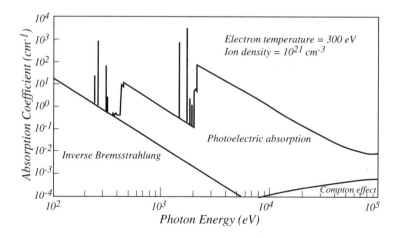

Figure 9.4 The relative importance of the various photoabsorption processes in the various spectral regions. (Reprinted with permission from Salzmann, D. and Krumbein, A., 1978, *J. Appl. Phys.*, **49**, 3229. Copyright 1978 American Institute of Physics.)

$$\kappa_{ib} = \frac{16\pi \bar{Z}^2 n_i n_e e^6 \ln \Lambda(\omega)}{3c(\omega/2\pi)^2 (2\pi m T_e)^{3/2}(1 - \omega_p^2/\omega^2)}$$

$$= 1.33 \times 10^{-4}\,\mathrm{s}^{-1}\,\frac{\bar{Z}(n_e/10^{21}\,\mathrm{cm}^{-3})^2 \ln \Lambda(\omega)}{(\hbar\omega/1\,\mathrm{keV})^2 (T_e/100\,\mathrm{eV})^{3/2}(1 - \omega_p^2/\omega^2)} \qquad (9.4.4)$$

Here $\Lambda = \min(v_T/\omega_p p_{min},\ v_T/\omega p_{min})$, v_T is the electrons thermal velocity, and $p_{min} = \max[Ze^2/T_e,\ \hbar/(mT_e)^{1/2}]$ is the minimum impact parameter for electron–ion collisions. In equation (9.4.4), $\omega_p = (4\pi e^2 n_e/m)^{1/2}$ is the plasma frequency, see equation (1.4.5). Equation (9.4.4) implies that only photons whose frequency is higher than the plasma frequency, $\omega > \omega_p$, can propagate in the plasma. Longer wavelength photons, $\omega \leq \omega_p$, are strongly absorbed on the *critical surface*, defined as the surface on which the electron density is such that $\omega = \omega_p$.

The applicability of this simple formula in laser-produced plasmas was revised by B. Langdon (1980). He found that due to the inverse bremsstrahlung absorption of the incident laser beam, a non-Maxwellian electron distribution is generated near the critical surface, where most of the absorption occurs. This non-Maxwellian distribution modifies the absorption coefficient, and may result in a reduction by as much as 50% in the absorption of the energy of the laser beam by the plasma, particularly for a high-Z absorbing material.

Compton Effect and Pair Production

The Compton effect and pair production are important only at very high photon energies, a few tens of keV and above for the Compton effect, and $2mc^2 = 1022\,\mathrm{keV}$ and above for pair production. Photons of such high energy are generally not generated in present-day plasmas, so that these absorption processes are mostly unimportant in the field of plasma physics.

An exception to this statement can be found in femtosecond laser-produced plasmas and some astrophysical ones, which are in a highly nonequilibrium state. These plasmas do emit photons in the MeV region during the initial phase of the plasma evolution. However, femtosecond laser produced plasmas have dimensions that are several orders of magnitude smaller than the mean free path of photons at such high energies and are therefore optically thin for such photons. The absorption by femtosecond plasmas of such high energy photons is not an important physical effect. In some astrophysical plasmas, however, the absorption of photons with energies of 1 MeV and above can be an interesting and important effect.

The cross sections for these two processes can be found in books on nuclear physics; for example, Siegbahn (1979).

9.5 The Line Photoabsorption Cross Section

The *bound-bound photoabsorption*, sometimes called *line photoabsorption* or *resonant photoabsorption*, occurs when an ion absorbs a photon whose energy exactly equals the energy difference between two atomic levels, to within the levels' widths. As a result of the absorption, the ion is excited to an upper state. Two properties of this process have promoted it to a subject of rather intensive research in recent years, particularly in high-Z plasmas. First, in high-Z plasmas the greater part of the radiative energy is concentrated in the line spectrum, which has a high probability of being resonantly absorbed. Second, the line absorption occurs mainly in the transmission windows below the absorption edges (see figure 9.4), where the line wings are the most important absorption mechanisms. This last property may substantially change the opacities of plasmas.

Consider two ionic states of an ion having charge ζ. Denote the ℓower one by ℓ, and the upper one by u. Let the average energy between these two levels be $\Delta E_{\ell,u} = E_u - E_\ell$. For brevity, we omit the ionic charge, ζ, from the notation. Let $\sigma_{ph-res}(\hbar\omega)$ be the cross section for *resonant photoabsorption*, namely, that a photon of energy $\hbar\omega$ be absorbed from the radiation field, accompanied by an upward excitation of the absorbing ion from state ℓ to state u. As all the photons have the same speed, c, in any inertial coordinate system, the rate coefficient for this process is simply

$$\langle v\sigma_{ph-res}(\hbar\omega)\rangle = c\sigma_{ph-res}(\hbar\omega) \tag{9.5.1}$$

The cross section $\sigma_{ph-res}(\hbar\omega)$ has a sharp maximum at the center of the line, that is, for photons whose energy is

$$\hbar\omega \cong \hbar\omega_{\ell,u} = \Delta E_{\ell,u} = E_u - E_\ell \tag{9.5.2}$$

and drops rapidly to zero at energies which are away from this average by more than the line width. We denote by $\mathcal{L}(\hbar\omega)$ the normalized line profile, $\int_0^\infty \mathcal{L}(\hbar\omega)\, d(\hbar\omega) = 1$ (see chapter 7), then

$$\sigma_{ph-res}(\hbar\omega) = \sigma_0 \mathcal{L}(\hbar\omega) \tag{9.5.3}$$

and

$$\sigma_0 = \int_0^\infty \sigma_{ph-res}(\hbar\omega) d(\hbar\omega) \tag{9.5.4}$$

The units of σ_0 are cm^2 eV. Assume that the plasma is uniform and isotropic, with no preferential direction. The density of photons whose energy is in an infinitesimal spectral region $[\hbar\omega, \hbar\omega + d(\hbar\omega)]$ is

$$N_{ph}(\hbar\omega)d(\hbar\omega) = d(\hbar\omega)\int_\Omega f_{ph}(\hbar\omega)\, d\Omega = 4\pi f_{ph}(\hbar\omega)\, d(\hbar\omega) \tag{9.5.5}$$

see equation (9.1.1). In equation (9.5.5) we have omitted from the argument of f_{ph} all the constant variables. As usual, the number of reactions per unit time per unit volume is the product of the densities of the interacting particles and the rate coefficient for the process,

$$\frac{\text{reactions}}{\text{cm}^3\,\text{s}} = \frac{\text{absorptions}}{\text{cm}^3\,\text{s}} = \frac{\text{photoexcitations}}{\text{cm}^3\,\text{s}}$$

$$= N_\ell N_{ph}\langle v\sigma_{ph-res}\rangle = 4\pi c N_\ell \int_0^\infty \sigma_{ph-res}(\hbar\omega)f_{ph}(\hbar\omega)\, d(\hbar\omega) \tag{9.5.6}$$

where N_ℓ is the density of the ions in state ℓ. Up to this point the treatment is suitable for the absorption of both a continuous and a sharply peaked radiation spectrum. From here on we have to split the discussion into two cases: (a) the radiation intensity is smooth within the range of the absorption line, and (b) the radiation intensity is sharply peaked within the absorption line.

Continuous Radiation Spectral Distribution

If $f_{ph}(\hbar\omega)$ is approximately constant within the line region, one can replace $f_{ph}(\hbar\omega)$ in equation (9.5.6) by its average value $f_{ph}(\Delta E_{u,\ell})$. Inserting this back into equation (9.5.6), one gets

$$\frac{\text{absorptions}}{\text{cm}^3\,\text{s}} = 4\pi c N_\ell \sigma_0 f_{ph}(\Delta E_{u,\ell}) \tag{9.5.7}$$

This rate of absorptions per unit volume can be derived also in a different way, using the definitions of the Einstein coefficients. The Einstein B-coefficient is defined originally such that

$$\frac{\text{absorptions}}{\text{cm}^3\,\text{s}} = N_\ell h J(\hbar\omega) B_{\ell\to u} \tag{9.5.8}$$

see equation (4.5.6). In equation (9.5.8), $J(\hbar\omega)$ is the specific intensity averaged over all solid angles,

$$J(\hbar\omega) = \int_{4\pi} I(\hbar\omega)\,\frac{d\Omega}{4\pi} \tag{9.5.9}$$

which for an isotropic plasma reduces simply to

$$J(\hbar\omega) = I(\hbar\omega) \tag{9.5.10}$$

The explicit appearance of the Planck constant, h, in equation (9.5.8) is due to the fact that in the original definition of $B_{\ell\to u}$ the specific intensity corresponds to a unit frequency spectral interval, whereas $I(\hbar\omega)$ in equation (9.1.2) is defined per unit photon energy interval. Note that the units of $J(\hbar\omega)$ are erg/(cm^2 s eV), which differ from those of $I(\hbar\omega)$ by the absence of the steradian in the denominator. Comparing equation (9.5.8) to equation (9.5.7), one finds

$$B_{\ell\to u} = \frac{2}{\hbar^2\omega}\sigma_0 \tag{9.5.11}$$

Equation (9.5.8) can also be expressed by means of the Einstein A-coefficient and the oscillator strength. The relations between the Einstein A- and B-coefficients were given in chapter 4, and are reproduced here for the reader's convenience,

$$g_u B_{u\to\ell} = g_\ell B_{\ell\to u} \tag{4.5.12}$$

$$A_{u\to\ell} = \frac{2h\nu^3}{c^2} B_{u\to\ell} \tag{4.5.13}$$

The relation between the Einstein A-coefficient and the oscillator strength is

$$A_{u \to \ell} = 2c \, \frac{e^2}{(\hbar c)^2 mc^2} \, (\Delta E_{u,\ell})^2 f_{u \to \ell} \qquad (4.5.4)$$

Using these relations, one can rewrite the photon absorption rate per unit volume, which equals also the upward transition rate per unit volume, as

$$\frac{\text{absorptions}}{\text{cm}^3 \, \text{s}} = \frac{\text{photoexcitations}}{\text{cm}^3 \, \text{s}} = \frac{1}{2} \frac{g_u}{g_\ell} \, N_\ell J(\Delta E_{u,\ell}) \, \frac{\lambda_{u,\ell}^3}{c} \, A_{u \to \ell} \qquad (9.5.12)$$

where $\lambda_{u,\ell} = 2\pi\hbar c / \Delta E_{u,\ell}$ is the wavelength of the absorbed photon. In terms of the oscillator strength, equation (9.5.12) yields the following form,

$$\frac{\text{absorptions}}{\text{cm}^3 \, \text{s}} = 4\pi^2 r_0 \, \frac{g_u}{g_\ell} \, N_\ell J(\Delta E_{u,\ell}) \lambda_{u,\ell} f_{u \to \ell} \qquad (9.5.13)$$

where $r_0 = e^2/mc^2$ is the electron electromagnetic radius, and $f_{u \to \ell}$ is the oscillator strength.

The *resonant photoabsorption coefficient* is related to the corresponding cross section by

$$\kappa(\hbar\omega) = N_\ell \sigma_{ph-res}(\hbar\omega) = N_\ell \sigma_0 \mathcal{L}(\hbar\omega) = N_\ell \frac{\hbar^2 \omega}{2} B_{\ell \to u} \mathcal{L}(\hbar\omega) \qquad (9.5.14)$$

see equation (9.3.4). In the second of these equations, we have used equation (9.5.3), and in the last we have used equation (9.5.11). By expressing the B-coefficient in terms of the Einstein A-coefficient and the oscillator strength, equations (4.5.13) and (4.5.4), equation (9.5.14) can be rewritten as

$$\kappa(\hbar\omega) = \frac{\lambda^2}{4} \frac{g_\ell}{g_u} \, N_\ell \hbar A_{u \to \ell} \mathcal{L}(\hbar\omega) = 2\pi^2 r_0 \, \hbar c \, \frac{g_\ell}{g_u} \, N_\ell f_{u \to \ell} \mathcal{L}(\hbar\omega) \qquad (9.5.15)$$

Line Radiation Spectral Distribution Function

The shape of the absorption spectrum is determined by the profile function $\mathcal{L}(\hbar\omega)$. This function has nonzero values in a very narrow part of the spectrum. In this part, however, it attains rather high values, orders of magnitude higher than the continuous absorption. Moreover, in optically thick plasmas, the positions of the absorption lines are approximately at the same energies as the emission lines. The strong lines emitted from the plasma are, therefore, also strongly absorbed by the plasma environment. The emission and the absorption line profiles are very close to each other, but are not exactly equal, for reasons that will be explained shortly, and the tiny difference between them plays an important role in the absorption properties of the plasma. The reabsorption of a line emitted by the plasma is calculated in the following way: denote by $\mathcal{L}_e(\hbar\omega)$ the profile of the emission line, and by $\mathcal{L}_a(\hbar\omega)$ the profile of the absorption line (see figure 9.5 for illustration). The specific spectral intensity of the line is then $I(\hbar\omega) = I_0 \mathcal{L}_e(\hbar\omega)$, and the

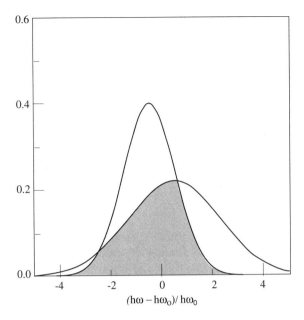

Figure 9.5 Schematic illustration of the overlap integral.

absorption cross section is $\sigma(\hbar w) = \sigma_0 \mathcal{L}_a(\hbar w)$. The absorption per unit propagation distance is, by equation (9.3.1),

$$dI(\hbar w) = -\kappa(\vec{r}; \hbar w) I(\vec{r}, t; \vec{e}, \hbar w)\, dx = -N_\ell \sigma(\hbar w) I(\hbar w)\, dx$$

$$= -N_\ell I_0 \sigma_0 \mathcal{L}_e(\hbar w) \mathcal{L}_a(\hbar w)\, dx = -N_\ell I_0 \sigma_0 \mathcal{L}_t(\hbar w)\, dx \qquad (9.5.16)$$

where

$$\mathcal{L}_t(\hbar w) = \mathcal{L}_e(\hbar w) \mathcal{L}_a(\hbar w) \qquad (9.5.17)$$

is the *overlap function* between the emission and the absorption line profiles. Note that $\mathcal{L}_t(\hbar w)$ is not normalized to unit integral. The overlap function has nonzero values only in that region of the spectrum where both functions, $\mathcal{L}_e(\hbar w)$ and $\mathcal{L}_a(\hbar w)$, are not too small; see the hatched area in figure 9.5. In practice, $\mathcal{L}_a(\hbar w)$ determines the amount of radiation that is reabsorbed by the medium. The total absorption of the spectral line is obtained by integrating both sides of equation (9.5.16) over $\hbar w$,

$$dI = -N_\ell I(x) \sigma_0 L_t\, dx \qquad (9.5.18)$$

In equation (9.5.18),

$$L_t = \int \mathcal{L}_t(\hbar w)\, d(\hbar w) = \int \mathcal{L}_e(\hbar w) \mathcal{L}_a(\hbar w)\, d(\hbar w) \qquad (9.5.19)$$

is the *overlap integral*, which is an important parameter in the reabsorption of lines in plasmas.

The difficulty in the calculations of the resonant photoabsorption coefficient in optically thick plasmas is mainly due to the calculation of the overlap integral. We have discussed the problem of the line profiles in chapter 7, but the question of the overlap integral is different and more complicated than that of the separate line shapes because in a real plasma the emission and absorption lines may have different widths and their centers may be slightly shifted. Even a relatively small shift may, under certain conditions, have a large effect on the overlap integral. Moreover, as the overlap function is an energy-dependent quantity, the radiation emerging from the plasma has a distorted shape, whose center may be shifted from that of the original emission line. Here we mention just a few factors that must be taken into account in the calculation of the overlap integral.

1. In rapidly expanding plasmas, there may be a velocity difference between the layer where the emission takes place and the absorbing layer. This produces a Doppler shift between the emission and the absorption lines. For instance, a relative velocity of the order of 10^6 cm/s produces a relative Doppler shift of $\delta(\hbar\omega)/\hbar\omega = v/c \approx 10^{-4}$. This shift is of the order of the line width and is large enough to substantially change the value of the overlap integral.

2. Density gradients in very dense plasmas have a similar effect. Assume that the emission takes place in the dense part of the plasma and the absorption in the lower density parts. Due to density effects, the emission lines will be slightly shifted relative to the absorption lines (see chapters 7 and 3), thereby altering the value of the overlap integral.

3. One of the greatest effects of the density gradients on the overlap integral is due to the fact that the emission and the absorption lines have different widths. This greatly modifies the absorption properties of the plasma medium for a given emission line. To simulate this effect accurately in a radiative transport problem, one has to divide the spectrum into very small energy groups. Computationally, this may be beyond the capacities of present-day computers.

4. Steep temperature gradients also influence the overlap between the absorption and the emission lines. If the gradients are large within a distance of one mean free path, the charge state distributions in two closely lying portions of the plasma can be different from each other. The line spectrum will be emitted by one kind of charge state, whereas in the absorbing region the photons will meet a different distribution of charge states whose absorption spectrum does not necessarily overlap the spectrum of the emitted photons. The photoabsorption in this case will proceed through continuous absorption or the wings of more distance lines.

5. Finally, the overlap integral is also affected by the shift between the absorption and the emission lines caused by the recoil energy of the emitting atom, which reduces the energy of the emitted photon. This shift between the emission and the absorption lines is $\delta(\hbar\omega)/\hbar\omega = \hbar\omega/2Mc^2$, where M is the atomic mass. For 1 keV photons, this ratio is of the order of 10^{-6}–10^{-7}, that is, it may have some influence in low density plasmas where the line broadening is small. This is generally not a large effect but, when combined with the other effects mentioned above, it may have a measurable influence.

9.6 The Basic Radiation Transport Equation

In the previous sections we have dealt with the individual processes by which ions and electrons absorb photons from the radiation field. This section is devoted to the problem of the interplay of successive radiation emission and absorption by the plasma medium. The sequence of these collective processes is the *radiation transfer* or *radiation transport* in the plasma. Obviously, this transport is important only in optically thick plasmas, in which absorption plays an important role in the energy balance.

A few words of apology are in order: Many books have been written on the problem of radiation transfer, particularly in the context of astrophysical applications. See, for example, the excellent books by Mihalas (1970) and Ivanov (1973). It is not the aim of this book to give a full account of this problem, but only a limited survey that can provide the reader with an intelligent insight into this very interesting subject. Readers who find interest in the material presented here, and would like to delve deeper into this subject, are encouraged to look into more specialized books.

The intensity emerging from an optically thin plasma is directly proportional to the emission coefficients, see chapter 6. In optically thick plasmas, the reabsorption of the radiation by the plasma particles lowers this proportionality and limits the intensity to that of a black body. As a first step, we derive the equation for the specific intensity that accounts for both the emission and the absorption of the plasma.

Assume that the radiation intensity at some point, $A(\vec{r})$, at a given time, t, is given by $I(\vec{r}, t; \vec{e}, \hbar\omega)\, d\Omega\, dS\, d(\hbar\omega)$. We recall that this is the radiative energy around the point $A(\vec{r})$, whose spectral distribution is within an infinitesimal spectral energy range $[\hbar\omega, \hbar\omega + d(\hbar\omega)]$, and which propagates into a solid angle $d\Omega$ around a direction defined by the unit vector $\vec{e}$, through an area dS perpendicular to the direction of propagation; see equation (9.1.2) and the explanation thereafter. What will be the intensity at a later time, $t + dt$, at some other point, $B(\vec{r} + d\vec{r})$, which is in the direction of the propagation, $d\vec{r} = dr\, \vec{e}$? An illustration of this problem is given in figure 9.6.

Since the radiation has the speed of light, the two quantities dt and $d\vec{r}$ are not independent but are connected by the relation $|d\vec{r}| = c\, dt$. Four processes contribute to the radiation intensity at $B(\vec{r} + d\vec{r})$ at $t + dt$: (a) the first is the intensity at $A(\vec{r})$ at the moment t, which arrives at $B(\vec{r} + d\vec{r})$ at time $t + dt$; (b) the second is the reduction of this intensity by the absorption along the path $d\vec{r}$; (c) third, there is a contribution to the radiant intensity by the ions between A and B, which emit radiation into the given direction during dt; (d) finally, the stimulated emission induced by the radiation field enhances the radiation intensity (chapter 6).

Denote by $\kappa(\vec{r}, t; \hbar\omega)$ the absorption coefficient of the plasma between A and B, and by $j(\vec{r}, t; \hbar\omega)$ the emissivity of this medium, which is the radiative energy emitted in the given spectral region, per unit volume, per unit time, per unit solid angle, (units $\mathrm{erg}/(\mathrm{cm}^3\, \mathrm{s}\, \mathrm{sr}\, \mathrm{eV})$). The radiative energy emitted into a solid angle $d\Omega$ equals $j(\vec{r}, t; \hbar\omega)\, d\Omega$. The emissivity is closely related to the emission rate, which was discussed in chapter 6. In fact, for isotropic emissivity,

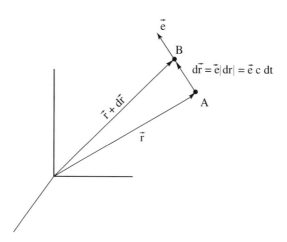

Figure 9.6 Illustration of the geometry and the
quantities in the basic radiation transfer equation.

$j(\hbar\omega) = \hbar\omega P(\hbar\omega)/4\pi$, where $P(\hbar\omega)$ is the emission rate of photons into the given
spectral region, per unit volume (see chapter 6). Applying these definitions, the
intensity at point $B(\vec{r} + d\vec{r})$ is given by

$$I(\vec{r} + d\vec{r}, \ t + dt; \vec{e}, \hbar\omega) \, d\Omega \, dS \, d(\hbar\omega) = I(\vec{r}, t; \vec{e}, \hbar\omega) \, d\Omega \, dS \, d(\hbar\omega)$$
$$- \kappa(\vec{r}, t; \hbar\omega) I(\vec{r}, t; \vec{e}, \hbar\omega) \, d\Omega \, dS \, d(\hbar\omega)|d\vec{r}| + j(\vec{r}, t; \hbar\omega) \, d\Omega \, dV \, d(\hbar\omega) \quad (9.6.1)$$

For brevity, we have omitted the stimulated emission term. This will be dealt with
later in this section. In the last term of equation (9.6.1), $dV = dS|d\vec{r}|$ is the volume
element between A and B. Figure 9.6 shows the geometry of the problem. For
infinitesimal $d\vec{r}$ and dt, the total change in the intensity can be expanded into a
two-dimensional Taylor series,

$$[I(\vec{r} + d\vec{r}, t + dt; \vec{e}, \hbar\omega) - I(\vec{r}, t; \vec{e}, \hbar\omega)] \, d\Omega \, dS \, d(\hbar\omega)$$

$$= \left[\frac{\partial I}{\partial t} dt + d\vec{r} \cdot \text{grad } I \right] d\Omega \, dS \, d(\hbar\omega) \quad (9.6.2)$$

$$= \left[\frac{\partial I}{\partial t} dt + c \, dt \, \vec{e} \cdot \text{grad } I \right] d\Omega \, dS \, d(\hbar\omega)$$

Combining equation (9.6.1) with equation (9.6.2) yields the *radiative transport
equation* in its most general form.

$$\frac{1}{c} \frac{\partial I}{\partial t} + \vec{e} \cdot \text{grad } I = -\kappa(\vec{r}, t; \hbar\omega) I(\vec{r}, t; \vec{e}, \hbar\omega) + j(\vec{r}, t; \hbar\omega) \quad (9.6.3)$$

In specific cases, equation (9.6.3) can be reduced to simpler forms. For steady
state plasmas that do not change too rapidly in time, the partial derivative with
respect to time drops out, and one gets a simpler form for equation (9.6.3), viz.,

$$\vec{e} \cdot \operatorname{grad} I = -\kappa(\vec{r}, \hbar\omega)I(\vec{r}, \vec{e}, \hbar\omega) + j(\vec{r}; \hbar\omega) \tag{9.6.4}$$

Moreover, if the plasma has a planar symmetry in the $x \cup y$ plane (figure 9.7) the intensity gradient is in the direction of the z-axis, $\operatorname{grad} I = (0, 0, \partial I/\partial z)$, and equation (9.6.4) reduces to

$$\mu \frac{\partial I(\mu; \hbar\omega)}{\partial z} = -\kappa(z; \hbar\omega)I(z, \mu; \hbar\omega) + j(z; \hbar\omega) \tag{9.6.5}$$

where $\mu = \cos\theta$ is the cosine of the angle between the directions of the radiation and the symmetry axis.

Equation (9.6.5) can be further simplified. For this purpose we define *the optical depth* in the spectral region $[\hbar\omega; \hbar\omega + d(\hbar\omega)]$ between two points, A and B, as the integral of the absorption coefficient between these two points,

$$\tau(A, B; \hbar\omega) = \int_A^B \kappa(\vec{r}; \hbar\omega)\, dl \tag{9.6.6}$$

where dl is the element of length along the line connecting A and B. The optical depth is a dimensionless quantity that defines how many times the distance between A and B is larger than the mean free path at the given photon energy. The *differential optical depth* is defined by

$$d\tau = \kappa(\vec{r}; \hbar\omega)\, dl \tag{9.6.7}$$

Dividing both sides of equation (9.6.5) by the absorption coefficient, $\kappa(z; \hbar\omega)$, equation (9.6.5) can be rewritten in the form

$$\mu \frac{\partial I}{\partial\tau} = -I(z, \mu; \hbar\omega) + S(z; \hbar\omega) \tag{9.6.8}$$

where the ratio of the emissivity to the absorption coefficient,

$$S(z; \hbar\omega) = \frac{j(z; \hbar\omega)}{\kappa(z; \hbar\omega)} \tag{9.6.9}$$

is called the *source function*.

Equation (9.6.8) has the structure of a simple ordinary linear differential equation of the first order, whose solution can be found in textbooks,

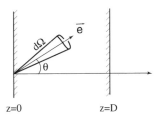

z=0   z=D

Figure 9.7 Radiation transport
in planar symmetry.

$$I(z, \mu; \hbar\omega) = e^{-\tau(0,z;\hbar\omega)/\mu} \left[I(0, \mu; \hbar\omega) + \frac{1}{\mu} \int_0^z d\tau'\, S(z'; \hbar\omega) e^{-\tau(0,z';\hbar\omega)/\mu} \right]$$

$$= I(0, \mu; \hbar\omega) e^{-\tau(0,z;\hbar\omega)/\mu} + \frac{1}{\mu} \int_0^z \kappa(z'; \hbar\omega)\, dz'\, S(z'; \hbar\omega) e^{-\tau(z',z;\hbar\omega)/\mu}$$

$$= I(0, \mu; \hbar\omega) e^{-\tau(0,z;\hbar\omega)/\mu} + \frac{1}{\mu} \int_0^z dz'\, j(z'; \hbar\omega) e^{-\tau(z',z;\hbar\omega)/\mu} \qquad (9.6.10)$$

$I(0, \mu, \hbar\omega)$ is the specific intensity of the radiation at $z = 0$, which propagates into a direction defined by μ. The meaning of equation (9.6.10) follows immediately: The intensity at any point z inside the plasma equals the incoming intensity attenuated by the exponential of the optical depth (first term), plus the intensity emitted along the path of propagation, which is also attenuated between the point of emission, z', and the point z (second term). Other boundary conditions are possible, such as the intensity at $z = \infty$, or more complicated ones. It should be emphasized again that equations (9.6.5–10) hold true for planar symmetry only.

To solve the equation of radiation transport explicitly, one first has to know the absorption coefficient, $\kappa(z; \hbar\omega)$, and the emissivity, $j(z; \hbar\omega)$, at every point inside the plasma. These are simple functions only for full thermodynamic equilibrium. In plasmas in LTE or corona equilibrium or for the steady-state collisional radiative model, these functions have to be computed according to the methods explained in chapter 6 and in the previous sections of this chapter. Such a computation is, in general, a major undertaking, and as yet only a few laboratories have the capability of obtaining results of reasonable accuracy.

The Source Function for Strong Lines in Thermodynamic Equilibrium and LTE

In thermodynamic equilibrium, the source function can be expressed in an explicit analytical form. To derive this function, we consider, for simplicity, the radiation transfer in a strong spectral line. For such a case, we include the stimulated emission into the equation of radiation transport. Using equation (9.5.14) for the relation between the absorption coefficient and the Einstein B-coefficient, the equation of radiation transfer, equation (9.6.5), reduces to the form

$$\mu \frac{\partial I(z, \mu; \hbar\omega)}{\partial z} = -I(z, \mu; \hbar\omega) \frac{\hbar^2\omega}{4\pi} [B_{\ell,u} N_\ell \mathcal{L}_a(\hbar\omega) - B_{u,\ell} N_u \mathcal{L}_e(\hbar\omega)]$$

$$+ \frac{\hbar\omega}{4\pi} N_u A_{u,\ell} \mathcal{L}_e(\hbar\omega) \qquad (9.6.11)$$

where the first term is the rate of resonant photoabsorption, the second is the rate of stimulated emission, and the last is the emission rate of radiation during spontaneous decay between these two levels, see equation (4.5.8). Assume, for simplicity, that the emission and absorption line profiles are equal, $\mathcal{L}_a(\hbar\omega) = \mathcal{L}_e(\hbar\omega)$, then, using equations (4.5.12–13), (see discussion following equation (9.5.11) in this chapter), one finds that

$$S(\hbar\omega) = \frac{j(\hbar\omega)}{\kappa(\hbar\omega)} = \frac{N_u A_{u,\ell}}{\hbar(B_{\ell,u} N_\ell - B_{u,\ell} N_u)}$$

$$= \frac{A_{u,\ell}}{\hbar B_{u,\ell}[N_\ell/N_u)(B_{\ell,u}/B_{u,\ell}) - 1]} = \frac{1}{\hbar} \frac{2h\nu^3}{c^2} \frac{1}{(N_\ell/N_u)(g_u/g_\ell) - 1} \qquad (9.6.12)$$

(We recall that in equations (9.6.11–12) there is an extra factor of $1/\hbar$, because we defined the radiation intensity per unit energy rather than unit frequency as used in other books. We shall continue to show this factor explicitly, to enable comparison with the other definitions.) In full thermodynamic equilibrium or in LTE, the ratio of the densities of the upper and lower states satisfies a Boltzmann distribution,

$$\frac{N_\ell}{N_u} = \frac{g_\ell}{g_u} e^{(E_u - E_\ell)/T} \qquad (9.6.13)$$

whence one obtains for the source function under equilibrium conditions,

$$S(\hbar\omega_{u,\ell}) = \frac{1}{\hbar} \frac{2h\nu_{u,\ell}^3}{c^2} \frac{1}{e^{(E_u - E_\ell)/T} - 1} = \frac{1}{\hbar} \frac{2h\nu_{u,\ell}^3}{c^2} \frac{1}{e^{\hbar\omega_{u,\ell}/T} - 1} = \mathbf{B}(\hbar\omega_{u,\ell}, T) \quad (9.6.14)$$

$\mathbf{B}(\hbar\omega_{u,\ell}, T)$ is the *Planck function*, defined by the last of these equations. $\mathbf{B}(\hbar\omega_{u,\ell}, T)$ depends on the photon energy as well as the plasma temperature. Although, strictly speaking, this simple analytical form of the source function was deduced here for spectral lines only, it holds true in general, by Kirchhoff's law, for all kinds of radiation, including the continuous.

9.7 Radiation Transport in Plasmas: Examples

A few solutions of equation (9.6.10) for several interesting cases are in order. The examples solved here have a rather general nature, and are important in several astrophysical and laboratory problems.

Example 1: The Spectrum Emerging from an Optically Thin Plasma

Assume that there is no radiation incident on the plasma at $z = 0$ (see figure 9.7), then $I(0, \mu; \hbar\omega) = 0$. What is the intensity emerging at $z = D$, perpendicular to the plasma surface ($\mu = 1$)? From the first of the equations in (9.6.10) we find that

$$I(D, \mu; \hbar\omega) = \int_0^{\tau(D)} d\tau' \, S(\tau'; \hbar\omega) e^{-\tau'(\hbar\omega)} \qquad (9.7.1)$$

where $\tau' = \int_0^{z'} \kappa(z'', \hbar\omega) \, dz''$. In optically thin plasmas, $\tau(D) \ll 1$ and $e^{-\tau'} \approx 1$, so that one can approximate

$$I(D, \mu; \hbar\omega) \approx \tau(D) S(\tau'; \hbar\omega) = \kappa(\hbar\omega) D S(\tau'; \hbar\omega) = j(\hbar\omega) D \qquad (9.7.2)$$

which means that in optically thin plasmas the emerging intensity depends only on the plasma emissivity and, as expected, the reabsorption does not play an important role.

Example 2: The Spectrum Emerging from a Semi-infinite Plasma

Assume now that the plasma fills the positive z-space. The radiation intensity into the negative half-space, $z < 0$, $\mu = -1$, can be obtained by integration by parts of equation (9.7.1),

$$I(0, \theta = 180°; \hbar\omega) = -\int_0^\infty S(z; \hbar\omega)e^{-\tau'} d\tau'$$

$$= S(0, \hbar\omega) + \frac{dS}{d\tau}\bigg|_{\tau=0} + \frac{d^2S}{d\tau^2}\bigg|_{\tau=0} + \cdots \qquad (9.7.3)$$

On the other hand, expanding S into a power series one gets

$$S(\tau = 1) = S(\tau = 0, \hbar\omega) + \frac{dS}{d\tau}\bigg|_{\tau=0} + \frac{1}{2}\frac{d^2S}{d\tau^2}\bigg|_{\tau=0} \cdots \qquad (9.7.4)$$

Equations (9.7.3) and (9.7.4) are identical for the first two orders. Deviations begin at the second and higher orders. So, to a good degree of accuracy one finds that

$$I(0, \theta = 180°; \hbar\omega) \cong S(\tau(\hbar\omega) = 1) \qquad (9.7.5)$$

In other words, the radiation emerging from the surface of a semi-infinite plasma is roughly equal to the source function at an optical depth of $\tau(\hbar\omega) = 1$. If the medium is in a steady state, the intensity at $z = 0$ is determined by the plasma temperature at depth $\tau(\hbar\omega) = 1$ from the surface. For instance, a measurement of the temperature of the sun by means of its outgoing radiation, which is the only means available for astrophysical plasmas, gives a temperature not on the sun's surface, but rather at some deeper distance, where the optical depth for the given spectral region is approximately 1.

Example 3: The Radiation Intensity Deep Inside a Homogeneous Plasma

If the plasma is homogeneous, more or less, with a source function that is almost constant throughout the plasma, equation (9.7.1) can be integrated analytically. In this case, the radiation intensity at any depth z is given by

$$I(z, \mu; \hbar\omega) = S(\hbar\omega)(1 - e^{-\tau(0,z;\hbar\omega)}) \qquad (9.7.6)$$

At great depths inside the plasma, far from the plasma boundaries, $\tau(0, z; \hbar\omega) \gg 1$, equation (9.7.6) reduces to

$$I(z \to \infty, \mu; \hbar\omega) = S(\hbar\omega) \qquad (9.7.7)$$

However, deep inside a homogeneous plasma, the influence of the boundaries is small, and therefore the conditions are presumably very close to those of thermodynamic equilibrium. This implies that the source function, $S(\hbar\omega)$, is close to the Planck function, $\mathbf{B}(\hbar\omega, T)$, which is the limiting value of $S(\hbar\omega)$ close to equilibrium conditions (equation 9.6.14). It turns out that, deep inside the plasma, the specific intensity, too, tends to the Planck function,

$$I(z \to \infty, \mu; \hbar\omega) \to \mathbf{B}(\hbar\omega, T) \qquad (9.7.8)$$

The result obtained here leads to a very important conclusion: *The specific intensity, in any spectral region $[\hbar\omega, \hbar\omega + d(\hbar\omega)]$, in a homogeneous plasma with temperature T, can never be larger than the Planck function, $\mathbf{B}(\hbar\omega, T)$.* In optically thin plasmas, the intensity is always less than this limiting value. In optically thick plasmas, it approaches this limit from below, but never crosses it. Although this conclusion was derived under the condition of a homogeneous plasma, it holds true for inhomogeneous plasmas as well.

Example 4: Escape Factors

An interesting approximate method for treating the radiation transfer of strong spectral lines was introduced by Holstein (1947). His method is very useful in some problems. Assume a spectral line of mean energy $\hbar\omega_{u,\ell}$ emitted by transitions in one of the charge states from an upper state, denoted by u, to a lower state, ℓ. By virtue of equation (9.6.11), the emission rate of photons per unit volume into the spectral region of this line can be written formally as

$$P(\hbar\omega_{u,\ell}, \vec{r}) = N_u \hbar\omega_{u,\ell} \Lambda_{u,\ell}(\vec{r}) A_{u\to\ell} \qquad (9.7.9)$$

The right hand side of this equation can be regarded as a shorthand for all three processes that take part in the emission and absorption of this spectral line, namely, spontaneous emission, stimulated emission, and resonant photoabsorption. The justification for such definitions is the fact that both $B_{u,\ell}$ and $B_{\ell,u}$ are proportional to $A_{u,\ell}$. The quantity $\Lambda_{u,\ell}(\vec{r})$, called the *escape factor*, takes into account the contributions of all these three processes, and it should be regarded as a multiplicative factor that modifies the Einstein A-coefficient in optically thick plasmas. $\Lambda_{u,\ell}(\vec{r})$ depends on the local plasma temperature and density, and thereby on the position vector $\vec{r}$. In an optically thin plasma,

$$\Lambda_{u,\ell} = 1 \qquad \text{optically thin plasmas} \qquad (9.7.10)$$

as can be seen by direct comparison with equation (6.2.1). In optically thick plasmas, the total emission rate depends on the plasma dimensions and geometry. Writing in full the three processes, the radiative energy emission rate becomes

$$P(\hbar\omega_{u,\ell}) = N_u[A_{u,\ell} + hJ(\hbar\omega)B_{u,\ell}] + N_\ell hJ(\hbar\omega)B_{\ell,u} \qquad (9.7.11)$$

where

$$J(\hbar\omega) = \int I(\vec{r}; \vec{e}, \hbar\omega) \frac{d\Omega}{4\pi} \qquad (9.7.12)$$

234 ATOMIC PHYSICS IN HOT PLASMAS

is the average radiation energy density in the spectral range of the line. Substituting equations (4.5.12–13) into (9.7.11) and comparing this result with equation (9.7.9), one obtains the relation

$$\Lambda_{u,\ell}(\vec{r}) = 1 - \left(\frac{g_u}{g_\ell}\frac{N_\ell}{N_u} - 1\right)\frac{\hbar c^2}{2h\nu_{u,\ell}^3} J(\hbar\omega) \tag{9.7.13}$$

or

$$\Lambda_{u,\ell}(\vec{r}) = 1 - \frac{J(\hbar\omega)}{S(\vec{r},\hbar\omega)} \tag{9.7.14}$$

where $S(\vec{r},\hbar\omega)$ is the local source function, see equation (9.6.12). On the other hand, inserting equation (9.7.6) for the specific intensity into equation (9.7.12), and assuming a source function constant in space, one obtains for the average radiation energy density,

$$J(\hbar\omega) = \int I(\vec{r};\vec{e},\hbar\omega)\frac{d\Omega}{4\pi} = \frac{1}{4\pi} S(\hbar\omega)\int_\Omega d\Omega\,(1 - e^{-\tau(\vec{r},\hbar\omega)}) \tag{9.7.15}$$

where $\tau(\vec{r},\hbar\omega)$ is to be understood as being taken between the boundaries of the plasma and $\vec{r}$. The integration over the whole sphere surrounding $\vec{r}$ can be carried out only when the geometry is known. To suppress the effects of the special geometry, we define a mean optical depth, $\bar{\tau}(\vec{r},\hbar\omega)$, at the position $\vec{r}$ by means of

$$J(\hbar\omega) = S(\hbar\omega)(1 - e^{-\bar{\tau}(\vec{r},\hbar\omega)}) \tag{9.7.16}$$

and write this mean optical depth in the form

$$\bar{\tau}(\vec{r},\hbar\omega) = \bar{N}_\ell(\vec{r})\sigma(\vec{r},\hbar\omega)l = \bar{N}_\ell(\vec{r})\sigma_0 \mathcal{L}(\hbar\omega)l \tag{9.7.17}$$

where $\sigma(\vec{r},\hbar\omega)$ is the cross section for an upward transition by resonant photoabsorption from level ℓ to level u, l is an effective or average distance, and $\bar{N}_\ell(\vec{r})$ denotes the average density along the path from the point $\vec{r}$ to the plasma boundaries. Inserting equations (9.5.3) and (9.7.17) into equation (9.7.16), one finds that

$$J(\hbar\omega) = S(\hbar\omega)(1 - e^{\bar{N}_\ell(\vec{r})\sigma_0 \mathcal{L}(\hbar\omega)l}) \tag{9.7.18}$$

When this last equation is inserted back into equation (9.7.14), one obtains

$$\Lambda_{u,\ell}(\vec{r}) = \int_0^\infty d(\hbar\omega)\mathcal{L}(\hbar\omega)\exp[-\bar{N}_\ell(\vec{r})\sigma_0 \mathcal{L}(\hbar\omega)l] \tag{9.7.19}$$

For optically thin plasmas, the exponent is small and one gets

$$\Lambda_{u,\ell}(\vec{r}) \approx \int_0^\infty d(\hbar\omega)\,\mathcal{L}(\hbar\omega) = 1 \tag{9.7.20}$$

in accordance with equation (9.7.10). For optically thick plasmas, the escape factor, $\Lambda_{u,\ell}(\vec{r})$, depends both on the line profile function and on the geometry of the plasma. For a Gaussian line profile, Holstein has shown that defining

$$\bar{\tau} = \bar{N}_\ell(\vec{r})l\frac{1}{\Delta\hbar\omega}\frac{\sqrt{\pi}he^2}{mc}f_{u,\ell} \tag{9.7.21}$$

($\Delta\hbar\omega$ is the width of the Gaussian line profile and $f_{u,\ell}$ is the oscillator strength), the escape factor for $\bar\tau \geq 3$ can be written to good accuracy as

$$\Lambda_{u,\ell}(\vec r) \approx \frac{1}{\bar\tau(\pi \log \bar\tau)^{1/2}} \tag{9.7.22}$$

regardless of the accurate shape of the plasma geometry. For a Lorentzian line profile, and $\bar\tau \geq 5$, the escape factor becomes

$$\Lambda_{u,\ell}(\vec r) \approx \frac{1}{(\pi\bar\tau)^{1/2}} \tag{9.7.23}$$

where $\bar\tau$ is defined in a manner similar to its definition in equation (9.7.21), with the Lorentzian width replacing the Gaussian one. For a Voigt profile, the escape factor is between the Gaussian and the Lorentzian profiles.

In optically thick plasmas, the escape factor approximation is a very convenient tool: One simply multiples the Einstein A-coefficient by the escape factor and gets a solution that is almost independent of the exact shape of the plasma. This is particularly helpful in the computations of the charge and excited state distributions in optically thick plasmas, which otherwise would be a major computational problem. For complicated geometries, however, the escape factor approximation is not always sufficiently accurate.

Example 5: The Radiative Heating of a Planar Slab

As a last example, we calculate the total radiative energy absorbed per unit volume, or per atom, at depth z inside a plasma having a planar symmetry, that is, the *radiative heating* of the slab.

Assume an absorbing volume of thickness dz, area dS (see figure 9.8) and volume $dV = dS\,dz$. Consider a layer of radiation-emitting ions, having thickness dz' at depth z'. As a first step we calculate how much of the radiation emitted by this layer is reabsorbed by the ions at z. Let $j(z';\hbar\omega)\,d(\hbar\omega)\,d\Omega\,dV'$ be the radiative energy rate emitted by this layer into the spectral region between $\hbar\omega$ and $\hbar\omega + d(\hbar\omega)$ and into the solid angle $d\Omega$. All the photons emitted by ions in a ring between radii r and $r + dr$ and angle θ will reach the small volume element of thickness dz and area dS. The emission volume is $dV' = 2\pi r\,dr\,dz'$. Expressing r in terms of θ, $r = (z - z')\tan\theta = Z\tan\theta$ ($Z = z - z'$) we find that

$$dV' = 2\pi Z^2\,\frac{\sin\theta}{\cos^3\theta}\,d\theta\,dz' \tag{9.7.24}$$

The total emission rate into the direction of the absorbing volume is therefore

$$J_1 = j(z';\hbar\omega)\,d(\hbar\omega)\,d\Omega\,2\pi Z^2\,\frac{\sin\theta}{\cos^3\theta}\,d\theta\,dz' \tag{9.7.25}$$

where $d\Omega$ is the solid angle subtended by the absorbing volume as viewed from the emitting volume. It is equal to

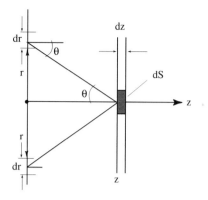

Figure 9.8 Schematic description of the geometry and the basic quantities of radiative heating of a planar slab.

$$d\Omega = \frac{dS \cos \theta}{R^2} = \frac{dS}{(Z/\cos \theta)^2} \cos \theta \qquad (9.7.26)$$

Substituting back into equation (9.7.25), one obtains for the emission rate,

$$J_1 = j(z'; \hbar\omega) \, d(\hbar\omega) \, 2\pi \, d\mu \, dS \, dz' \qquad (9.7.27)$$

where $\mu = \cos \theta$. The radiation *arriving* at the absorbing layer, J_2, is the emission rate attenuated along the path of propagation,

$$J_2 = J_1 \exp\left(-\frac{1}{\cos \theta} \int_{z'}^{z} \kappa(z''; \hbar\omega) \, dz''\right) \qquad (9.7.28)$$

Finally, the absorbed energy in layer dz is

$$dE(z) = J_2 n_i(z) \sigma(z; \hbar\omega) \frac{dz}{\cos \theta}$$

$$= 2\pi n_i(z) \, dS \, \sigma(z; \hbar\omega) \, dz \int_{-\infty}^{z} dz' \int_0^\infty d(\hbar\omega) j(z'; \hbar\omega) \int_0^1 d\mu \, e^{-\tau/\mu} \qquad (9.7.29)$$

where the optical path is measured between z' and z, $\tau = \tau(z', z, \hbar\omega)$. The angular integration can be carried out analytically. It equals $\tau E_2(\tau)$, where $E_2(x) = \int_x^\infty dt \, e^{-t}/t^2$ is the exponential integral of the second order. Taking into account that the number of ions in the absorbing volume is $n_i(z) \, dS \, dz$, one finds the total energy absorbed per atom as

$$\frac{dE(z)}{n_i(z) \, dS \, dz} = 2\pi \int_0^\infty d(\hbar\omega) \, \sigma(z; \hbar\omega) \int_{-\infty}^{z} dz' \, j(z'; \hbar\omega) \tau(z', z; \hbar\omega) E_2[\tau(z', z; \hbar\omega)]$$

$$(9.7.30)$$

This last formula corresponds to absorption from the left half-space. For the total absorbed energy per atom, a similar formula for the absorption from the right half-space has to be added. Formally, this is equivalent to extending the upper limit of the integral over z' to infinity.

9.8 Diffusion Approximation, Radiative Heat Conduction, and Rosseland Mean Free Path

A problem of radiation transport of practical interest arises deep inside an optically thick plasma, where the radiation is close to, but is not quite, in thermodynamic equilibrium with the material. These are the conditions, for instance, deep inside star interiors and other astrophysical plasmas, where the radiation transport plays an important role in the local energy balance. In general, in such plasmas there is a slow temperature gradient, whose scale length, $L = T/|\nabla T|$, is much longer than the average mean free path of the photons. This gradient gives rise to an excess flux of the radiation, such that more photons move from the high temperature part of the plasma to the lower temperature portion than the other way. If the gradient is not too steep, the photons' directional distribution is almost, although not exactly, isotropic. Under such conditions, the flow of the radiation propagating from the high temperature regions heats the material in the lower temperature portions by means of a process that resembles the process of heat conduction by material particles. In fact, the propagation of the radiation in the direction of the temperature gradient goes through a series of absorption, reemission, and scattering processes, which are similar to the processes that take part in the diffusion of heat by material particles. The computational method most frequently used to handle the radiation transport in such cases is, therefore, called the *diffusion approximation*.

If the basic steady state equation of radiation transfer (equation 9.6.4) is integrated over all directions, we get a relationship between the radiation flux and the radiation energy density. Using the definitions in equations (9.1.4) and (9.1.6) for the radiant flux and energy density, the integration over the angular variables yields

$$\nabla \cdot \vec{F}(\vec{r}; \hbar\omega) = -c\kappa(\vec{r}; \hbar\omega)\left(U(\vec{r}; \hbar\omega) - \frac{4\pi}{c}\mathbf{B}(\hbar\omega; T) \right) \qquad (9.8.1)$$

where we have used the fact that the local conditions are close to thermodynamic equilibrium, and thus the source function can be approximated by the Planck function, equation (9.6.14). A second approximate equation relating the radiant energy and flux can be obtained when the directional distribution of the photons' propagation is almost isotropic. This equation is derived as follows: Multiply equation (9.6.4) by the unit vector in the direction of the photon propagation, $\vec{e}$, and integrate again over all angles. The ith component of the integral on the left hand side is

$$\left(\int \vec{e}(\vec{e}\cdot\nabla I)\,d\Omega\right)_i = \sum_{k=1}^{3}\int d\Omega\, e_i e_k\,\frac{\partial I(\vec{r},\vec{e};\hbar\omega)}{\partial x_k} \tag{9.8.2}$$

If the intensity is almost isotropic, its partial derivative is approximately constant and can be taken out of the integration. The polar form of the unit vector in the direction (θ,ϕ) has the form

$$\vec{e} = \begin{pmatrix} \sin\theta\cos\phi \\ \sin\theta\sin\phi \\ \cos\theta \end{pmatrix} \tag{9.8.3}$$

and direct integration shows that

$$\int e_i e_k\,d\Omega = \frac{4\pi}{3}\delta_{i,k} \tag{9.8.4}$$

where $\delta_{i,k}$ is the *Kronecker* δ, which equals 1 for $i=k$ and is zero otherwise. Inserting this last result into equation (9.8.2), one obtains

$$\int \vec{e}(\vec{e}\cdot\nabla I)\,d\Omega = \frac{4\pi}{3}\nabla I(\vec{r};\hbar\omega) = \frac{c}{3}\nabla U(\vec{r};\hbar\omega) \tag{9.8.5}$$

As already mentioned, this equation holds true only when the radiation field has low anisotropy. Carrying out the same operation on the right hand side of equation (9.6.4), the emission term, $j(\vec{r};\hbar\omega)$, being independent of the photon direction, gives a zero integral, $\int j(\vec{r};\hbar\omega)\vec{e}\,d\Omega = 0$. The first term on the right hand side of equation (9.6.4) can be expanded as follows:

$$\int d\Omega\,\kappa(\hbar\omega)I(\vec{e};\hbar\omega)\vec{e} = \kappa(\hbar\omega)\int d\Omega\,I(\vec{e};\hbar\omega)\vec{e} = -\kappa(\hbar\omega)\vec{F}(\hbar\omega) \tag{9.8.6}$$

Finally, combining the last two equations, we obtain

$$\vec{F}(\hbar\omega) = \frac{cl(\hbar\omega)}{3}\nabla U(\hbar\omega) \tag{9.8.7}$$

In this equation, $l(\hbar\omega) = 1/\kappa(\hbar\omega)$ is the *mean free path* of the radiation.

Equation (9.8.7) connects the radiation flux to the gradients of the local radiation density and it resembles the diffusion equations of particles. It is called, therefore, *the radiation diffusion equation*, and it is useful when the anisotropy and the gradients of the radiation density are not too large.

In optically thick plasmas, when the plasma is close to thermodynamic equilibrium, the radiation density and the plasma temperature are related by equation (9.2.4). Even if the plasma is not in perfect equilibrium but close to it, we can approximate the radiation density by equation (9.2.4) and write

$$\nabla U \approx \nabla U_{TE}(T) = \frac{16}{c}\sigma_{SB}T^3\nabla T \tag{9.8.8}$$

Integrating both sides of equation (9.8.7) over the whole spectrum, and inserting equation (9.8.8) into the result, one obtains

$$\vec{F} = -\frac{16}{3} l_R \sigma_{SB} T^3 \nabla T \qquad (9.8.9)$$

where l_R is the average mean free path, with $\nabla U_{TE}(\hbar\omega)$ as the weighting factor. l_R is called the *Rosseland mean free path*, and its explicit value is given by

$$l_R = \int_0^\infty \frac{1}{\kappa(\hbar\omega)} G(u)\, du \qquad (9.8.10)$$

where $u = \hbar\omega/T$ and

$$G(u) = \frac{dU_{TE}/dT}{\int (dU_{TE}/dT)\, d(\hbar\omega)} = \frac{15}{4\pi^4} \frac{u^4 e^{-u}}{(1 - e^{-u})^3} \qquad (9.8.11)$$

The energy regions where $\kappa(\hbar\omega)$ is small have the largest contribution to the Rosseland mean free path. These are, in general, the relatively transparent regions slightly below the absorption edges. In these spectral regions, most of the opacity is produced by the collective absorption of the spectra lines' wings. The main difficulty in a reliable computation of the Rosseland mean free path stems from the necessity for an accurate modeling of these line wings.

Equation (9.8.9) shows that the radiation diffusion from the high to the low temperature regions of the plasma is proportional to the product of the third power of the temperature and the temperature gradient, which is the well known $T^3 \nabla T$ law. This strong dependence on the plasma temperature indicates the special importance of the radiation heat conduction in high temperature plasmas, such as astrophysical and fusion-related laboratory plasmas.

A similar but conceptually different average value is the *Planck mean free path*, defined as

$$\kappa_P = \frac{\int \kappa(\hbar\omega) \mathbf{B}(\hbar\omega; T)\, d\hbar\omega}{\int \mathbf{B}(\hbar\omega; T)\, d\hbar\omega} \qquad (9.8.12)$$

where $\mathbf{B}(\hbar\omega; T)$ is the Planck function defined in equation (9.6.14). In contrast to the Rosseland m.f.p., the Planck m.f.p. emphasizes the regions of large $\kappa(\hbar\omega)$, that is, the more opaque spectral regions.

10

Applications

This final chapter is devoted to a short review of the potential applications of radiation from hot plasmas for commercial purposes. X-rays emitted from hot plasmas have remarkable advantages over the use of visible light in several applications. In particular, the short wavelength, which is capable of providing orders of magnitude better spatial resolution than visible rays, is of paramount importance in several industrial and high-tech applications. The conversion efficiency of visible into x-ray intensity in laser-produced plasmas may attain several tens percent, thereby providing an x-ray source in the range of 10^9–10^{10} watts, which is equivalent to a few joules in $\sim 10^{-9}$ s. This is a very high radiant intensity that may have industrial or high-tech applications.

Applications of plasma-generated x-rays are, however, still scarce. This can be explained by the great difficulties in the fabrication and application of x-ray sources, mainly because optical components for x-rays are not yet available commercially and are very expensive. Optical devices for focusing or deflecting a beam, such as lenses or mirrors, which are simple and inexpensive for visible light, require special development or are nonexistent for the x-ray domain. It is, therefore, no surprise that this field has taken its first steps only in the last decade. During this period it has, however, made enormous progress that is continuing at an accelerated pace, and it would appear that its main accomplishments still lie ahead. A review of the present status of the applications of x-rays from hot plasmas is, therefore, necessarily condemned to be out of date shortly after its publication. With this in mind, we nevertheless make an attempt to give an updated list of topics that are being investigated with an intention of applying them in other high-tech fields and that have already reached some degree of maturity.

10.1 X-Ray Lasers

Lasing of x-rays from plasmas was demonstrated for the first time in 1985 by an experimental group at the Lawrence Livermore National Laboratory (Matthews et al., 1985) and somewhat later by a group at the Plasma Physics Laboratory at

Princeton University (Suckewer et al., 1985). Since then, population inversion and lasing have been demonstrated in several other laboratories in a variety of plasma materials. Most of these experiments have used large plasma-generating machines to produce the lasing of x-rays. Recently, however, in a very important step forward in this field, a compact discharge-pumped "table-top" soft x-ray laser was built by a group at Colorado State University (Rocca et al., 1994).

Population Inversion, Stimulated Emission, Gain, and
Other Basic Quantities of X-Ray Lasers

The status of x-ray lasers up to 1990 is summarized by R. C. Elton (1990). The principles of x-ray lasers are similar to those in the visible, and are based on the generation of population inversion in a material medium ("pumping"). Stimulated emission then generates a well-defined beam of coherent radiation.

As shown in chapter 9, passage of radiation in a material medium is affected by three processes, namely, spontaneous emission, stimulated emission, and absorption by the medium. The equation of radiation transport is given by equations (9.6.3–8) in its most general form. For the transport of a spectral line, the relevant equation is (9.6.11), reproduced here for the reader's convenience,

$$\mu \frac{\partial I(z, \mu; \hbar\omega)}{\partial z} = -I(z, \mu; \hbar\omega) \frac{\hbar^2 \omega}{4\pi} [B_{\ell,u} N_\ell \mathcal{L}_a(\hbar\omega) - B_{u,\ell} N_u \mathcal{L}_e(\hbar\omega)]$$

$$+ \frac{\hbar\omega}{4\pi} N_u A_{u,\ell} \mathcal{L}_e(\hbar\omega) \qquad (9.6.11)$$

We recall that in equation (9.6.11) $A_{u,\ell}$, $B_{u,\ell}$, and $B_{\ell,u}$ are the Einstein coefficients for spontaneous decay, stimulated emission, and resonant photoabsorption, respectively. If the radiation field in the spectral region of the given line (i.e., within the region defined by the line width) is strong and the density of the upper state is sufficiently large, this equation reduces to the approximate form

$$\frac{\partial I(z)}{\partial z} = G(\hbar\omega)I(z) \qquad (10.1.1)$$

where $I(z) = I(z, \mu = 1; \hbar\omega_{u,\ell})$ is the specific intensity of the beam in the forward direction, $G(\hbar\omega)$ is the *gain coefficient*,

$$G(\hbar\omega) = \frac{\hbar^2 \omega}{4\pi} [B_{u,\ell} N_u \mathcal{L}_e(\hbar\omega) - B_{\ell,u} N_\ell \mathcal{L}_a(\hbar\omega)] \qquad (10.1.2)$$

and we assume that the radiation field within the spectral range of the line is already strong enough that the last term of equation (9.6.11), the spontaneous emission, is negligibly small relative to the stimulated emission. The gain coefficient, $G(\hbar\omega)$, here plays the role of a negative absorption coefficient, $G(\hbar\omega) = -\kappa(\hbar\omega)$.

Let L be the length of a homogeneus plasma (see figure 10.1). If an external beam of intensity I_0 impinges on the plasma at $z = 0$, then by equation (10.1.1) the radiation emerging from the back surface is simply

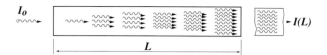

Figure 10.1 Illustration of the quantities in equation (10.1.3).

$$I(L) = I_0 e^{G(\hbar\omega)L} \tag{10.1.3}$$

Thus, the outgoing intensity, $I(L)$, may be larger or smaller than the incident intensity, I_0, depending on whether the gain coefficient, $G(\hbar\omega)$, is positive or negative, and this in turn depends on which of the two terms in equation (10.1.2) is the larger. If $G(\hbar\omega)$ is positive, one gets an amplification of the beam, that is, a laser.

The above process can proceed without the need for an incident external beam, but rather by the plasma's own radiation. Denote by $J(\hbar\omega)$ the emissivity of the plasma per unit length (figure 10.1), which is a constant for a homogeneous plasma. This is exactly the problem solved in section 9.7, Example 1. Replacing $\kappa(\hbar\omega)$ by $-G(\hbar\omega)$ in equation (9.7.1), one obtains

$$I(L; \hbar\omega) = \int_0^{G(\hbar\omega)L} G(\hbar\omega)\, dz' \, \frac{J(\hbar\omega)}{-G(\hbar\omega)} \, e^{G(\hbar\omega)z'}$$

$$= -J(\hbar\omega) \int_0^{G(\hbar\omega)L} dz' \, e^{G(\hbar\omega)z'} = J(\hbar\omega) \frac{(e^{G(\hbar\omega)L} - 1)}{G(\hbar\omega)} \tag{10.1.4}$$

The small-gain limit of the specific intensity, I_s, is obtained when $G(\hbar\omega)L \ll 1$,

$$I_s \approx \frac{J(\hbar\omega)}{G(\hbar\omega)} G(\hbar\omega)L = J(\hbar\omega)L \tag{10.1.5}$$

Equations (10.1.4) and (10.1.5) hold for every energy interval within the line profile. To obtain the overall amplification of the beam, equation (10.1.4) has to be integrated over the whole line profile. Such an integration depends, of course, on the details of the profile function. A widely used approximate expression that gives good results for every sharply peaked line profile, including the Doppler, is (Elton, 1990)

$$I_0 = \frac{J_0}{G_0} \frac{(e^{G_0 L} - 1)^{3/2}}{(G_0 L e^{G_0 L})^{1/2}} \tag{10.1.6}$$

where $J_0 = \int J(\hbar\omega)\, d(\hbar\omega)$ is the total emissivity integrated over the line profile, and $G_0 = G(\hbar\omega_{u\ell})$ is the gain coefficient at the line center.

A quantity closely related to the gain coefficient is the *inversion factor*, F, defined by

$$G_0 = \frac{\hbar^2 \omega_{u\ell}}{4\pi} B_{u,\ell} N_u \mathcal{L}_e(\hbar\omega) F \tag{10.1.7}$$

Using equation (10.1.2), F can be rewritten as

$$F = 1 - \frac{N_\ell B_{\ell,u}}{N_u B_{u\ell}} = 1 - \frac{g_u N_\ell}{g_\ell N_u} \tag{10.1.8}$$

where in the second equation we have used equation (4.5.12) to replace the ratio of the Einstein coefficients by the ratio of the statistical weights. As already mentioned, lasing occurs when $G_0 > 0$, that is, when $F > 0$. Under conditions of LTE, the ratio of the populations of the upper to that of the lower level is given by the Boltzmann factor,

$$\frac{N_u}{N_\ell} = \frac{g_u}{g_\ell} \exp\left(-\frac{E_u - E_\ell}{T}\right) \tag{10.1.9}$$

where $E_u > E_\ell$. Therefore, for an LTE-type population distribution of the excited states,

$$F = 1 - \exp\left(-\frac{E_u - E_\ell}{T}\right) < 0 \tag{10.1.10}$$

This means that no laser action is possible under conditions of equilibrium. To get lasing, the inversion factor has to be positive, which by equation (10.1.8) occurs only when

$$\frac{N_u}{N_\ell} > \frac{g_u}{g_\ell} \tag{10.1.11}$$

In other words, lasing between two ionic levels can occur only if the upper level is *overpopulated*, which means that the ratio of its population to that of the lower one is larger than the ratio of their statistical weights. This condition is also called *population inversion*. Such population inversion can be produced only by inserting energy ("pumping") into the plasma through a mechanism that can efficiently and preferentially overpopulate the upper level.

Schemes for Population Inversion

The characteristics of population inversion in plasmas are similar in most aspects to lasing in the visible. Population inversion can be induced by two methods (Elton, 1990). The first is selective population of an upper state by collisional or photon-induced mixing from a nearby highly populated state. The second possible scheme includes a nonequilibrium population of the ionic states. As most lower levels decay more rapidly than the upper ones, population inversion develops within the time scale of the lifetime of the lower level.

In the most general scheme, four ionic states take part in a lasing system (see figure 10.2). The upper state u is populated by some mechanism from a ground state ion, labeled i. A lower level, designated by ℓ, which in general is populated by means of the same mechanism, is rapidly depopulated into a deeper final state f (which may or may not be identical to i). Population inversion is generated between u and ℓ, such that the condition in equation (10.1.11) is fulfilled. A

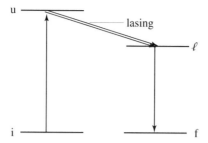

Figure 10.2 General scheme of ionic states for a laser.

beam of photons of energy $\hbar\omega_{u\ell}$ that propagates through the plasma stimulates emission from the ions along its path, gradually amplifying the beam intensity.

The description up to this point is suitable for lasers in both the visible and x-ray regions. There are, however, problems that are specific to lasing in plasmas. In a plasma, for instance, one has to look for a level scheme that is based on closed-shell ions. Being stable configurations, closed-shell ions are the abundant charge states and their population probabilities are relatively constant for a wide range of plasma temperatures and densities. This may reduce the sensitivity of the laser to the local plasma conditions. On the other hand, open-shell ions undergo rapid ionization and recombination processes, which may smear out any momentary population inversion.

Another problem specific to x-ray lasers is the very short lifetime of the upper level, which means that any population inversion holds for a short time only. This implies that all the energy must be emptied out within a short period, and only single pass amplifications can be expected. This is in sharp contrast to the visible, where multipass and even continuous lasing can be generated.

One of the most attractive schemes for population inversion, which has been the basis of several successful experiments, is the level structure of neonlike ions. The energy levels of the excited $2p^5 3s$ and $2p^5 3p$ electronic configurations are shown in figure 10.3. These levels in a plasma are excited from the ground state by electron impact with rather high probability. Population inversion occurs after the rapid spontaneous decay of the $3s$ electrons to the $2p^6$ closed-shell ground state. The wavelengths of the transitions, and the gain coefficients, as measured in experiments are given in table 10.1. As can be seen from the table, as the ion charge increases, the wavelength of the transition decreases, and there also seems to be a slow increase of the gain coefficient. On the other hand, more energy is required to ionize high-Z materials into a neonlike configuration. X-ray laser energy beams of the order of 10^{-5} J have been measured (Matthews et al., 1985), corresponding to a conversion efficiency of the order of 10^{-7}–10^{-8} from the energy in the plasma into the energy of the coherent x-ray beam.

Similar ideas led to the search for lasing in hydrogenlike ions. Suckewer et al. (1985) have found lasing in the 182 Å $3 \rightarrow 2$ transition in H-like carbon. Since the Einstein A-coefficient of the $n = 2$ state is much larger than that of the $n = 3$ state,

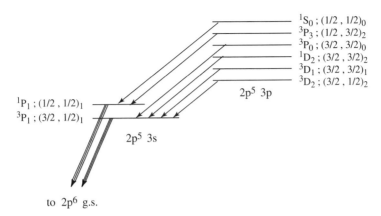

Figure 10.3 The laser transitions from the $2p^5 3p$ to the $2p^5 3s$ excited configurations.

the $n = 2$ state depopulates much faster than the $n = 3$ state, thereby producing a population inversion between these two states.

A series of experiments has also been carried out on the $4d$–$4p$ transitions in nickellike ions. Lasing has been found in Eu^{35+}, Yb^{42+}, Ta^{45+}, W^{46+}, and Re^{47+} plasmas (MacGowan et al., 1987b). The wavelengths of these spectral lines are much shorter—between 40 and 65 Å (depending on the ion charge)—but the gain coefficients are similar to those of neonlike ions (1–$5\,cm^{-1}$). A review of neonlike and nickellike x-ray lasers, generated by applying a prepulse technique, has been published by Li et al. (1997).

Experimental Demonstration of Stimulated X-Ray
Emission

Experimental demonstration of an x-ray laser is based on the exponential increase of the intensity of the lasing line as a function of the plasma length, see equations (10.1.4–6). The standard experimental arrangement consists of a cylindrical plasma and a high resolution spectrometer that measures the outgoing radiation along the long axis of the plasma (see figure 10.4). The emission spectrum is then measured as a function of the plasma length. The intensities of lines that undergo stimulated amplification increase with the increase of the plasma length, whereas the intensities of all the other lines diminish. As lasing occurs for a limited period only, until the population inversion is extinguished, a time-resolved spectrometer is used to obtain a better identification of the stimulated x-ray amplification.

From the large body of experiments that have been carried out to investigate the possibilities of x-ray lasers, we have selected only two: an experiment carried out in Lawrence Livermore National Laboratory (LLNL) with neonlike selenium, which was actually the first demonstration of lasing of x-rays in a laser-produced plasma, and a more recent successful experiment with a discharge pumped "table-top" small dimension soft x-ray laser.

Table 10.1 Wavelengths and Measured Gain Coefficients (cm⁻¹) of the 3p–3s Lasing Lines in Ne-like Systems

Ion	$^1S_0-^1P_1$ λ (Å)	$^1S_0-^1P_1$ G (cm⁻¹)	$^1D_2-^3P_1$ λ (Å)	$^1D_2-^3P_1$ G (cm⁻¹)	$^3P_2-^1P_1$ λ (Å)	$^3P_2-^1P_1$ G (cm⁻¹)	$^3D_1-^3P_1$ λ (Å)	$^3D_1-^3P_1$ G (cm⁻¹)	$^3D_2-^3P_1$ λ (Å)	$^3D_2-^3P_1$ G (cm⁻¹)	$^3P_0-^3P_1$ λ (Å)	$^3P_0-^3P_1$ G (cm⁻¹)	Reference
Ar⁸⁺	468.75	0.6											(a)
Cu¹⁹⁺	221.11	2.0	279.31	1.7	284.67	1.7							(b)
Zn²⁰⁺	212.17	2.3	262.32	2.0	267.23	2.0							(c)
Ge²²⁺	196.06	3.1	232.24	4.1	236.26	4.1	247.32	2.7	286.46	4.1			(b,d)
As²³⁺			218.84	4.1	222.56	5.4							(c)
Se²⁴⁺	182.43	≤1–2.7	206.38	4.0	209.78	3.8	220.28	2.3	262.94	3.5	169.29	≤1.0	(e,f)
Sr²⁸⁺	159.8		164.1	4.4	166.5	4.0	165.0		218.0				(f)
Y²⁹⁺			155.0	4.5	157.1	4.5							(f,g)
Mo³²⁺	141.6	0	131.0	4.1	132.7	4.2	139.4	2.9			106.4	2.2	(f,h)

(a) Rocca et al. (1994), (b) Lee et al. (1987), (c) Lee et al. (1988), (d) Elton (1987), (e) Matthews et al. (1985), (f) Keane et al. (1989), (g) Matthews (1986), (h) MacGowan et al. (1987a).

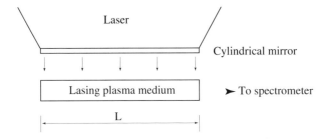

Figure 10.4 Schematic arrangement for an experiment to demonstrate and measure x-ray lasing.

Neonlike Selenium

The first demonstration of stimulated emission in the x-ray region was by a group of 19 researchers in LLNL (Matthews et al., 1985). As the lasing material they chose laser-produced neonlike selenium. The target was composed of a thin layer of approximately 750 Å, deposited on one side of a 1500 Å formvar substrate. The foil length was 1.1 cm. The Novette laser-target facility was used to irradiate the target. The wavelength of the Novette laser is $\lambda = 0.532 \, \mu$m, and its pulse length is 4.5×10^{-10} s. The laser beam was focused by means of a cylindrical lens to a line focus on the selenium side of the target, which had dimensions of $0.02 \times 1.12 \, cm^2$. In some of the experiments the foils were irradiated from both sides. In addition, the longitudinal dimension could be varied, so that the line integrated gain, G_0L, could be measured for a series of plasma lengths. Computer simulations show that the conditions in the selenium plasma are fairly constant for a significant portion of the laser pulse, with temperature around $1.0 \, keV$ and ion density of about $1 \times 10^{21} \, cm^{-3}$. The spectrum emitted in the longitudinal direction was measured by an x-ray streak camera (see section 8.5), and the results are shown in figure 10.5. The exponential increase of the line integrated gain is clearly seen. From the curves, a gain coefficient of $G = 5.5 \pm 1.0 \, cm^{-1}$ was deduced.

A Discharge Pumped "Table-Top" Soft X-Ray Laser

Most of the x-ray lasers mentioned above have been obtained by using high power laser-produced plasmas as the gain medium. The use of such huge machines automatically limits the applicability of the resulting coherent x-ray beam to these laboratories. Moreover, as the cost of running such lasers is in the range of $\$10^4$–$10^5$ per shot, the generated x-ray laser pulse is also extremely expensive.

J. J. Rocca and his collaborators (Rocca et al., 1994) succeeded in generating a soft x-ray laser by means of a fast capillary discharge. The lasing material is neonlike argon. The experiment consisted of a fast discharge (half period of 60 ns) of a $\sim 40 \, kA$ pulse through 3, 6, and 12 cm long capillary channels, filled with pure argon or a 1:2 mixture of Ar and H_2. Stimulated amplification has been

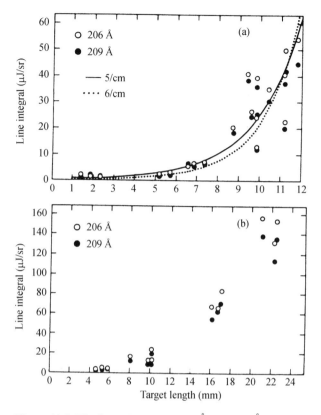

Figure 10.5 The intensity of the 206 Å and 209 Å lines on the plasma axis, as a function of the plasma length. Taken from figure 4 in Matthews et al. (1985). (a) Double sided and (b) single sided laser irradiation conditions.

found for the 468.75 Å $J = 0$–1 transition in neon-like argon (figure 10.3). The measured gain coefficient is $0.6\,\mathrm{cm}^{-1}$.

The wavelength of this x-ray laser is very long, deep in the XUV domain, and is easily absorbed in any small amount of material. Nevertheless, the fact that such a relatively "small" and inexpensive x-ray laser could be built has great importance to the development of the field. This group has reported lasing in neonlike sulfur at $\lambda = 608$ Å, using the same technique. It seems that the search for such small, easily operated lasers will be the target in this field in the coming years.

10.2 Applications of High Intensity X-Ray Sources

This last section is devoted to the description of the state of art of several potential applications of the spectra emitted from hot plasmas at the time of writing. The list is definitely not complete and, as these techniques are still in the stage of experimentation, the material in this section refers to the present status only and may change rapidly in the near future.

X-Ray Lithography

Soft x-ray projection lithography in the wavelength range 50–200 Å has recently emerged as a leading candidate for the fabrication of integrated circuits and optoelectronic devices. The ultimate aim is the design of devices with sizes reaching below 0.1 μm (Silfvast and Ceglio, 1993). Such an achievement will undoubtedly induce major improvements in the fabrication of components for electronics, computers, robotics, and electrooptics.

The recent advance in the field of lithography was made possible by the rapid progress in x-ray optics. This involved the development of high-reflecting multilayer coatings and efficient x-ray sources. Multilayer coatings with reflectivity up to 65% have been developed. Although these components are still very expensive, costs are expected to come down in the future.

Soft x-ray projection lithography has several advantages. These include large depth of field and good spatial resolution. Laser-plasma sources cost less than those from a synchrotron, but are presently still much more expensive than sources in the visible or the ultraviolet. The greater problem is still the development of low cost optics, and the optimization of the source to get higher emission efficiency in the required spectral range.

Several experiments have already demonstrated the potential applicability of this technique. By using 140 Å radiation, Early et al. (1993) have imaged 0.1 μm wide lines and spaces. The borders of the lines are defined to an accuracy of $\sim$140 Å. Such resolution is hard to obtain with visible or UV light. Similar results were also reported by Tichenor et al. (1993) and Macdowell et al. (1993).

X-Ray Microscopy of Biological Specimens

High resolution analysis of biological specimens are at present carried out routinely by electron miroscopy. While this technique can provide spatial resolution of the order of a few angstroms, it suffers from several drawbacks, such as small depth of field and relatively long irradiation time. X-ray microscopy could, in principle, avoid these limitations. So far, two techniques have been used to produce the x-ray source: synchrotron and laser-plasma radiation. In accordance with the general topic of this book, we review here briefly the laser-plasma source only. It has the advantage over the synchrotron source of having much smaller dimensions (particularly the "table-top" femtosecond laser plasma facilities) and is comparable in cost to the electron microscope.

X-ray microscopy has several advantages over electron microscopy. First, the longer mean free path of the x-rays in the specimen facilitates a deeper depth of field and the imaging of thicker samples. Second, the short burst of the x-rays from a pulsed laser-plasma enables the probing of the internal structure of in-vitro assemblies, thereby providing the opportunity of observing complex biological features in their natural even live, state.

The highest contrast in x-ray images of in-vitro biological specimen is expected in the so-called *water window*, between 2.3 and 4.4 nm (282–540 eV). Careful design of the plasma material, laser intensity, and pulse length can provide a high conversion efficiency of the laser energy into x-ray intensity in the water window. Magnification of the image is obtained by means of a pinhole camera.

A commercial x-ray microscope should consist of a laser-plasma point x-ray source, an electrooptical image magnifier with a converter into the visible, an image intensifier, and a CCD detector. At the present time, however, the conventional method consists of a thin layer of polymethylmethacrylate (PMMA) photoresist, which is developed after the exposure to obtain the image of the specimen (Richardson et al., 1992; Kinjo et al., 1994). The method has been successfully demonstrated on human chromosomes (Richardson et al., 1992; Kinjo et al., 1994) and live bacteria (Richardson et al., 1992).

References

Aaron, F. D., Costescu, A., and Dinu, C., 1993, *J. Phys. II France*, **3**, 1227.

Abramowitz, M. and Stegun, I., 1965, *Handbook of Mathematical Functions*, Dover, New York.

Aguilera-Navarro, V. C., Estevez, G. A., and Kostecki, A., 1988, *J. Appl. Phys*, **63**, 2848.

Arpigny, C., 1963, *Astrophys. J.*, **A138**, 607.

Audebert, P., Geindre, J. P., Gauthier, J. C., and Popovics, C., 1984, *Phys. Rev.*, **A30**, 768.

Bar-Shalom, A., Oreg, J., Goldstein, W. H., Shvarts, D., and Zigler, A., 1989, *Phys. Rev.*, **A40**, 3183.

Bauche-Arnoult, C., Bauche, J., and Klapisch, M., 1979, *Phys. Rev.*, **A20**, 2424.

Bauche, J. and Bauche-Arnoult, C., 1990, *Comp. Phys. Rep.*, **12**, 1.

Beigman, I. L. and Chichkov, B. N., 1980, *J. Phys. B.*, **13**, 565.

Beigman, I. L., Vainshtein, L. A., and Sunyaev, R. A., 1968, *Sov. Phys. Uspekhi*, **11**, 411.

Beigman, I. L., Vainshtein, L. A., and Chichkov, B. N., 1981, *Sov. Phys. JETP*, **53**, 490.

Belic, D. S., Dunn, G. J., Morgan, T. J., Mueller, D. W., and Timmer, C., 1983, *Phys. Rev. Lett.*, **50**, 339.

Bely-Dubau, F. and Volonte, S., 1980, *Rep. Prog. Phys.*, **43**, 199.

Bely-Dubau, F. Gabriel, A. H., and Volonte, S., 1979, *Mon. Not. R. Astron. Soc.*, **189**, 801.

Berg, H. F, Ali, A. W., Lincke, R., and Griem, H. R., 1962, *Phys. Rev.*, **125**, 199.

Bethe, H. A. and Salpeter, E. E., 1957, *Quantum Mechanics of One and Two-Electron Atoms*, Springer, Berlin.

Bloom, S. D. and Goldberg, A., 1986, *Phys. Rev.*, **A34**, 2865.

Bödecker, St., Günter, S., Könies, A., Hitzschke, L., and Kunze, H.-J., 1995, *Phys. Rev.*, **E47**, 2785.

Brachman, M. K., 1951, *Phys. Rev.*, **83**, 1263.

Bradley, D. K., Kilkenny, J., Rose, S. J., and Hares, J. D., 1987, *Phys. Rev. Lett.*, **59**, 2995.

Brooks, R. L., Datla, R. U., Krumbein, A. D., and Griem, H. R., 1978, *Phys. Rev. Lett.*, **41**, 107.

Brooks, R. L., Datla, R. U., Krumbein, A. D., and Griem, H. R., 1980, *Phys. Rev.*, **A21**, 1387.

Brush, S. and Armstrong, B., 1965, in *Proceedings of Workshop Conference on Lowering of the Ionization Potential*, JILA-Report 79, University of Colorado, Boulder, Colo.

Burgess, A., 1965, *Astrophys. J.,* **A141**, 1588.

Büscher, S., Glenzer, S., Wrubel, Th., and Kunze, H.-J., 1995, *J. Quant. Spectrosc. Rad. Trans.*, **54**, 73.

Callaway, J., 1982, *Phys. Rev.*, **A26**, 199.

Chang, T. Y. and Izabelle, A., 1989, *J. Appl. Phys.*, **65**, 2162.

Chichkov, B. N., Mazing, M. A., Shevelko, A. P., and Urnov, A. M., 1981, *Phys. Lett.*, **83A**, 401.

Chihara, J., 1991, *Phys. Rev.*, **A44**, 1247.

Condon, E. U. and Shortley, G. H., 1987, *The Theory of Atomic Spectra*, Cambridge University Press, Cambridge, UK.

Cowan, R. D., 1981, *The Theory of Atomic Structure and Spectra*, University of California, Berkeley.

Cullen, D. E. et al., 1989, *Tables and Graphs of Photon-Interaction Cross Sections from 10 eV to 100 GeV*, UCRL-50400, vol. 6, Lawrence Livermore National Laboratory.

Davis, J. and Blaha, M., 1982, *J. Quant. Spectrosc. Rad. Trans.*, **27**, 307.

DeWitt, H. E. and Hubbard, W. B., 1976, *Astrophys. J.*, **A205**, 295.

Dharma-wardana, M. W. C. and Perrot, F., 1982, *Phys. Rev.*, **A26**, 2096.

Dharma-wardana, M. W. C. and Taylor, R., 1981, *J. Phys. C*, **14**, 629.

Drawin, H. W., 1968, in *Plasma Diagnostics*, edited by W. Lochte-Holtgreven, North Holland, Amsterdam.

Drawin, H. W. and Felenbok, P., 1965, *Data for Plasmas in LTE*, Gauthier-Villars, Paris.

Early, K., et al., 1993, *Appl. Opt.*, **32**, 7044.

Eidmann, K., Schwanda, W., Földes, I. B, Sigel, R., and Tsakiris, G. D., 1994, *J. Qaunt. Spectrosc. Rad. Trans.*, **51**, 77.

Eliezer, S., Ghatak, A., and Hora, H., 1986, *An Introduction to Equations of State*, Cambridge University Press, Cambridge, UK.

Elton, R. C., 1987, *J. de Phys.*, **C9**, 359.

Elton, R. C., *X-ray Lasers*, Academic Press, San Diego, 1990.

Faussurier, G., Blancard, C., and Decoster, A., 1997, *J. Quant. Spectrosc. Radiat. Trans.*, **58**, 233.

Feldman, U., Doschek, G. A., and Seely, J. F., 1988, *J. Opt. Soc. Am. B*, **5**, 2237.

Feldman, U., 1992, *Physica Scripta*, **46**, 202.

Fermi, E., 1930, *Mem. Acad. Italia*, **1**, 1.

Fernandez-Velicia, F. J., 1986, *Phys. Rev.*, **A34**, 4387.

Feynman, R. P., 1974, *Statistical Mechanics*, W. A. Benjamin, Reading, Mass.

Fischer, C. F., 1991, *Comp. Phys. Comm.*, **A64**, 399.

Förster, E., Gabel, K., and Uschmann, I., 1991, *Laser Part. Beams*, **9**, 135.

Förster, E., Fill, E. E., Gäbel, K., He, H., Missalla, T., Renner, O., Uschmann, I., and Wark, J., 1994, *J. Quant. Spectrosc. Rad. Trans.*, **51**, 101.

Furutani, Y., Ohashi, K., Shimizu, M., and Fukuyama, A., 1993, *J. Phys. Soc. Japan*, **62**, 3413.

Gabriel, A. H., 1972, *Mon. Not. R. Astron. Soc.*, **160**, 99.

Gabriel, A. H. and Jordan, C., 1972, *Case Studies Atom. Coll. Phys.*, **2**, 209.

Gabriel, A. H. and Paget, T. M., 1972, *J. Phys. B.*, **5**, 673.

Gabriel, A. H. and Volonte, S., 1973, *J. Phys. B*, **6**, 2684.

Galanti, M. and Peacock, N. J., 1975, *J. Phys. B.*, **8**, 2427.

Gavrila, M., 1959, *Phys. Rev.*, **113**, 5.

Gauthier, J. C., Monier, P., Audebert, P., Chenais-Popovics, C., and Geindre, J. P., 1986, *Laser Part. Beams*, **4**, 421.

Glenzer, S., Wrubel, Th., Büscher, S., Kunze, H-J, Godbert, L., Calisti, A., Stamm, R., Talin, B., Nash, J., Lee, R. W., and Klein, L., 1994, *J. Phys. B*, **27**, 5507.

Goldberg, A., Rozsnyai, B. F., and Thompson, P., 1986, *Phys. Rev.*, **A34**, 421.

Golden, L. B. and Sampson, D. H., 1977, *J. Phys. B*, **10**, 2229.

Golden, L. B. and Sampson, D. H., 1978, *J. Phys. B*, **11**, 541.

Golden, L. B. and Sampson, D. H., 1980, *J. Phys. B*, **13**, 2645.

Goldsmith, S., Griem, H. R., and Cohen, L., 1984, *Phys. Rev.*, **A30**, 2775.

Gombas, P., 1956, in *Handbuch der Physik*, vol. 36, p. 109, Springer, Berlin.

Gordon, W., 1929, *Ann. Phys.*, **2**, 1031.

Goto, T. and Burgess, D. D., 1974, *J. Phys. B.*, **7**, 857.

Greig, I. G., Griem, H. R., Jones, L. A., and Oda, T., 1970, *Phys. Rev. Lett.*, **24**, 3.

Griem, H. R., 1963, *Phys. Rev.*, **131**, 1170.

Griem, H. R., 1964, *Plasma Spectroscopy*, McGraw-Hill, New York.

Griem, H. R., 1974, *Spectral Line Broadening by Plasmas*, Academic, New York and London.

Griem, H. R., 1988, *Phys. Rev.*, **A38**, 2943.

Grützmacher, K. and Wende, B., 1978, *Phys. Rev.*, **A18**, 2140.

Gupta, U. and Rajagopal, A. K., 1980, *Phys. Rev.*, **A22**, 2792.

Gupta, P. D., Popil, R., Fedosejevs, R., Offenberger, A. A., Salzmann, D., and Capjack, C. E., 1986, *Appl. Phys. Lett.* **48**, 103.

Haan, S. L. and Jacobs, V. L., 1989, *Phys. Rev.*, **A40**, 80.

Hahn, Y., 1993, *J. Quant. Spectrosc. Rad. Trans.*, **49**, 81.

Hammel, B. A., Keane, C. J., Kilkenny, J. D., Landen, O. L., Ress, D. B., Bell, P., Pasha, R., Wallace, R. J., and Bradley, D. K., 1992, *Plasma Physics and Controlled Nuclear Fusion Research*, IAEA-CN-56/B-4-9, p. 215.

Hammel, B. A., Keane, C. J., Cable, M. D., Kania, D. R., Kilkenny, J. D., Lee, R. W., and Pasha, R., 1993, *Phys. Rev. Lett.*, **70**, 1263.

Hammel, B. A., Keane, C. J., Dittrich, T. R., Kania, D. R., Kilkenny, J. D., Lee, R. W., and Levedahl, W. K., 1994, *J. Quant. Spectrosc. Rad. Trans.*, **51**, 113.

Hansen, J. P., 1973, *Phys. Rev.*, **A8**, 3096.

Hansen, J. P. and McDonald, I. R., 1986, *Theory of Simple Liquids*, Academic Press, London.

Heisenberg, W., 1926, *Zeits. f. Phys.*, **39**, 499.

Heitler, W., 1954, *The Quantum Theory of Radiation*, Oxford University Press, London, UK.

Holstein, T., 1947, *Phys. Rev.*, **72**, 1212.

Holstein, T., 1951, *Phys. Rev.*, **83**, 1159.

Holtsmark, J., 1919, *Ann. Phys. (Leipzig)* **58**, 577.

Hooper, C. F., 1966, *Phys. Rev.*, **149**, 77.

Hooper, C. F., 1968, *Phys. Rev.*, **169**, 193.

Hubbard, W. B. and Slattery, W. L., 1971, *Astrophys. J.*, **A168**, 131.

Hulten, L., 1935, *Z. Physik*, **95**, 789.

Ichimaru, S., 1982, *Rev. Mod. Phys.*, **54**, 1017.

Ichimaru, S., Mitake, S., Tanaka, S., and Yan, X., 1985, *Phys. Rev.*, **A32**, 1768 [This is the first of a series of papers published by this group in *Phys. Rev. A* during 1985–86 under the title "Theory of interparticle correlations in dense high-temperature plasmas".

Inglis, D. R. and Teller, E., 1939, *Astrophys. J.*, **A90**, 439.

Ivanov, V. V., 1973, *Transfer of Radiation in Spectral Lines*, NTIS, U.S. Department of Commerce, Springfield, VA, USA.

Jacobs, V. L., 1985, *Astrophys. J.*, **A296**, 121.

Jacobs, V. L., Cooper, J., and Haan, S. L., 1987, *Phys. Rev.*, **A36**, 1093.

Jamelot, G., Jaegle, P., Lemaire, P., and Carillon, A., 1990, *J. Quant. Spectrosc. Rad. Trans.*, **44**, 71.

Johnston, T. W. and Dawson, J., 1973, *Phys. Fluids*, **16**, 722.

Kalitkin, N., 1960, *Sov. Phys. JETP*, **11**, 1106.

Kalitkin, N. and Kuzmina, L., 1960, *Sov. Phys. Solid State*, **7**, 287.

Karzas, W. J. and Latter, R., 1961, *Astrophys. J. Suppl. VI*, **55**, 167.

Kato, T. and Janev, R. K., 1992, *Nucl. Fusion, Special Suppl.*, **3**, 33.

Keane, C. J., Ceglio, N. M., MacGowan, B. J., Matthews, D. L., Nilson, N. G., Trebes, J. E., and Whelan, D. A., 1989, *J. Phys. B.*, **22**, 3343.

Kinjo, Y., Shinohara, K., Ito, A., Nakano, H., Watanabe, M., Horiike, Y., Kikuchi, Y., Richardson, M. C., and Tanaka, K. A., 1994, *J. Microscopy*, **176**, 63.

Kirzhnitz, D., 1959, *Sov. Phys. JETP*, **8**, 1081.

Kohn, W. and Sham, L. J., 1965, *Phys. Rev.*, **137**, A1697.

Kondo, K., 1995, *ILE Annual Report, 1995*, Institute of Laser Engineering, University of Osaka, Osaka, Japan.

Koniges, A. E., Eder, D. C., Wan, A. S., Scott, H. A., Dalhed, H. E., Mayle, R. W., and Post, D. E., 1995, *J. Nucl. Mat.*, **220**, 1116.

Landshoff, R. K. and Perez, J. D., 1976, *Phys. Rev.*, **A13**, 1619.

Landau, L. D. and Lifshitz, E. M., 1959, *Statistical Mechanics*, Pergamon Press, London.

Langdon, A. B., 1980, *Phys. Rev. Lett.*, **44**, 575.

Latter, R., 1955, *Phys. Rev.*, **99**, 1854.

Leboucher,-Dalimier, E., Poquerusse, A., and Angelo, P., 1993, *Phys. Rev.*, **E17**, R1467.

Lee, T. N., McLean, E. A., and Elton, R. C., 1987, *Phys. Rev. Lett.*, **39**, 1185.

Lee, T. N., McLean, E. A., Stamper, J. A., Griem, H. R., and Manka, C. K., 1988, *Bull. Am. Phys. Soc.*, **33**, 1920.

Li, Y., Pretzler, G., Lu, P., and Fill, E. E., 1997, *Phys. Plasmas*, **4**, 479.

Liberman, D. A., 1979, *Phys. Rev.*, **A20**, 4981.

Lisitsa, V. S., 1994, *Atoms in Plasmas*, Springer, Berlin.

Lotz, W., 1968, *Z. Phys.*, **216**, 241.

Macdowell, A. A. et al., 1993, *Appl. Opt.*, **32**, 7072.

MacGowan, B. J. et al., 1987a, *J. Appl. Phys.*, **61**, 5243.

MacGowan, B. J. et al., 1987b, *Phys. Rev. Lett.*, **59**, 2157.

Malnoult, P., d'Etat, B., and Nguyen, H., 1989, *Phys. Rev.*, **A40**, 1983.

March, N. H., 1953, *Phil. Mag.* **44**, 346.

March, N. H., 1957, *Adv. Phys.* **6**, 1.

Matsushita, T., Kodama, R., Chen, Y. W., Nakai, M., Shimada, K., Saito, M., and Kato, Y., 1995, *ILE Annual Report, 1995*, Institute of Laser Engineering, University of Osaka, Osaka, Japan.

Matthews, D. L., 1986, *J. de Phys.*, **C6**, 1.

Matthews, D. L. et al., 1985, *Phys. Rev. Lett.*, **54**, 110.

McWhirter, R. W. P., 1965, in *Plasma Diagnostic Techniques*, vol. 2, edited by R. H. Huddlestone and S. L. Leonard, Academic Press, New York.

Meeron, E., 1960, *J. Math. Phys.*, **1**, 192.

Meng, H. C., Greve, P., Kunze, H.-J., and Schmidt, T., 1985, *Phys. Rev.*, **A31**, 3276.

Merts, A. L., Cowan, R. D., and Magee, N. H., 1976, LASL Report LA-6220-MS, Los Alamos, Calif.

Metropolis, N. A., Rosenbluth, A. W., Rosenbluth, M. N., Teller, A. H., and Teller, E., 1953, *J. Chem. Phys.*, **21**, 1087.

Mihalas, D., 1970, *Stellar Atmospheres*, W. H. Freeman, San Francisco.

Minguez, E., 1990, *Laser Part. Beams*, **8**, 709.

Moore, C. E., 1949, 1952, 1958, *Atomic Energy Level Tables*, NBS Circular 467, vols. I, II and III, U.S. Government Printing Office, Washington, D.C.

Moores, D. L., Golden, L. B., and Sampson, D. H., 1980, *J. Phys. B.*, **13**, 385.

More, R. M., 1979, *Phys. Rev.*, **A19**, 1234.

More, R. M., 1981, *Atomic Physics in Inertial Confinement Fusion*, LLNL-Report, no. UCRL-84991.

More, R. M., 1982, *J. Quant. Spectros. Rad. Trans.*, **27**, 345.

More, R. M., 1983, In *Atomic and Molecular Physics of Thermonuclear Fusion*, edited by C. J. Joachin and D. E. Post, p. 399, Plenum, New York.

More, R. M., 1985, *Adv. At. Mol. Phys.*, **21**, 305.

Neiger, M. and Griem, H. R., 1976, *Phys. Rev.*, **A14**, 291.

Nguyen Hoe, Koenig, M., Benderjen, D., Caby, M., and Coulaud, C., 1986, *Phys. Rev.*, **A33**, 1279.

Ochi, Y., Fujita, K., Miki, H., and Nishimura, H., 1995, *ILE Annual Report*, Institute of Laser Enginering, University of Osaka, Osaka, Japan.

Ornstein, L. S. and Zernicke, F., 1914, *Proc. Acad. Sci. Amsterdam*, **17**, 793.

Pankratov, P. and Meyer-ter-Vehn, 1992a, *Phys. Rev.*, **A46**, 5497.

Pankratov, P. and Meyer-ter-Vehn, 1992b, *Phys. Rev.*, **A46**, 5500.

Parr, R. G. and Ghosh, S. K., 1986, *Proc. Natl. Acad. Sci. USA*, **83**, 3577.

Perrot, F., 1979, *Phys. Rev.*, **A20**, 586.

Perrot, F., 1989, *Phys. Scr.*, **39**, 332.

Pittman, T. L. and Fleurier, C., 1986, *Phys. Rev.*, **A33**, 1291.

Poquerusse, A., 1993, *J. Phys. II France*, **3**, 197.

Post, D. E., 1995, *J. Nucl. Mat.*, **220–222**, 143.

Pratt, R. H., 1960, *Phys. Rev.*, **117**, 1017.

Pratt, R. H., Ron, A., and Tseng, H. K., 1973, *Rev. Mod. Phys.*, **45**, 273. [Among the many other papers published by this group on the subject of photoionization we mention here only the following: Botto, D. J., McEnnan, J., and Pratt, R. H., 1978, *Phys. Rev.*, **A18**, 580; Oh, S. D., McEnnan, J., and Pratt, R. H., 1976, *Phys. Rev.*, **A14**, 1428].

Richardson, M. C., Shinohara, K., Tanaka, K. A., Kinjo, Y., Ikeda, N., and Kado, M., 1992, *Proc. SPIE*, 1741.

Rickert, A. and Meyer-ter-Vehn, J., 1990, *Laser Part. Beams*, **8**, 715.

Riley, D., Willi, O., Rose, S. J., and Afshar-Rad, T., 1989, *Europhys. Lett.*, **10**, 135.

Rocca, J. J., Shlyaptsev, V., Tomasel, F. G., Cortazar, O. D., Hartshorn, D., and Chilla, J. L. A., 1994, *Phys. Rev. Lett.*, **73**, 2192.

Rose, S. J., 1983, *J. de Physique Suppl.*, **C8**, 159.

Rosen, A. and Lindgren, I., 1972, *Phys. Scr.*, **6**, 109.

Rousse, A., Audebert, P., Geindre, J. P., Fallies, F., Gauthier, J. C., Mysyrowicz, A., Grillon, G., and Antonetti, A., 1994, *Phys. Rev. E.*, **50**, 2200.

Rozsnyai, B., 1972, *Phys. Rev.*, **A5**, 1137.

Rozsnyai, B., 1975, *J. Quant. Spectrosc. Rad. Trans.*, **15**, 695.

Rozsnyai, B., 1982, *J. Quant. Spectrosc. Rad. Trans.*, **27**, 211.

Rozsnyai, B., 1991, *Phys. Rev.*, **A43**, 3035.

Safronova, U. I., Shlyapzeva, A. S., Vainshtein, L. A., Kato, T., and Masai, K., 1992, *Phys. Scr.*, **46**, 409.

Sahal-Brechot, S., 1969, *Astron. Astrophys.*, **1**, 91.

Salzmann, D., 1979, *Phys. Rev.*, **A20**, 1704.

Salzmann, D., 1980, *Phys. Rev.*, **A22**, 2245.

Salzmann, D., 1994, *Phys. Rev.*, **A49**, 3729.

Salzmann, D. and Krumbein, A., 1978, *J. Appl. Phys.*, **49**, 3229.

Salzmann, D. and Szichman, H., 1987, *Phys. Rev.*, **A35**, 807.

Salzmann, D. and Wendin, G., 1978, *Phys. Rev.*, **A18**, 2695.

Salzmann, D., Stein, J., Goldberg, I. B., and Pratt, R. H., 1991, *Phys. Rev.*, **A44**, 1270.

Sampson, D. H., 1985, *Excitation and Ionization of Highly Charged Ions by Electron Impact*, DOE/ET/53056-07. [This author in collaboration with L. B. Golden, R. E. H. Clark, C. J. Fontes, S. J. Goett, H. Zhang, and G. V. Petrou published series of papers on the subject of electron impact excitation. Among many others these are the following: (i) 1983, *At. Dat. Nucl. Dat. Tabl.*, **A28**, 279; (ii) 1983, *At. Dat. Nucl. Dat. Tabl.*, **A28**, 299; (iii) 1983, *At. Dat. Nucl. Dat. Tabl.*, **A29**, 467; (iv) 1983, *At. Dat. Nucl. Dat. Tabl.*, **A29**, 535; (v) 1984, *At. Dat. Nucl. Dat. Tabl.*, **A30**, 125; (vi) 1990, *At. Dat. Nucl. Dat. Tabl.*, **A44**, 31; (vii) 1990, *At. Dat. Nucl. Dat. Tabl.*, **A44**, 209; (viii) 1990, *At. Dat. Nucl. Dat. Tabl.*, **A44**, 273; (ix) 1991, *At. Dat. Nucl. Dat. Tabl.*, **A48**, 25; (x) 1991, *At. Dat. Nucl. Dat. Tabl.*, **A48**, 91].

Sampson, D. H., and Zhang, H., 1992, *Phys. Rev.*, **A45**, 1556.

Sarfaty, M., Maron, Y., Krasik, Ya.E., Weingarten, A., Arad, R., Shpitalnik, R., Fruchtman, A., and Alexiou, S., 1995, *Phys. Plasmas*, **2**, 2122.

Sauter, F., 1931, *Ann. Physik*, **11**, 454.

Schwanda, W. and Eidmann, K., 1992, *Phys. Rev. Lett.*, **69**, 3507.

Seaton, M. J., 1964, *Planet. Space Sci.*, **12**, 55.

Siegbahn, K., 1979, *Alpha-, Beta-, and Gamma Ray Spectroscopy*, North-Holland, Amsterdam.

Silfvast, W. T. and Ceglio, N. M., 1993, *Appl. Opt.*, **32**, 6895.

Skupsky, S., 1980, *Phys. Rev.*, **A21**, 1316.

Slattery, W. L., Doolen, G. D., and DeWitt, H. E., 1980, *Phys. Rev.*, **A21**, 2087.

Slattery, W. L., Doolen, G. D., and DeWitt, H. E., 1982, *Phys. Rev.*, **A26**, 2255.

Sobelman, I. I., 1972, *Introduction to the Theory of Atomic Spectra*, Pergamon Press, Oxford, UK.

Sobelman, I. I., Vainshtein, L. A., and Yukov, E. A., 1981, *Excitation of Atoms and Broadening of Spectral Lines*, Springer, Berlin.

Spitzer, L., 1962, *Physics of Fully Ionized Gases*, Interscience Publishers, New York.

Spruch, L., 1991, *Rev. Mod. Phys.*, **63**, 151.

Stein, J. and Salzmann, D., 1992, *Phys. Rev.*, **A45**, 3943.

Stein, J., Goldberg, I. B., Shalitin, D., and Salzmann, D., 1989, *Phys. Rev.*, **A39**, 2078.

Stone, S. R. and Weisheit, J. C., 1984, LLNL report, UCID-20262.

Suckewer, S., Skinner, C. H., Milchberg, H., Keane, C., and Voorhees, D., 1985, *Phys. Rev. Lett.*, **55**, 1753.

Tawara, H. and Kato, T., 1987, *At. Dat. Nucl. Dat. Tabl.*, **A36**, 167.

Tichenor, D. A., et al., 1993, *Appl. Opt.*, **32**, 7068.

Trees, R. E., 1951a, *Phys. Rev.*, **83**, 756.

Trees, R. E., 1951b, *Phys. Rev.*, **84**, 1089.

van Regemorter, H., 1962, *Astrophys. J.*, **A132**, 906.

Vinogradov, A. V. and Shevelko, V. P., 1976, *Sov. Phys. JETP*, **4**, 542.

Vollbrecht, M., Uschmann, I., Förster, E., Fujita, K., Ochi, Y., Nishimura, H., and Mima, K., 1998, *J. Quant. Spectrosc. Rad. Trans.*, in press.

Voronov, G. S., 1997, *At. Dat. Nucl. Dat. Tabl.*, **A65**, 1.

Weisheit, J. C., 1988, *Adv. At. Mol. Phys.*, **25**, 101.

Weisheit, J. C. and Rozsnyai, B., 1976, *J. Phys. B.*, **9**, L63.

Wiese, W. L., Smith, M. W., and Glennon, B. M., 1966, *Atomic Transition Probabilities*, NSRDS-NBS 4, National Bureau of Standards.

Wiese, W. L., Smith, M. W., and Miles, B. M., 1969, *Atomic Transition Probabilities*, NSRDS-NBS 22, National Bureau of Standards.

Williams, J. F., 1988, *J. Phys. B.,* **21**, 2107.

Younger, S. M., 1980a, *Phys. Rev.,* **A22**, 111.

Younger, S. M., 1980b, *Phys. Rev.,* **A22**, 1425.

Younger, S. M., 1981a, *Phys. Rev.,* **A24,** 1278.

Younger, S. M., 1981b, *Phys. Rev.,* **A23**, 1138.

Younger, S. M., 1981c, *Phys. Rev.,* **A24**, 1272.

Younger, S. M., 1982a, *Phys. Rev.,* **A25**, 3396.

Younger, S. M., 1982b, *Phys. Rev.,* **A26**, 3177.

Younger, S. M., 1986, *Phys. Rev.,* **A34**, 1952.

Younger, S. M., 1988, *Phys. Rev.,* **A37**, 4125.

Younger, S. M., Harrison, A. K., Fujima, K., and Griswold, D., 1988, *Phys. Rev. Lett.,* **61**, 962.

Younger, S. M., Harrison, A. K., and Sukiyama, G., 1989, *Phys. Rev.,* **A40**, 5256.

Zeldovitch, Y. and Raizer, Y., 1966, *Physics of Shock Waves and High Temperature Hydrodynamic Phenomena*, Academic Press, New York.

Zimmerman, G. B. and More, R. M., 1980, *J. Quant. Spectrosc. Rad. Trans.,* **23**, 517.

Index

259